Hellenic Institute of Nuclear Physics; 6th International Workshop: Perspectives on Nuclear Physics; From Fundamentals to Applications (HINPw6)

EPJ Web of Conferences Volume 252 (2021)

Online
14 - 16 May 2021

Editors:

Athena Pakou
Dennis Bonatsos
George Lalazissis
George Souliotis

ISBN: 978-1-7138-3644-5

Printed from e-media with permission by:

Curran Associates, Inc.
57 Morehouse Lane
Red Hook, NY 12571

Some format issues inherent in the e-media version may also appear in this print version.

For additional information, please contact EDP Sciences – Web of Conferences
at the address below.

EDP Sciences – Web of Conferences
17, Avenue du Hoggar
Parc d'Activité de Courtabœuf
BP 112
F-91944 Les Ulis Cedex A
France

Phone: +33 (0) 1 69 18 75 75
Fax: +33 (0) 1 69 28 84 91

contact-edps@webofconferences.org

Additional copies of this publication are available from:

Curran Associates, Inc.
57 Morehouse Lane
Red Hook, NY 12571 USA
Phone: 845-758-0400
Fax: 845-758-2633
Email: curran@proceedings.com
Web: www.proceedings.com

TABLE OF CONTENTS

Statement of Peer review

In submitting conference proceedings to *Web of Conferences*, the editors of the proceedings certify to the Publisher that

1. They adhere to its **Policy on Publishing Integrity** in order to safeguard good scientific practice in publishing.
2. All articles have been subjected to peer review administered by the proceedings editors.
3. Reviews have been conducted by expert referees, who have been requested to provide unbiased and constructive comments aimed, whenever possible, at improving the work.
4. Proceedings editors have taken all reasonable steps to ensure the quality of the materials they publish and their decision to accept or reject a paper for publication has been based only on the merits of the work and the relevance to the journal.

Title, date and place of the conference

HINPw6-Hellenic Institute of Nuclear Physics; 6th International workshop; Perspectives on Nuclear Physics; from Fundamentals to Applications
14-16 May 2021- zoom confernce organized by the National and Kapodistrian University of Athens and the University of Ioannina

Proceedings editor(s):

Athena Pakou
Dennis Bonatsos
George Lalazissis
George Souliotis

Date and editor's signature

2-9-2021

EPJ Web of Conferences **252**, 00001 (2021) https://doi.org/10.1051/epjconf/202125200001
HINPw6

Preface: HINPw6 - Hellenic Institute of Nuclear Physics; 6th International workshop; Perspectives on Nuclear Physics; from fundamentals to applications

Athena Pakou[1], *Dennis* Bonatsos[2], *Georgios A.*Lalazissis[3] and *George* A. Souliotis[4]

[1] Department of Physics and HINP, University of Ioannina, 45110 Ioannina, Greece
[2] Institute of Nuclear and Particle Physics, National Centre for Scientific Research ``Demokritos", GR-15310 Aghia Paraskevi, Attiki, Greece
[3] Department of Physics, Aristotle University of Thessaloniki, Thessaloniki GR-54124, Greece
[4] Department of Chemistry, National and Kapodistrian University of Athens and HINP, 15771 Athens, Greece

The sixth International workshp - HINPw6 - of the Hellenic Institute of Nuclear Physics (HINP) was held on May 14 to 16, 2021, by means of a zoom platform, due to Covid-19 restrictions. HINP is a virtual Institute of non-profitable character that constitutes a form of collaboration between three Nuclear Physics Laboratories, the University of Ioannina, the Aristotle University of Thessaloniki, and the National & Kapodistrian University of Athens. Members of the National Centre for Scientific Research " Demokritos" are also participating. Details on the Institute can be found in http://hinp.physics.uoi.gr/

One of the main goals of HINP is the collaboration between Nuclear Physics Laboratories in Greece with strong international links, promoting exchange of ideas among senior researchers, postdocs and graduate students with experimental and theoretical background. The HINP series of workshops are devoted to this goal.

The present web-based conference was co-organized by the National and Kapodistrian University of Athens (NKUA) and the University of Ioannina (UoI). While this is the 6th workshop of HINP, it is the third one with a clearly international character, following HINPw4, held at the U. of Ioannina in May 2017, and HINPw5, organized at the Aristotle U. of Thessaloniki in April 2019, on which further information can be found at http://hinp.physics.uoi.gr/Workshops.htm

Over 70 participants attended the workshop, welcomed by the Dean of the School of Natural Sciences of NKUA, Professor Ioannis Emmanouil, and the Director of the Physical Chemistry Laboratory of NKUA, Professor Andreas Koutselos. We warmly thank both of them. The workshop included scientific talks by 25 invited speakers from several parts of the world, including Europe, USA, and China. In total 47 speakers reported their research on various topics in specific sessions of

1. Nuclear Structure
2. Advances in Nuclear Facilities
3. Nuclear Reactions with weakly bound nuclei
4. Nuclear Astrophysics
5. Heavy-Ion Reactions and Rare Isotope Production
6. Applications

Posters were replaced by 5 min flash talks, a novelty highly appreciated by the participants.

While most of the talks were focused on fundamental open problems, an effort has been made by the organizing committee to demonstrate the unified character of fundamental and applied research. Thus the opening talk, given by Nicolas Alamanos from CEA Paris-Saclay, was on "Nuclear Physics; The Societal Impact – A project for Greece?" Highlights in the same direction were the talks by Sherry Yennello from Texas A&M on "Advancing research in Texas through Experiments in Medical Isotope Science", and by Luis Acosta from UNAM on "The AMS technique as an important tool for the measurement of astrophysical cross sections", as well as a flash talk by Vasileios Soukeras from INFN-LNS, Catania on "Low energy proton induced reactions for application purposes".

Important new results were outlined in various subfields. We warmly thank all speakers for their high level talks. We especially thank our invited speakers with intriguing subjects on Nuclear Structure (relativistic Brueckner-Hartree-Fock and relativistic Energy Density Functional theories for finite nuclei; chirality and wobbling phenomena; octupole deformation in the "nuclear clock" metastable state in ^{229}Th), Nuclear Astrophysics (^{12}C+^{12}C sub-barrier fusion; ternary fission in low density matter ; astrophysics at the n-TOF facility) and Nuclear Reactions from light weakly bound halo nuclei (^{6,8}He, ^{11}Be, ^{15}C) to heavy neutron rich drip line elements and the connection of the nuclear equation of state to the interplay between fusion and quasi-fission processes in low-energy nuclear reactions. We further warmly thank our invited speakers presenting special advances in their laboratories for weak interactions, including the *TwinSol* facility at the Notre Dame U., the IEAP at the Czech Technical U. in Prague, and the NUMEN project - MAGNEX facility in the INFN-LNS, Catania.

Once more we feel that we would like to warmly thank all our colleagues and collaborators from abroad, and especially those, who constantly support with their presence and their inspiring talks our workshops : Nicolas Alamanos from CEA Paris-Saclay, Francesco Cappuzzello from INFN-LNS and the U. of Catania, and Peter Ring from the Technical U. of Munich.

Nuclear Physics. The Societal Impact – A project for Greece?

Nicolas Alamanos[1*]

[1]Université Paris-Saclay, IRFU, CEA, F-91191 Gif-sur-Yvette, France

Abstract. Answers to some of the most important questions that our world is facing out, will come from interdisciplinary efforts in medicine, energy and climate. These are involving contributions from fundamental research and in particular from nuclear physics and associated techniques. From the different types of radiation used in hospitals to Magnetic Resonance Imaging (MRI), nuclear physics and its associated technologies that is accelerators and superconducting magnets are omnipresent. The development of new radioisotope production techniques, therapy of certain cancers with ions and hadron therapy or high field MRI are among the subjects undergoing rapid development. Furthermore, archaeometry and many other societal applications are also benefited from the techniques of nuclear physics. My presentation in the 6th workshop of the Hellenic Institute of Nuclear Physics (HINP) was focused on the links between fundamental research and society and was partially inspired by an article I have published in 2020 on this subject [1]. In the last part of my presentation, I have brushed up the situation concerning the construction of low energy accelerator facilities worldwide. The construction of new accelerator installations is going through a flourishing period in particular in Europe with the construction of new accelerators dedicated to fundamental research in Nuclear Physics, to neutron production and to societal applications. This favourable climate could motivate the Hellenic nuclear physics community to design a new accelerator facility dedicated to fundamental research, neutron production and to the multiple applications of nuclear physics techniques to societal problems. Its construction could benefit from European funds and technical contributions from many European countries like France or Italy.

1 Applications of nuclear physics techniques in medicine

The medical imaging science started out in the beginning of last century following the discovery of X-rays by Wilhelm Röntgen. Radiography exploits the absorption of the ionizing X radiation when crossing the human body: the denser the tissues and the higher its average atomic number, the stronger the absorption. In radiography, the different colours or shades correspond to different levels of absorption, i.e. density and thickness.

Nearly two weeks after his discovery Wilhelm Röntgen immobilized for some moments his wife Anna Bertha's hand in front of a photographic plate in the path of the rays. After development, he observed an image of the bones in the hand. This was the first "Röntgenogram" ever taken. This first medical application of X rays represents one of the most obvious benefits of nuclear physics in medicine.

A century ago, the living body, like most of the material world, was opaque. Then Wilhelm Röntgen captured an X-ray image of his wife's finger - her wedding ring 'floating' around a white bone - and our vision changed forever.

1.1 Imaging

Positron Emission Tomography and computer Tomography (PET/CT): In order to function and reproduce cells need energy in the form of glucose. Cancer cells require a lot of energy and therefore have a very high glucose consumption. PET consists of administering intravenously to the patient a molecule marked with a radioactive isotope of a relatively short lifetime, which decays, by positron emission. An electron immediately captures the positron and the two particles annihilate each other emitting two photons each of 511keV [1]. The electron–positron annihilations results in gamma photons emitted at almost 180 degrees to each other; hence, it is possible, with an appropriated detection system to localize their source along a straight line. In 1969, Godfrey Hounsfield improved this technique by inventing the Computer Tomography (CT). The basic principle of a CT scanner is to take a set of X-ray images from a large number of different directions, and then to apply special algorithms to perform the so-called back-projection in order to obtain a 3D representation of the density profile of the object under observation. The large majority of PET scanners installed in hospitals are hybrid systems where a PET detector is combined with X-ray Computer Tomography (CT). According to medical specialists, the technical evolution, which led to the PET/CT technique, led to a

* Corresponding author: nicolas.alamanos@cea.fr

real medical revolution. In 1979 Godfrey Hounsfield was awarded for this discovery the Nobel Prize in Physiology and Medicine.

Producing radioisotopes for medical treatment: Medical imaging is currently used for obtaining anatomical information, such as bone displacement, the condition of organs or the presence of tumours. In Europe alone there are over 12 million medical procedures per year (diagnostic and therapy) using radioisotopes, i.e. more than 30,000 procedures per day. About 90% of radioisotopes are used for diagnostic purposes and 10% for therapy. The short-lived isotopes used in medicine require accelerators, in particular cyclotrons, nuclear reactors and possibly dedicated nuclear physics techniques for their production.

Currently, research into new and even more effective elements for medical treatment is actively being pursued in many physics laboratories. What are the reaction mechanisms for producing new radioelements? The main goal is to find efficient production routes for novel radiopharmaceuticals (theranostics, multi-modal imaging …), with special considerations concerning purity and yield. The reaction mechanisms, the underlying theory and the most commonly used nuclear reaction codes for simulations and cross section predictions are important inputs in this field [2]. This is a good example of the application of nuclear physics techniques to other fields; codes developed for the needs of fundamental nuclear physics research are today applied to calculating the production of radioelements.

Another widely used radioisotope is ^{99}Mo/^{99m}Tc. Technetium-99m itself is used in more than 80% of diagnostic nuclear imaging procedures, with over 30,000 scans performed with this isotope in the United States each day and global consumption exceeding 40 million scans per year. In 2013 eight research reactors were involved in the large-scale production (>95% of world supply) of ^{99}Mo. Today three of them are shut down. Others are expected to stop irradiating targets for Technetium-99m production within the decade [3]. These supply shortages also raised interest in alternative production routes. Studies to find alternative methods for producing Technetium-99m require strong collaboration between nuclear physicists, accelerator specialists and the medical world. Several nuclear reactions have been investigated for producing Technetium-99m. Bombarding a Mo-100 target with a 22-MeV-proton beam currently seems to be one of the most effective ways for its direct production. However, the very short half-life of Technetium-99m requires that the isotope must be used quickly after production, limiting its use geographically to an area close to the cyclotron facility. Even so, for some countries, cyclotron type "methods" stand as a unique and invaluable alternative for the production of Technetium-99m, even if it seems necessary to pave the country with medical cyclotrons close to hospital complexes [4, 5]. This requires innovation to provide compact, easy-to-use and cost-effective cyclotrons.

When I was writing this text, I was informed that for the production of molybdenum-99 (Mo-99) other techniques, based on electron accelerators, are also under investigation [6].

Using radiations for therapy - hadron-therapy: Today, cancer is among the highest causes of death in developed countries and its treatment presents a challenge at the global level. The first to suggest the use of proton beams for radiation therapy, in 1946, was Robert Wilson, an accelerator pioneer. In conventional radiation therapy, beams of X rays (high-energy photons) or hadrons are produced and then projected at the patient to destroy tumour cells. Using crossing beams from many angles, radiation oncologists irradiate the tumour cells while trying to spare the surrounding normal tissues.

The strength of hadron-therapy lies in the unique physical and radiobiological properties of these particles. In particular, carbon ion beams have physical properties enabling the radiation damage to be concentrated intensively at the cancer site. These beams present a ballistic advantage with the maximum energy deposition at the end of the path (i.e. the Bragg peak) [1]. Furthermore, from the biological point of view, hadrons are more efficient in killing cancer cells and the treatment period is shorter compared to other radiotherapy methods [7, 8]. Carbon therapy has grown up only very recently. The leading country is Japan where the first treatment centre was built in Chiba in 1994. In Germany, this type of treatment is available at the Heidelberg Ion-Beam Therapy Centre (HIT) and at the Marburg Ion-Beam Therapy Centre (MIT). In Italy, the National Centre of Oncological Hadron therapy (CNAO) also provides this treatment. Austria opened a Carbon-ion radiotherapy centre in 2017. Centres in France, South Korea, Taiwan, and China will open soon.

One of the dreams of scientists working in this field is to develop low-cost laser driven accelerators for hadron therapy. Charged particle acceleration using ultra-intense and ultra-short laser pulses has gathered strong interest in the scientific community and it is now one of the most attractive topics in relativistic laser-plasma interaction research [1, 9]. Laser driven acceleration of electrons and particles may offer new possibilities for medical applications. The peculiarities of laser-driven beams led to the development of new strategies and advanced techniques for the transport, diagnostics and dosimetry of the accelerated particles, due to the wide energy spread, the angular divergence and the extremely intense pulses. In this framework, INFN-LNS (Italian Institute of Nuclear Physics, Catania) in collaboration with the ELI-Beamline Institute (Dolny Brezany, CZ) is building the ELIMED (ELI-Beamlines MEDical and multidisciplinary applications) beamline. ELIMED will be the first user addressed transport beamline dedicated to medical and multidisciplinary studies with laser-accelerated ion beams and completely open to the scientific community wishing to perform experiments with these new beams [9].

Among the open issues in this field is the compactness of the gantries [10]. Compared with protons, heavy ions are much more difficult to bend and direct to the target. Currently, HIT (Heidelberg Ion-Beam Therapy Centre) is the world's only carbon-ion facility with a beam that can be rotated to 360°.

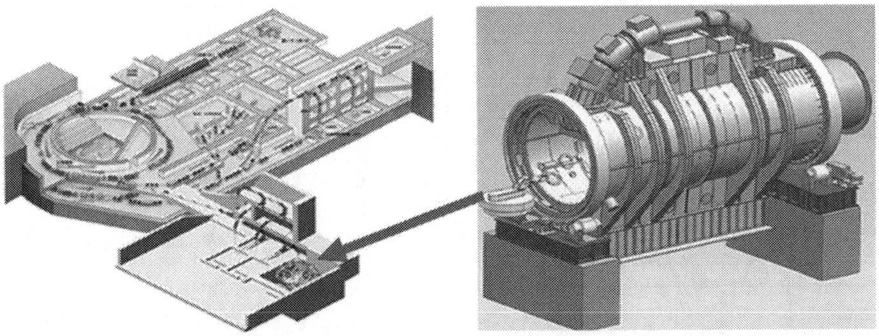

Fig. 1. Left: The Heavy-Ion Medical Accelerator in the Chiba (HIMAC) complex consists of three ion sources, one linear accelerator, two synchrotron rings, and three treatment rooms. The new treatment facility also has three treatment rooms. Right: A three-dimensional image of the iso-centric rotating gantry under development for the new treatment facility.

To achieve this goal and bend the carbon beams, HIT has deployed an enormous gantry weighing 670 tons. Engineers in Nuclear and Elementary particle physics laboratories have developed techniques for building high field superconducting magnets such as the ISEULT magnet (Fig. 2). These techniques could be applied in the construction of superconducting gantries whose weight and size should be more than one-third lower than those of conventional magnets and will offer the possibility to apply higher fields.

Magnetic Resonance Imaging (MRI): To perform a study, the subject is positioned within an MRI scanner that forms a strong magnetic field around the area to be imaged. An oscillating magnetic field is temporarily applied to the patient at the appropriate resonance frequency. The only nucleus in abundance in the human body with non-zero spin is hydrogen. The excited hydrogen atoms emit a radio frequency signal, which can be measured. Hydrogen contributes two thirds to the water molecule (H_2O), which forms 80% of the brain mass. MRI is a "water story" [11]. Neither oxygen nor carbon, which are key nuclei for living organisms, have spin. The larger the magnetic field, the better the image quality. For example, a field of 1.5 Tesla allows details of the brain of ~2 mm to be imaged. A field of 11.7 Tesla would increase this resolution by a factor of 10 down to ~0.2 mm. With this resolution, high accuracy segmentation of the hippocampus becomes possible. Furthermore, highly interesting information for clinical research into Alzheimer's disease, epilepsy and schizophrenia may become available. The future of scientific research in this domain also relies on high field magnets.

In the race to increase the strength of the fields, the 11.7 Tesla ISEULT whole body magnet installed at NeuroSpin (CEA Paris-Saclay), is the magnet with the highest field existing today, see Fig. 2. It has a 900 mm diameter opening, capable of scanning a patient's entire body. During the tests carried out on July 18, 2019, it reached a magnetic field of 11.70 Tesla [12]. It is quite different from conventional MRI magnets. The niobium titanium superconducting wires are maintained at a very low temperature (1.8 K above absolute zero) and are powered by an excessively stable current of 1400 Amperes. The main winding of the magnet, weighing 45 tons, must be positioned as precisely as possible around the brain (a few tenths of a millimetre away).

In the coming months, many new devices will be added to the magnet to transform it into an MRI device capable of producing images of the brain. After the installation and testing of the whole system, the first images are expected to be produced in the end of the year.

Fig. 2. Top: With the new record of 11.7 Tesla, ISEULT exceeds the most powerful MRI magnet currently in service in the world, based in Minneapolis, reaching 10.5 Tesla, but with a smaller central opening of 880 mm, compared to 900 mm for ISEULT. Bottom: The magnetic field must be homogeneous at 0.5 PPM (parts per million) around the patient's brain. 5904 pieces of shim (small iron pellets) were thus screwed onto their rails and installed inside the magnet tunnel.

2 Other societal applications

If the medical field is the one where the contributions of nuclear physics are probably the most remarkable, it is not the only beneficiary. Archaeometry and in particular dating, air pollution measurements, and "neutron" techniques for safety investigations are among the beneficiaries of the instrumentation developed in nuclear physics laboratories.

Cultural heritage – Archaeometry: The purpose of archaeology is to examine, interpret and date archaeological remains. Many nuclear physics methods are used to analyse the composition of archeometrical samples. Among these, the most popular are probably the PIXE and PIGE (Particle Induced X-ray Emission) and (Particle Induced γ-ray Emission) methods, providing information on the "elementary" composition (in the sense of chemical element) of a sample (Z), and the RBS (Rutherford Back Scattering) method providing information on the thickness and composition (A) of the sample. These techniques allow the state of preservation of artworks and finds to be studied, the support of restoration (to find for instance the most appropriate materials which are compatible with the original ones), and the study of artistic techniques and the dating of samples [1, 13].

Dating, and in particular the ^{14}C method, is probably among the greatest contributions of Nuclear Physics to Archaeology. How does it work? Cosmic rays collide with atoms in the atmosphere creating energetic neutrons. These neutrons collide with nitrogen atoms creating ^{14}C via the ($n+^{14}N \rightarrow p+^{14}C$) reaction. ^{14}C is radioactive, decaying by β emission, with a half-life of about 5,700 years. The ^{14}C atoms combine with oxygen to form carbon dioxide (CO_2), which is absorbed and incorporated into plant fibres by photosynthesis. Animals and people eating the plants absorb ^{14}C too. The ratio of carbon ^{12}C to ^{14}C in the air and in all living organisms is nearly constant at any time.

The ^{14}C atoms are always decaying, but new ^{14}C atoms are replacing them at a constant rate. Our bodies have a certain percentage of ^{14}C atoms in them, and all living plants and animals have the same percentage. When an organism dies, carbon exchange with the biosphere ceases, it stops absorbing ^{14}C and the ^{14}C isotope content decreases at a rate set by the laws of radioactive decay. For this reason, the ratio between ^{12}C and ^{14}C gives precise information on the date of death.

Dating is based on a comparison of the ratio of ^{12}C and ^{14}C contained in a sample with that of a standard reference sample. From this comparison, the "^{14}C age" of the sample is deduced, Fig. 3. This "^{14}C age" is then translated into real age (or "calendar age"). There is a possibility that the $^{14}C/^{12}C$ ratio might have varied over time and that a correction will need to be applied to radiocarbon ages to obtain calendar dates [13]. The most accurate method for obtaining the $^{14}C/^{12}C$ ratio is accelerator mass spectroscopy (AMS). The ions are accelerated before mass analysis [13]. Their special strength relies on the power to separate a rare isotope from an abundant neighbouring mass ("abundance sensitivity", e.g. ^{14}C from ^{12}C). AMS makes possible the detection of naturally occurring, long-lived radioisotopes such as ^{10}Be, ^{36}Cl, ^{26}Al and ^{14}C with typical isotopic abundancies ranging from 10^{-12} to 10^{-18}.

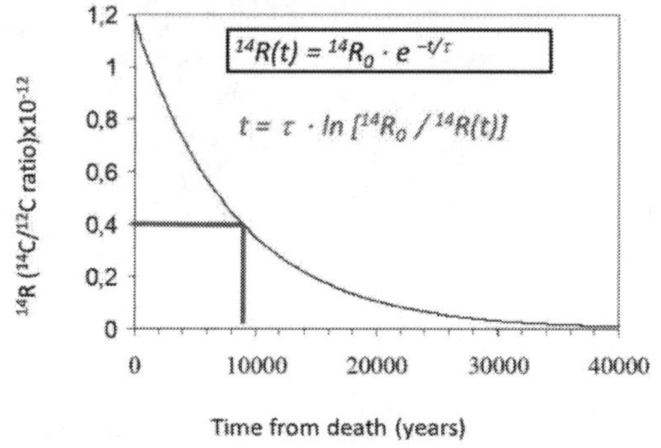

Fig. 3. ^{14}C is used as a natural chronometer to date organic materials, those systems that have been directly (or indirectly) in equilibrium with the atmosphere, acquiring ^{14}C from it, and that have ceased that equilibrium condition at a well-defined moment.

A few years ago, the ^{14}C dating method was brought to public prominence thanks to the dating of the shroud of Turin. The shroud of Turin is a linen cloth that tradition associates with the crucifixion and burial of Jesus. It has undergone numerous scientific tests, the most notable of which is radioactive dating. In 1988, scientists at three separate laboratories dated samples from the Shroud to a range between 1260 and 1390 AD, which coincides with the first certain appearance of the shroud in the 1350s much later than the burial of Jesus in 30 or 33 AD.

The ^{14}C dating method is not just applied to old (or very old!) objects. Indeed, nuclear test explosions performed after the Second World War injected a large amount of neutrons into the atmosphere that induced a huge increase in atmospheric ^{14}C concentration. The effect was so considerable that after 1955, in less than ten years radiocarbon concentration almost doubled, reaching its maximum value in 1963–1965 [13]. This effect helped to date a supposed Fernand Léger painting (Contraste de Formes) allegedly painted in 1913-14. It was bought by Peggy Guggenheim in the late 1960s but never displayed to the public because of early suspicions that it was a forgery. From the data shown in Fig. 4 it appears obvious that the painting is a fake [14, 15]. This method is also used to date great Bordeaux wines.

The accuracy and precision of radiocarbon (^{14}C) dating depend on several factors affecting the main steps of the dating process, from the selection of the samples to their preparation and measurement, involving a systematic evaluation of anomalies potentially introduced during each stage of the process and attempting to minimize the uncertainties. Archeometrical methods are at the crossroads of several scientific fields. Supported by sophisticated technologies, they refine our vision of world Cultural Heritage. This results in renewed interest in analytical pipelines, data management and preservation plans for cultural remains.

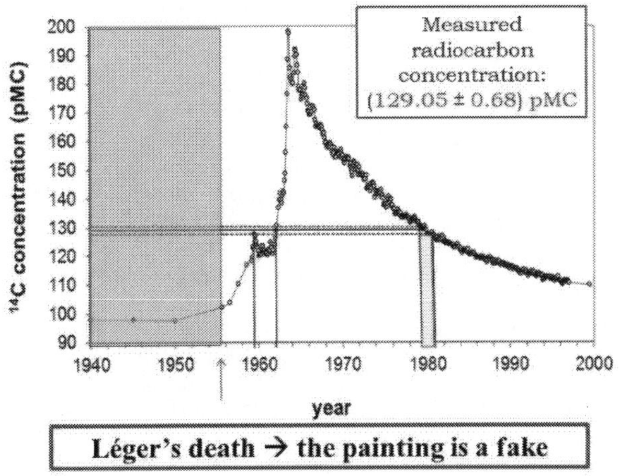

Fig. 4. The black curve represents the ^{14}C concentration increase until 1963-1965 due to nuclear test explosions. From the ^{14}C concentration, it appears obvious that the canvas was produced using cotton plants harvested in 1959, 1962, or 1979-80, I.e. after Fernand Léger's death on August 17, 1955 in Gif-sur-Yvette.

3 The worldwide landscape concerning emerging accelerator facilities

Several accelerator facilities are under construction in the world for satisfying the needs of fundamental research, neutrons production or demands associated to societal applications.

Fundamental research: The international scientific landscape concerning the production of radioactive beams has been profoundly enriched with the commissioning of new facilities such as FAIR (Facility for Antiproton and Ion Research) in Germany, FRIB (Facility for Rare Isotope Beams) at Michigan State University or ARIEL (Advanced Rare Isotope Laboratory) at TRIUMF. FAIR is aiming to accelerate a large variety of particle beams of a previously unparalleled intensity and quality, as well as antiprotons. FRIB is a truly next generation projectile-fragmentation facility with a 200MeVA 400kW LINAG followed by a fragmentation separator. Few months ago, April 16th, FRIB reached an important milestone with first acceleration of beam in ReA6. A $^{14}N^{6+}$ beam was accelerated to 10.2 MeV/u using the new ReA6 beta=0.085 cryomodule with all cavities at full field. ARIEL at TRIUMF with the use of a super-conducting electron LINAG, a new 30MeV electron linear accelerator, is aiming to obtain ~10^{14} fission/s. Other countries have decided to build their own radioactive beam facilities such as RISP (Radioactive Isotope Science Project) in KOREA, Beijing ISOL (Isotope Separation On Line) in CHINA or DERICA (Dubna Electron-Radioactive Ion Collider fAcility) at DUBNA in RUSSIA. A new project is also under consideration at GANIL France. Furthermore, intermediate scale facilities as the HIE ISOLDE at CERN a natural extension of the present facility by the addition of a 10MeVA post-accelerator has been constructed or SPES at INFN-Legnaro a national 8kW-ISOL is under construction.

Compact accelerators for neutron production: Since several decades, nuclear reactor based neutrons sources, have proven their benefit in supporting science, industry, agriculture and health. However, the neutrons landscape worldwide and in particular in Europe will change profoundly over the next few decades [16]. The ORPHEE reactor in France was recently shut down and others like the BER2 in Germany, REZ in the Czech Republic, BRR in Hungary or Democritus in Greece and eventually ILL in Grenoble are expected to be shut down in the very near future. Therefore, two-thirds of currently operating neutron sources in Europe will close within the next few decades. Even if an increase in the capacity of spallation sources will take place with the increase of ISIS capacities and the commissioning of ESS, this situation is excessively penalizing for the 6000 users of neutrons in Europe. One hope for coping with this worldwide shortage of neutron sources comes from a possible massive deployment of Compact Accelerator Neutron Sources (CANS) [17]. What is it all about? A CANS is accelerating a very high intensity beam of protons or deuterons, typically from 5 to 100mA, at energies between a few MeV, typically from 3MeV to 5MeV up to ~40MeV. After acceleration, the beam impinges on a dedicated neutron production target, which may be surrounded by a moderator for providing epithermal and thermal neutron beams. Low energy CANS, using deuteron and proton beams of energy between 3MeV and 5MeV are attractive because they are relatively inexpensive, (~10M€) and can be installed and operated relatively easily, for instance in University environments. This type of CANS can be used for a variety of applications such as radiography, radioelements production for nuclear medicine; etc. but also for educational purposes and less demanding in terms of neutron flux experiments (see the JCANS network in Japan [18]). Their construction uses techniques perfectly mastered nowadays, i.e. a proton or deuteron source of ~5mA up to 100mA and a radiofrequency quadrupole (RFQ) to reach energies between 3MeV and 5MeV. Nevertheless, the blocking point for this type of development is the neutron production target. For proton energies below 3MeV, lithium should be preferred due to the low stripping reaction threshold and a higher neutron production cross section. For proton energies higher than 3MeV, beryllium and lithium are roughly equivalent in terms of neutron yield [19].

Demands associated to societal applications: Supporting the knowledge of artworks past times and civilizations by nuclear physics technics is a very valuable intellectual contribution to the historical and archaeological studies. Every important museum, like the Acropolis museum, has to dispose or to be associated with an accelerator facility and eventually participate in collaboration with Nuclear Physics Laboratories to the construction of experimental setups aiming to perform in situ measurements. This is for instance the goal of MACHINA, the first transportable proton accelerator facility based on RFQ cavities, jointly developed by INFN and CERN and specifically designed to perform in situ cultural heritage studies [13].

4 Outlook

Recent developments in accelerator technologies have led several countries to consider the possibility of building medium-size facilities with the objective of promoting for large number of disciplines national research. An example of this type of initiatives is the SARAF project in Israel. SARAF will be a user facility for basic and applied nuclear physics. The research plans for this facility span several disciplines: Precision studies beyond the Standard-Model. Extended nuclear astrophysics research with high-energy neutrons. Studies of exotic neutron-rich isotopes relevant to the r-process. Nuclear structure of exotic isotopes. High-energy neutron cross section measurements for basic nuclear physics and material science research. Neutron based imaging, therapy, novel radiopharmaceuticals development, and production [20].

A SARAF type facility, eventually financed by European funds, could be also consider for Greece. I am convinced that many European countries will be eager to participle in its contraction. I would like to remind that the prize of such a facility is a fraction, between 25 and 50%, of the prize of a football player like the French player Kylian Mbappé. Into this very favourable context, the Hellenic nuclear physics community could contribute by nominating a scientific and technical group aiming to design and propose a national project. This project, if accepted by the national authorities, should boost not only nuclear physics but also the ensemble of the scientific fields using its techniques. It would also help to create a multicultural centre and increase the number of young scientists in Greece.

References

1. Nicolas Alamanos, "An introduction to the school "Rewriting Nuclear Physics Textbooks: one-step forward" and future perspectives", Eur. Phys. J. Plus 135, 417 (2020).
2. A. Fontana and L. Canton, "Nuclear Physics applied to the production of Innovative Radio-Pharmaceuticals", Eur. Phys. J. Plus (to be published).
3. R. Van Noorden, The medical testing crisis, Nature 504, 202.
4. P. Schaffer et al., "Direct Production of 99mTc via ^{100}Mo(p,2n) on Small Medical Cyclotrons", 23rd International Conference on the Application of Accelerators in Research and Industry (CAARI).
5. J. Tanguay et al., "A fast and simple dose-calibrator-based quality control test for the radionuclide purity of cyclotron-produced 99mTc", Phys. Med. Biol. 60, 8229–8247 (2015) and "Cyclotron based production of 99mTc", IAEA report.
6. Press Releases, May 11th, 2021, https://www.northstarnm.com/northstar-medical-radioisotopes-receives-electron-beam-accelerators-for-first-of-its-kind-advanced-medical-radioisotope-production/
7. H. Tsujii, Overview of Carbon-ion Radiotherapy, IOP Conf. Series: Journal of Physics: Conf. Series 777, 012032 (2017).
8. D. Kramer, Carbon-ion cancer therapy shows promise, Physics Today 68, 6, 24 (2015).
9. D. Margarone, G.A.P. Cirrone et al., "ELIMAIA: A Laser-Driven Ion Accelerator for Multidisciplinary Applications", Quantum Beam Sci. 2, 8 (2018).
10. Y. Iwata et al., Superconducting gantry and other developments at HIMAC, Proceedings of PAC2013, Pasadena, CA USA.
11. D. Le Bihan, « Le cerveau de cristal : Ce que nous révèle la neuro-imagerie ». Odile Jacob. Paris, 2012.
12. L. Quettier et al., IEEE Transactions on applied superconductivity, Vol. 28, No. 3, April 2018.
13. M. Fedi, "How a small accelerator can be useful for interdisciplinary applications part II: cultural heritage studies", Eur. Phys. J. Plus 136, 411 (2021)
14. L. Caforio, et al, Eur. Phys. J. Plus 129, 6 (2014)
15. M. Fedi et al, Nuclear Instruments and Meth. B 294, 662 (2013).
16. ESFRI Physical Sciences and Engineering Strategy Working Group Neutron Landscape Group. ISBN: 978-88-901562-5-0
17. I.S. Anderson, C. Andreani, J.M. Carpenter, G. Festa, G. Gorini, C.-K. Loong, and R. Senesi, « Research opportunities with compact accelerator-driven neutron sources», Physics Reports 654, 1–58 (2016).

https://doi.org/10.1051/epjconf/202125201001

18. http://www.jcans.net/
19. O. Sgouros, V. Soukeras and A. Pakou, "Low energy proton induced reactions with weakly bound nuclei for application purposes", Eur. Phys. J. A, 57, 125 (2021).
20. I. Mardor et al., Eur. Phys. J. A 54, 91 (2018)

Relativistic Brueckner-Hartree-Fock Theory in Infinite Nuclear Matter

Peter Ring[1,2,*], *Sibo* Wang[2], *Qiang* Zhao[2], and *Jie* Meng[2]

[1]Fakultät für Physik, Technische Universität München, D-85747 Garching, Germany
[2]State Key Laboratory of Nuclear Physics and Technology, School of Physics, Peking University, Beijing 100871, China

Abstract. On the way of a microscopic derivation of covariant density functionals, the first complete solution of the relativistic Brueckner-Hartree-Fock (RBHF) equations is presented for symmetric nuclear matter. In most of the earlier investigations, the G-matrix is calculated only in the space of positive energy solutions. On the other side, for the solution of the relativistic Hartree-Fock (RHF) equations, also the elements of this matrix connecting positive and negative energy solutions are required. So far, in the literature, these matrix elements are derived in various approximations. We discuss solutions of the Thompson equation for the full Dirac space and compare the resulting equation of state with those of earlier attempts in this direction.

1 Introduction

In recent years, microscopic theories of the nuclear many-body problem, starting with the bare nucleon-nucleon forces, showed considerable progress in describing the properties of light and even specific medium-heavy nuclei [1–6]. However, the largest part of the nuclear chart is still only accessible in energy density functional (EDF) theories [7] which are based on the mean-field concept. In Coulombic systems, where density functional theory has been originally introduced by Kohn and Sham [8, 9] such functionals can, nowadays, be derived, in a fully microscopic way, from the bare Coulomb force [10], but in nuclei the situation is much more complex. First, the extreme strength of the bare nucleon-nucleon force with a strongly repulsive core at short distances and an attractive Yukawa tail caused by pion exchange forbids any mean-field approximation. On the other side, the existence of shell effects [11, 12] and the success of the phenomenological mean-field models [13] show that nuclei can be well described within an averaged field. Therefore, already in the fifties the concept of an effective force has been introduced by Brueckner [14, 15]: nucleons in the medium do not feel the strong bare force. The Pauli principle forbids the scattering into occupied orbits and reduces this force. Taking this effect into account in the scattering theory, Brueckner replaced the Lippman-Schwinger equation [16] for the free scattering by the Bethe-Goldstone equation [17] for the scattering in the nuclear medium and introduced the G-matrix that depends on the density. In the fifties, neither were the bare nucleon forces known well enough, nor the computer facilities were strong enough to solve the corresponding self-consistent equations successfully. In the sixties and seventies, when the bare forces were better known and when computers became available to solve the Brueckner-Hartree-Fock (BHF) equations, it turned out that this concept failed in describing the proper saturation properties of nuclear matter [18]. On the other side, nuclear density functionals with phenomenological, effective, density-dependent, Skyrme [19] and Gogny [20] interactions turned out to be very successful. Covariant density functional theories (CDFTs) [21–26] are of particular interest because Lorentz invariance is one of the underlying symmetries of QCD. This symmetry not only allows to describe the spin-orbit coupling, which has an essential influence on the underlying single-particle structure, but it also puts stringent restrictions on the number of parameters in the corresponding functionals without reducing the quality of the agreement with experimental data [27].

Brueckner theory is only an approximation, but other methods, for instance variational methods, also failed [28]. In this situation, two new concepts have been introduced in the following years: (a) The concept of deriving the properties of finite nuclei only from bare two-nucleon forces has been given up and three-body forces have been introduced in various microscopic theories [29–31]. (b) Lorentz invariance was taken into account. This leads to relativistic Brueckner theory. In fact, the Brooklyn group [32] could show that an approximate solution of the relativistic Brueckner-Hartree-Fock equations can reproduce the saturation properties of nuclear matter rather well. Over the years many groups worked in this direction [33–41] and solved the relativistic BHF problem in various approximations.

In this contribution we discuss the relativistic Brueckner-Hartree-Fock problem in detail and show results of new full solutions, where the positive energy states (PES) and the negative energy states (NES) of the relativistic HF problem are fully taken into account. We also compare with approximate treatments of the NES.

*e-mail: peter.ring@tum.de

2 Relativistic Brueckner-Hartree-Fock Theory

The full solution of the Brueckner-Hartree-Fock equations is an iterative process. For each density characterized by the Fermi momentum k_F an effective interaction, the G-matrix, is determined by the solution of the Bethe-Goldstone equation [17]:

$$\hat{G}(W) = \hat{V} + \hat{V}\frac{\hat{Q}}{W - \hat{H}_{HF}}\hat{G}(W). \tag{1}$$

Here W is the starting energy. $\hat{Q}$ is the Pauli operator, which allows only scattering processes into intermediate states above the Fermi surface in the nuclear system, and $\hat{H}_{HF}$ is the Hartree-Fock single particle operator for the motion of the two particles in the intermediate states.

$$\hat{H}_{HF} = \hat{T} + \hat{U}, \tag{2}$$

where the the matrix elements of the single-particle potential are given by

$$\langle n|\hat{U}|n'\rangle = \sum_{k<k_F}\langle nk|\hat{G}|n'k\rangle. \tag{3}$$

The index k runs over all the states in the Fermi sea below the Fermi level.

In the relativistic case, the free scattering process is usually described by the Thompson equation [42], one of the relativistic three-dimensional reductions of the Bethe-Salpeter equation [43]. In particular, this equation has been used for the adjustment of the relativistic bare nucleon-nucleon potential Bonn A [44] to the experimental phase shifts for scattering processes of two particles with positive energies.

In RBHF theory, i.e. for the scattering process in infinite nuclear matter, the Thompson equation has the form

$$G(\boldsymbol{q}', \boldsymbol{q}|\boldsymbol{P}, W) = V(\boldsymbol{q}', \boldsymbol{q}|\boldsymbol{P}) + \int \frac{d^3k}{(2\pi)^3} V(\boldsymbol{q}', \boldsymbol{k}|\boldsymbol{P}) \frac{Q(\boldsymbol{k}, \boldsymbol{P})}{W - E_{\boldsymbol{P}+\boldsymbol{k}} - E_{\boldsymbol{P}-\boldsymbol{k}}} G(\boldsymbol{k}, \boldsymbol{q}|\boldsymbol{P}, W), \tag{4}$$

where $\boldsymbol{P} = \frac{1}{2}(\boldsymbol{k}_1 + \boldsymbol{k}_2)$ is the center-of-mass (c.m.) and $\boldsymbol{k} = \frac{1}{2}(\boldsymbol{k}_1 - \boldsymbol{k}_2)$ is the relative momentum of the two nucleons in nuclear matter with momenta $\boldsymbol{k}_1$ and $\boldsymbol{k}_2$. $\boldsymbol{q}, \boldsymbol{k}$, and $\boldsymbol{q}'$ are the initial, intermediate, and final relative momenta. W is the energy of the two scattered particles, usually denoted as *starting energy*. The Pauli operator Q allows only the scattering to unoccupied states above the Fermi momentum, i.e. $Q(\boldsymbol{k}, \boldsymbol{P}) = 1$ for $|\boldsymbol{k}_1| = |\boldsymbol{P} + \boldsymbol{k}| > k_F$ and $|\boldsymbol{k}_2| = |\boldsymbol{P} - \boldsymbol{k}| > k_F$ and it vanishes for all other cases.

$E_{\boldsymbol{P}\pm\boldsymbol{k}}$ are the single-particle energies for nucleons with momenta $\boldsymbol{P} \pm \boldsymbol{k}$, i.e. the eigenvalues of the Dirac equation in the nuclear medium:

$$[\boldsymbol{\alpha} \cdot \boldsymbol{p} + \beta M + \mathcal{U}(\boldsymbol{p})]\,\psi(\boldsymbol{p}, s) = E_{\boldsymbol{p}}\psi(\boldsymbol{p}, s). \tag{5}$$

Here $\psi(\boldsymbol{p}, s)$ is a Dirac-spinor and $\mathcal{U}$ is the relativistic single particle potential

$$\langle a|\mathcal{U}|b\rangle = \sum_c \langle ac|\hat{G}(W)|bc\rangle. \tag{6}$$

where $|a\rangle$, $|b\rangle$, and $|c\rangle$ are eigenstates of the Dirac operator (5) and the index c runs over all the states in the Fermi sea, in analogy to the no-sea approximation of Ref. [21].

Due to symmetries, the single-particle potential $\mathcal{U}$ in the rest frame of nuclear matter can be decomposed as

$$\mathcal{U}(\boldsymbol{p}) = \beta U_S(p) + U_0(p) + \boldsymbol{\alpha} \cdot \hat{\boldsymbol{p}} U_V(p), \tag{7}$$

with the scalar potential U_S and the Lorentz vector potentials $(U_0, \hat{\boldsymbol{p}}U_V)$.

As in the non-relativistic case, we have a self-consistent problem that is solved by iteration. The relativistic Hartree-Fock equations are based on an effective (density-dependent) two-body interaction $\hat{G}$ in Eq.(5) which is given by the solution of the Thompson equation (4) depending in turn on the eigenvalues E_p of the Dirac operator (5).

Besides the problem of the starting energy W in Eq. (6), which occurs also in the non-relativistic case (see for instance Refs. [6, 45]), there are eigenstates of the Dirac operator in Eq. (5) with positive energies (PES) and those with negative energies (NES) and it is evident that one needs for the solution of the relativistic Hartree-Fock problem in Eqs. (5) and (6), in each step of the iteration, not only the matrix elements $\mathcal{U}^{++}$ of the potential (6) with positive energies, but also the matrix elements $\mathcal{U}^{+-}$ and $\mathcal{U}^{--}$ with negative energies, i.e. the matrix elements of the G-matrix in the full Dirac space. This means, one has to solve the Thompson equation (4) in full Dirac space too.

The computer code for the solution of the Thompson equation is rather complicated. It has been developed in the helicity scheme [46, 47] for the relativistic scattering process of particles with positive energy to determine the relativistic matrix elements of the bare nucleon-nucleon potential from the scattering phase shifts [44], and it has been used in the literature also in the medium for the calculation of the G-matrix elements for states with positive energy: G^{++++}. The solution of Eq. (4) leads immediately to the matrix elements $\mathcal{U}^{++}$ of the Dirac potential. In most of the applications of RBHF theory, the remaining matrix elements $\mathcal{U}^{+-}$ and $\mathcal{U}^{--}$ have been derived in various approximations. As examples we mention the Green's function method of Weigel and collaborators [34, 48], the momentum-independence approximation [35] and the projection methods [33, 37]. Solutions including PESs and NESs have been discussed in Refs. [36, 41, 49]

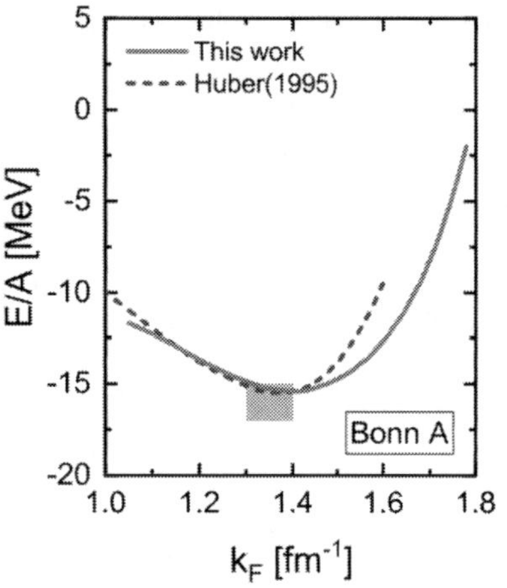

Figure 1. Binding energy per nucleon E/A in symmetric nuclear matter as a function of the Fermi momentum k_F. Treatment with NESs (red) is compared with an approximate treatment of Ref. [48](dashed blue).

Today we report on the full solution of the Thompson equation (4) in full Dirac space for symmetric nuclear matter using the relativistic potential Bonn A [44]. This leads to coupled integral equations for the various channels G^{++++}, G^{+-++}, G^{+-+-}, etc. Details can be found in Ref. [50]. An additional problem occurs through the fact that the Thompson equation (4) is usually solved in the c.m. frame of the scattering process. On the other side the RHF-problem in Eqs.(5) and (6) is solved in the rest frame of nuclear matter. To avoid this problem, we also solved the Thompson equation in the rest frame of nuclear matter using approximations discussed in detail in Ref. [50]. In our calculations, we used for the starting energy W the procedure discussed in Ref. [45], which means that we consider the matrix elements for negative energy states in Eq. (6) in the same way as for occupied states.

3 Results

In Fig. 1, we show the binding energy per nucleon E/A for symmetric nuclear matter as a function of the Fermi momentum k_F. The red line is obtained by RBHF calculations in the full Dirac space with the relativistic potential Bonn A [44]. The shaded area indicates the empirical saturation region of symmetric nuclear matter. We find that the nuclear matter saturation point is reasonably described, much better than in a non-relativistic calculation with Bonn A [35], where the saturation energy is 23.55 MeV and the saturation density corresponds to k_F = 1.85 fm^{-1} (see Table 1). The blue line shows the results obtained by the relativistic Green's function approach in the full Dirac space [48]. They agree with our results below the saturation density. The discrepancy above the saturation density is found mainly arising from the different schemes for the starting energies W in Eq. (6): in this work the NESs are treated as occupied, whereas in Ref. [48] they were treated as unoccupied states.

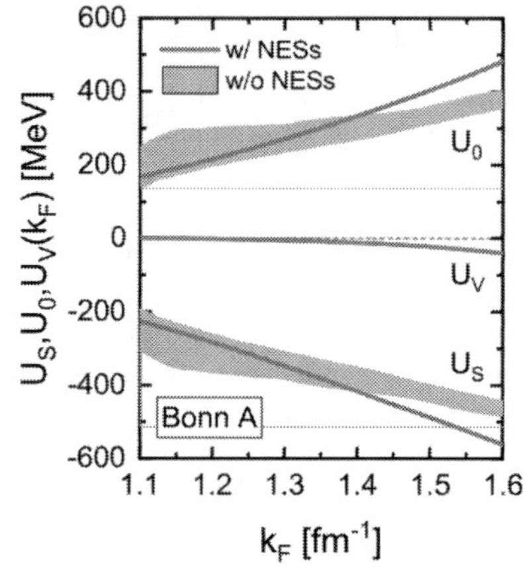

Figure 2. The different contributions to the single particle potential in Eq. (7). Results in full Dirac space (red) are compared with those, where the NES are treated approximately (shaded gray). Figures taken from Ref. [50]

Fig. 2 shows the different contributions of the relativistic potential in Eq. (7) as a function of the Fermi momentum k_F. The calculations in the full Dirac space (red) are compared with results of the momentum-independence approximation of Ref. [35]. In this approximation, U_V is neglected, and the momentum-independent potentials U_S and U_0 are extracted from the values of the single-particle potential energy at two different momenta. The results of this procedure depend on the values of these two momenta and the corresponding uncertainties are indicated by the shaded areas in Fig. 2. Although these uncertainties are reduced with the increasing density, the strengths of U_S and U_0 are both underestimated above the saturation density, as compared to the results obtained in the full Dirac space. As it is known that the total binding energy depends on the strong cancelation of these two quantities, it is important to solve the RBHF problem in the full Dirac space.

In Table 1 we show the saturation properties of symmetric nuclear matter calculated by RBHF theory in the full Dirac space using the interactions Bonn A, B and C [44]: the saturation density ρ_0, the binding energy per nucleon E/A, the compression modulus K_∞, and the Dirac mass M_D^*/M at saturation density. These quantities are compared with the corresponding values from the non-relativistic BHF calculations and the phenomenological covariant density functionals NL3 [51], DD-ME2 [52], DD-PC1 [53], PC-PK1 [54] and PKO1 [55] and empirical values for ρ_0, E/A [35].

In Fig. 3, we show the single-particle potentials $U_S(p)$, $U_0(p)$, and $U_V(p)$ of Eq. (7) at saturation density k_F = 1.34 fm^{-3} as a function of the momentum p. The momentum-independence approximation seems to be well justified. The space-like component $U_V(p)$ is also found

https://doi.org/10.1051/epjconf/202125202001

Table 1. Saturation properties of symmetric nuclear matter. See text for details.

Potential	ρ_0 [fm^{-3}]	E/A [MeV]	K_∞ [MeV]	M_D^*/M
RBHF Bonn A	0.188	-15.40	258	0.55
RBHF Bonn B	0.164	-13.36	206	0.61
RBHF Bonn C	0.144	-12.09	150	0.65
BHF Bonn A	0.428	-23.55	204	
BHF Bonn B	0.309	-18.30	160	
BHF Bonn C	0.247	-15.75	103	
NL3	0.148	-16.30	272	0.60
DD-ME2	0.152	-16.14	251	0.57
DD-PC1	0.152	-16.06	230	0.58
PC-PK1	0.154	-16.12	238	0.59
PKO1	0.152	-16.00	250	0.59
Empirical	0.16±0.01	-16±1	240±20	

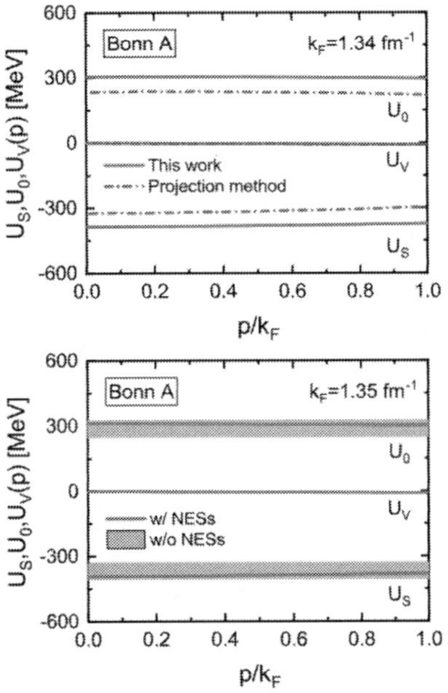

Figure 3. Momentum dependence of the single particle potential. Treatment with NESs (red) is compared with two approximations: the projection method of Ref. [37] (upper part) and the momentum-independence approximation of Ref. [35] (lower part). The uncertainties of this approximation are shown by the gray shaded area. Figures taken from Ref. [50].

to be rather small. In the upper panel, the calculations in full Dirac space (red) are compared with the projection method of Ref. [37] (blue dashed-dotted). In this method, the G matrix elements are projected onto a complete set of five Lorentz invariant amplitudes [33], and the single-particle potentials are calculated analytically. However, the choice of these Lorentz invariant amplitudes is not unique. Different schemes of projections have been used [33, 37, 56, 57] which differ mainly in the effect of the pseudoscalar meson exchange. In the lower panel the calculations in full Dirac space (red) are compared with the momentum-independence approximation and its uncertainties (gray shaded areas).

4 Conclusions

The RBHF equations have been solved for symmetric nuclear matter in the full Dirac space for potential Bonn A. In this way, one avoids all the uncertainties of previous RBHF calculations without the negative energy states. The Thompson equation is chosen as the scattering equation, and the matrix elements of the Bonn potential are treated in the c.m. frame. The saturation properties of symmetric nuclear matter obtained in this way are in good agreement with the empirical values. They agree with the results based on the relativistic Green's function approach up to the saturation density. The discrepancy above the saturation density is found mainly arising from the different schemes for the starting energies. Since there is no clear-cut prescription for the starting energy in the Brueckner-Hartree-Fock method, one has to go beyond the Brueckner approximation for such cases, i.e., one has to include more than two hole lines in the approximation.

An analysis of the uncertainties of the RBHF calculation in the Dirac space with PESs only shows that these uncertainties can reach more than 100 MeV for the single-particle potentials. It also turns out that the equation of state is less bound above the saturation density. These analyses demonstrate the significance of the RBHF calculations in the full Dirac space. Therefore further investigations for asymmetric nuclear matter with the RBHF calculations in the full Dirac space are necessary to clarify the isospin dependence of the single-particle potentials.

5 Acknowledgements

This work was supported in part by the National Key R&D Program of China (No. 2017YFE0116700 and No. 2018YFA0404400), the National Natural Science Foundation of China (No. 11935003, No. 11975031, No. 11875075, and No. 12070131001), and the Deutsche Forschungsgemeinschaft (DFG, German Research Foundation) under Germany's Excellence Strategy EXC-2094-390783311, ORIGINS. Part of this work was achieved by using the High-performance Computing Platform of Peking University, and the supercomputer OCTOPUS at the Cybermedia Center, Osaka University under the support of Research Center for Nuclear Physics of Osaka University.

References

[1] M. Baldo, C. Maieron, J. Phys. G **34**, R243 (2007)
[2] B.R. Barrett, P. Navratil, J.P. Vary, Prog. Part. Nucl. Phys. **69**, 131 (2013)
[3] G. Hagen, et al., Rep. Prog. Phys. **77**, 096302 (2014)
[4] J. Carlson, et al., Rev. Mod. Phys. **87**, 1067 (2015)
[5] H. Hergert, S. Bogner, T. Morris, A. Schwenk, K. Tsukiyama, Phys. Rep. **621**, 165 (2016)
[6] S. Shen, H. Liang, W. Long, J. Meng, P. Ring, Prog. Part. Nucl. Phys. **109**, 103713 (2019)
[7] M. Bender, P.H. Heenen, P.G. Reinhard, Rev. Mod. Phys. **75**, 121 (2003)
[8] W. Kohn, L.J. Sham, Phys. Rev. **137**, A1697 (1965)
[9] W. Kohn, L.J. Sham, Phys. Rev. **140**, A1133 (1965)
[10] J.P. Perdew, S. Kurth, Lecture Notes in Physics (Springer Verlag Heidelberg) **620**, 1 (2003)
[11] O. Haxel, J.H.D. Jensen, H.E. Suess, Phys. Rev. **75**, 1766 (1949)
[12] M. Goeppert-Mayer, Phys. Rev. **78**, 16 (1950)
[13] S.G. Nilsson, Mat. Fys. Medd. Dan. Vid. Selsk. **29**, No. 16 (1955)
[14] K.A. Brueckner, Phys. Rev. **97**, 1353 (1955)
[15] B.D. Day, Rev. Mod. Phys. **39**, 719 (1967)
[16] B.A. Lippmann, J. Schwinger, Phys. Rev. **79**, 469 (1950)
[17] H.A. Bethe, J. Goldstone, Proc. Roy. Soc. A **238**, 551 (1957)
[18] F. Coester, S. Cohen, B. Day, C.M. Vincent, Phys. Rev. C **1**, 769 (1970)
[19] D. Vautherin, D.M. Brink, Phys. Rev. C **5**, 626 (1972)
[20] J. Dechargé, D. Gogny, Phys. Rev. C **21**, 1568 (1980)
[21] J.D. Walecka, Ann. Phys. (NY) **83**, 491 (1974)
[22] J. Boguta, A.R. Bodmer, Nucl. Phys. A **292**, 413 (1977)
[23] B.D. Serot, J.D. Walecka, Adv. Nucl. Phys. **16**, 1 (1986)
[24] P.G. Reinhard, Rep. Prog. Phys. **52**, 439 (1989)
[25] P. Ring, Prog. Part. Nucl. Phys. **37**, 193 (1996)
[26] J. Meng, H. Toki, S.G. Zhou, S.Q. Zhang, W.H. Long, L.S. Geng, Prog. Part. Nucl. Phys. **57**, 470 (2006)
[27] J. Meng Ed., *Relativistic Density Functional for Nuclear Structure*, Int. Rev. Mod. Phys. **10**, (2016)
[28] V.R. Pandharipande, R.B. Wiringa, Rev. Mod. Phys. **51**, 821 (1979)
[29] J. Fujita, H. Miyazawa, Prog. Theor. Phys. **17**, 360 (1957)
[30] W. Zuo, A. Lejeune, U. Lombardo, J. Mathiot, Nucl. Phys. A **706**, 418 (2002)
[31] H.W. Hammer, A. Nogga, A. Schwenk, Rev. Mod. Phys. **85**, 197 (2013)
[32] M.R. Anastasio, L.S. Celenza, C.M. Shakin, Phys. Rev. Lett. **45**, 2096 (1980)
[33] C.J. Horowitz, B.D. Serot, Nucl. Phys. A **464**, 613 (1987)
[34] P. Poschenrieder, M.K. Weigel, Phys. Rev. C **38**, 471 (1988)
[35] R. Brockmann, R. Machleidt, Phys. Rev. C **42**, 1965 (1990)
[36] F. de Jong, H. Lenske, Phys. Rev. C **58**, 890 (1998)
[37] T. Gross-Boelting, C. Fuchs, A. Faessler, Nucl. Phys. A **648**, 105 (1999)
[38] E. Schiller, H. Müther, Eur. Phys. J. A **11**, 15 (2001)
[39] Z. Ma, L. Liu, Phys. Rev. C **66**, 024321 (2002), 0202721
[40] C. Fuchs, Lecture Notes in Physics (Springer Verlag Heidelberg) **641**, 111 (2004)
[41] T. Katayama, K. Saito, Phys. Lett. B **747**, 43 (2015)
[42] R.H. Thompson, Phys. Rev. D **1**, 110 (1970)
[43] E.E. Salpeter, H.A. Bethe, Phys. Rev. **84**, 1232 (1951)
[44] R. Machleidt, Adv. Nucl. Phys. **19**, 189 (1989)
[45] S. Shen, H. Liang, J. Meng, P. Ring, S. Zhang, Phys. Lett. B **781**, 227 (2018)
[46] K. Erkelenz, R. Alzetta, K. Holinde, Nucl. Phys. A **176**, 413 (1971)
[47] K. Erkelenz, Phys. Rep. **13**, 191 (1974)
[48] H. Huber, F. Weber, M.K. Weigel, Phys. Rev. C **51**, 1790 (1995)
[49] T. Katayama, K. Saito, arXiv[nucl-th]:1410.7166 (2014)
[50] S. Wang, Q. Zhao, P. Ring, J. Meng, Phys. Rev. C **103**, 054319 (2021)
[51] G.A. Lalazissis, J. König, P. Ring, Phys. Rev. C **55**, 540 (1997)
[52] G.A. Lalazissis, T. Nikšić, D. Vretenar, P. Ring, Phys. Rev. C **71**, 024312 (2005)
[53] T. Nikšić, D. Vretenar, P. Ring, Phys. Rev. C **78**, 034318 (2008)
[54] P.W. Zhao, Z.P. Li, J.M. Yao, J. Meng, Phys. Rev. C **82**, 054319 (2010)
[55] W.H. Long, N. Van Giai, J. Meng, Phys. Lett. B **40**, 150 (2006)
[56] L. Sehn, C. Fuchs, A. Faessler, Phys. Rev. C **56**, 216 (1997)
[57] C. Fuchs, T. Waindzoch, A. Faessler, D.S. Kosov, Phys. Rev. C **58**, 2022 (1998)

EPJ Web of Conferences **252**, 02002 (2021)
HINPw6
https://doi.org/10.1051/epjconf/202125202002

Nuclear magnetic transitions in the relativistic energy density functional approach

Nils Paar[1,*], *Goran* Kružić[2], and *Tomohiro* Oishi[1]

[1]Department of Physics, Faculty of Science, University of Zagreb, Bijenička c. 32, HR-10000, Zagreb, Croatia
[2]Research department, Ericsson - Nikola Tesla, Krapinska 45, HR - 10000, Zagreb, Croatia

Abstract. Recently a novel theory framework has been established for description of magnetic dipole (M1) transitions in finite nuclei, based on relativistic nuclear energy density functional with point coupling interactions. The properties of M1 transitions have been studied, including the sum rules, spin, orbital, isoscalar and isovector M1 transition strengths in magic and open shell nuclei. It is shown that pairing correlations and spin-orbit interaction plays an important role in the description of M1 transition strength distributions. The analysis of the evolution of M1 transition properties in the isotope chain $^{100-140}$Sn shows the interplay between single and double-peak structures, determined by the evolution of single-particle states, their occupations governed by the pairing correlations, and two-quasiparticle transitions involved. Comparison of the calculated B(M1) transition strength with recent data from inelastic proton scattering on $^{112-124}$Sn, shows that quenching of the g factors g_{eff}/g_{free} =0.80-0.93 is required to reproduce the experimental data. Further experimental investigations are needed to determine accurately the quenching factor.

1 Introduction

Magnetic dipole (M1) transitions represent fundamental excitation phenomena in finite nuclei. Because of their relevance for various nuclear properties and applications, M1 transitions have been extensively studied both theoretically and experimentally [1–6]. The isovector spin-flip M1 response is relevant for the understanding of single-particle properties in nuclei, spin-orbit interaction, and shell closures from stable nuclei toward limits of stability [7–9]. It is also relevant for applications related to the design of nuclear reactors [10]. The M1 transitions are also important for resolving the problem of quenching of the spin-isospin response in nuclei that is necessary for reliable description of double beta decay matrix elements [11]. The M1 spin-flip excitations can also be used as a probe for inelastic neutrino-nucleus cross sections [12], and they are relevant in modelling the r-process nucleosynthesis [6, 13]. In deformed nuclei, M1 scissors mode has been extensively studied, where the orbital part of M1 operator plays a dominant role, with protons and neutrons oscillating with opposite phase around the core [2, 5, 14, 15]. Experimental studies of M1 transitions are rather challenging, due to simultaneous presence of electric dipole (E1) and electric quadrupole (E2) transitions, which dominate over the M1 response [2, 16–19].

Various theoretical approaches, mainly in the non-relativistic framework, have been employed in studies of M1 transitions - for more details see Refs. [4, 9, 14, 20] and references therein. While Skyrme functionals successfully reproduce electric excitations, there are difficulties to describe magnetic transitions, and further improvements in the the spin channel are required [20]. A study of M1 strength functions based on D1M Gogny force showed that the available experimental data could be reproduced if the calculated strength is shifted globally by about 2 MeV and increased by an empirical factor of 2 [6]. Clearly, further theoretical studies of M1 transitions are required to resolve open questions on their properties. To respond to some of the open challenges, recently a novel theory framework has been established for studies of M1 excitations, based on the relativistic nuclear energy density functional [21, 22]. This framework has been employed in several studies, including the analysis of basic M1 properties, their isotopic dependence and relationship with spin-orbit interaction, the role of pairing properties, constraining the quenching of the g factors for the free nucleons, and others. One of the important advantages to employ the relativistic framework is the natural description of the spin-orbit splitting, which plays an essential role in the M1 transitions. Here we give only a brief overview of recent studies, for more details see Refs.[21–24].

2 Magnetic dipole transitions in the relativistic quasiparticle random phase approximation

The relativistic quasiparticle random phase approximation (RQRPA) for description of M1 transition strength, based on relativistic point coupling interactions, was introduced in Ref.[21]. The nuclear ground state properties are described using the relativistic Hartree-Bogoliubov (RHB)

*e-mail: npaar@phy.hr

model with DD-PC1 interaction, that provides single-particle states and wave functions obtained in the canonical basis for the implementation in the RQRPA [21]. In addition to the point coupling interaction DD-PC1 used in the RHB model [21], for the description of unnatural parity excitations of the M1 type ($J^\pi = 1^+$), the RQRPA residual interaction is supplemented with the relativistic isovector-pseudovector contact interaction,

$$\mathcal{L}_{IV-PV} = -\frac{1}{2}\alpha_{IV-PV}[\bar{\Psi}_N\gamma^5\gamma^\mu\vec{\tau}\Psi_N]\cdot[\bar{\Psi}_N\gamma^5\gamma_\mu\vec{\tau}\Psi_N]. \quad (1)$$

Since this term does not contribute to the ground state due to parity arguments, its strength parameter α_{IV-PV} is constrained by the experimental data on M1 excitation energies of ^{48}Ca and ^{208}Pb [21]. By solving the RQRPA equations in the matrix formulation, the M1 transition strengths are obtained and further analyzed. The M1 sum rule for core-plus-two-nucleon systems [25] has been used to validate the model calculations. The spin, orbital, isoscalar and isovector M1 transition strengths have been studied in detail in magic nuclei ^{48}Ca and ^{208}Pb, and open shell nuclei ^{42}Ca and ^{50}Ti [21]. In these systems, the isovector spin-flip M1 transition is dominant, mainly between one or two spin-orbit partner states. As an illustrative example, in Fig.1 the M1 transition strength distributions are shown for ^{208}Pb, including the full response R(M1), spin response $R^{\sigma}_{M1}(E)$, and orbital response $R^{\ell}_{M1}(E)$ functions. For comparison, the unperturbed response without the RQRPA residual interaction (Hartree response) is also shown. The spin M1 response dominates in the transition strength, and the residual RQRPA interaction clearly separates the M1 response from the unperturbed one. Two main peaks are identified in the spectra, the low one at 6.11 MeV, with the major contribution from the transitions between spin-orbit partner states for neutrons, $(\nu 1i^{-1}_{13/2} \rightarrow \nu 1i_{11/2})$ and protons, $(\pi 1h^{-1}_{11/2} \rightarrow \pi 1h_{9/2})$, that interfere destructively. The second state at 7.51 MeV is composed of coherent transitions $(\nu 1i^{-1}_{13/2} \rightarrow \nu 1i_{11/2})$ and $(\pi 1h^{-1}_{11/2} \rightarrow \pi 1h_{9/2})$ (for more details see Ref.[21]).

In further studies of open shell nuclei, it is shown that pairing correlations have a significant impact on the centroid energy and major peak position of the M1 mode [21, 22]. From the analysis of M1 mode in the Ca isotope chain, the role of the isovector-pseudovector (IV-PV) residual interaction has been discussed [22]. It has been shown that the experimental data on M1 mode could improve and optimize the theoretical aspects to describe the residual interactions in the RQRPA. In Ref. [24] a relation between the M1 excitation of nuclei and the pairing modes used in model calculations has been explored. For nucleons, two modes of the Cooper-pair coupling exist, namely of $S_{12} = 0$ with $L_{12} = 0$ (spin-singlet s-wave) and $S_{12} = 1$ with $L_{12} = 1$ (spin-triplet p-wave). It has been shown that the model calculation of M1 transitions, combined with accurate experimental data, allow to discern the role of the spin singlet and spin triplet pairing correlations [24].

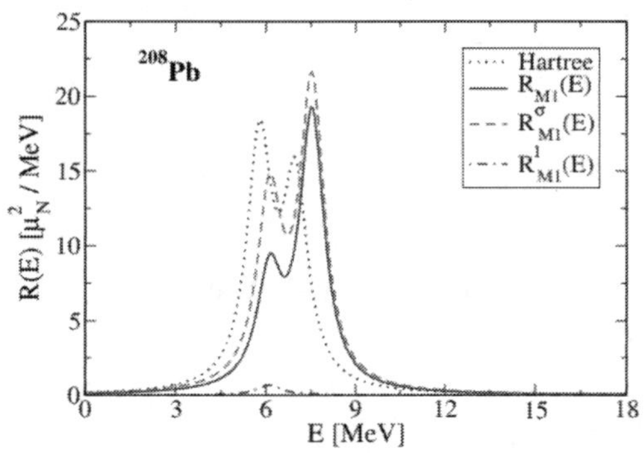

Figure 1. The M1 transition strength distributions for ^{208}Pb, including full response $R_{M1}(E)$, spin response $R^{\sigma}_{M1}(E)$, orbital response $R^{\ell}_{M1}(E)$, and unperturbed (Hartree) response functions.

3 Isotopic dependence of the M1 transition strength in $^{100-140}$Sn and quenching of the g-factors

Recent experimental study using inelastic proton scattering on Sn isotopes provided novel data on E1 and M1 strength distributions along the even-even $^{112-124}$Sn nuclei [26]. Since the resulting photoabsorption cross sections derived from the E1 and M1 strength distributions showed significant differences when compared to those from previous (γ, xn) experiments [27, 28], it is interesting to explore how the new experimental data compare to the model calculations based on the relativistic nuclear energy density functional. Resolving the properties of M1 modes of excitation using new experimental data is also essential for our understanding of the role of M1 transitions in modeling radiative neutron capture cross sections of relevance for nucleosynthesis. Therefore, in Ref. [23] the M1 transitions have been investigated in even-even $^{100-140}$Sn isotopes using the RQRPA. It has been shown that the M1 transition strength distribution is characterized by an interplay between single and double-peak structures, that can be understood from the evolution of single-particle states, their occupations governed by the pairing correlations, and two-quasiparticle transitions involved [23].

Figure 2 shows the evolution of transition strengths for two dominant M1 peaks along Sn isotope chain. Starting from ^{102}Sn, the strength of the low- (high-) energy peak is increasing (decreasing), until its maximal (minimal) value is obtained at ^{116}Sn. Further, the strength continues decreasing (increasing) until doubly magic ^{132}Sn, where the trend reverses again. From the analysis of the relevant transitions, one can understand the evolution of the strengths of two main peaks. The low-energy state is characterizd by destructive interference between dominating nearly constant proton contributions and smaller neutron contributions that display variation along the isotope chain. The evolution of high-energy state is a result of dominating neutron contributions followed by the same behavior of smaller proton contributions. Different structures of the low- and high-energy M1 state govern different

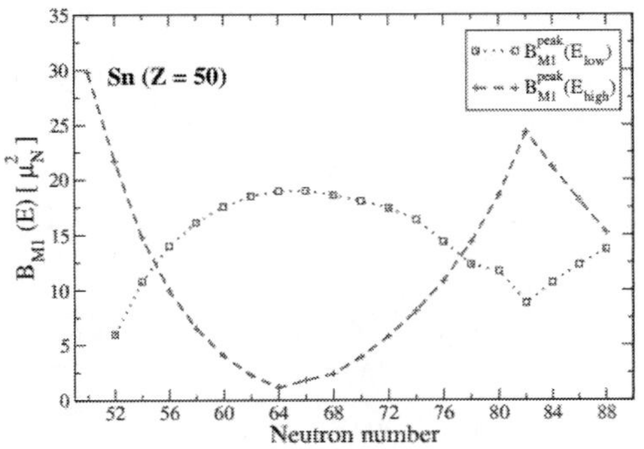

Figure 2. The M1 transition strengths for the two main peaks in even-even $^{100-140}$Sn isotopes, where E_{low} and E_{high} denote energies of the low- and high-energy states.

evolution of their strengths along the Sn isotope chain. The M1 excitations are composed from transitions between the spin-orbit (SO) partner states. Therefore, it is interesting to explore the relationship between the SO energy splittings and R(Q)RPA excitation energies along the Sn isotope chain, as shown in Fig. 3. The M1 excitations induced by spin-flip transitions hold orbital quantum numbers unchanged. As shown in Fig. 3, the lower M1 state is very close to the proton SO splitting energy $(1g_{9/2})$-$(1g_{7/2})$, and the residual R(Q)RPA interaction only slightly reduces the M1 excitation energies with respect to the relevant SO splittings. However, in the case of second M1 state, the respective transitions are dominated by $\nu(1g_{9/2} - 1g_{7/2})$ configuration until ^{116}Sn and $\nu(1h_{11/2} - 1h_{9/2})$ configuration in heavier Sn isotopes. We note that their R(Q)RPA-excitation energies are different than the SO splitting energies of neutrons, thus showing the relevance of the IV-PV and pairing residual interactions in the RQRPA.

In the following the overall calculated B(M1) strength is compared to the recent experimental data from inelastic proton scattering [26]. The RQRPA summation of the B(M1) strength amounts 22.81, 22.61, 22.56, 22.76, 23.34, and 25.55 μ_N^2 for even-even isotopes $^{112-124}$Sn, respectively. The corresponding experimental values amount 14.7(1.4), 19.6 (1.9), 15.6 (1.3), 18.4 (2.4), 15.4(1.4), and 19.1(1.7) μ_N^2, respectively. From the comparison of the RQRPA results with the experimental data, the effective g factor can be obtained, that is needed to reproduce the experimental B(M1) transition strength. In this way one can determine the quenching factor as the ratio of the effective g factor including quenching (g_{eff}) with respect to the one for free nucleons (g_{free}). With the assumption that the general quenching applies to all gyromagnetic factors in the M1 transition operator, the quenching factor is obtained, g_{eff}/g_{free}=0.80-0.93, that is higher value than previously reported, $g_{eff}/g_{free} \approx$ 0.6-0.75 [2, 29–31]. Therefore, smaller quenching of the g_{free} factor is needed to reproduce recent inelastic proton scattering data on M1 transitions. Since part of the B(M1) strength above the neutron threshold may be missing in the experiment, the present results show that the actual quenching of the g factors in nuclei may be rather small in comparison to the ones for the free nucleons. Additional experimental studies are required to support this conclusion.

4 Conclusion

Recently established theory framework for description of M1 transitions, based on the relativistic nuclear energy density functional, provides the insight into spin, orbital, isoscalar and isovector M1 transition strengths in magic and open shell nuclei. Since in open shell nuclei pairing correlations are essential to reproduce the M1 transition strength distributions, this mode could be used as an additional constraint for the nuclear pairing interactions. The evolution of M1 transition properties in $^{100-140}$Sn isotopes demonstrate interesting interplay between single and double-peak structures, determined by the evolution of single-particle states, their occupations governed by the pairing correlations, and involvement of various two-quasiparticle configurations. Because M1 excitations originate from transitions between spin-orbit partner states, their properties also represent useful constraint for the spin-orbit interaction channel. From the comparison with the inelastic proton scattering data for $^{112-124}$Sn, it is shown that quenching in the g factors needed to reproduce the experimental data amounts g_{eff}/g_{free}=0.80-0.93, that is larger value than known from previous studies, $g_{eff}/g_{free} \approx$ 0.6-0.75. Since some measured B(M1) strength may be missing around the neutron threshold, further investigations are required to determine the quenching factor accurately.

Acknowledgments

This work is supported by the "QuantiXLie Centre of Excellence", a project co-financed by the Croatian Government and European Union through the European Regional Development Fund, the Competitiveness and Cohesion Operational Programme (KK.01.1.1.01).

https://doi.org/10.1051/epjconf/202125202002

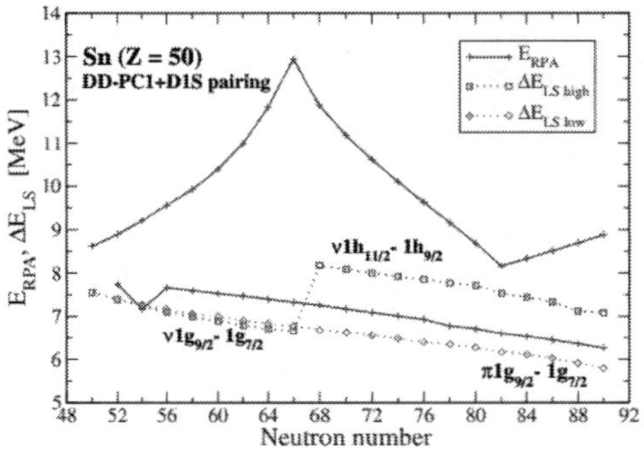

Figure 3. The M1-excitation energies in Sn isotopes and the corresponding SO-splitting energies ΔE_{LS} ((nlj) quantum numbers are denoted for the SO-partner levels). $\Delta E_{LS_{low}}$ and $\Delta E_{LS_{high}}$ represent the SO splitting contributions at lower and higher RPA energies, respectively.

References

[1] A. Richter, Progress in Particle and Nuclear Physics **34**, 261 (1995)

[2] K. Heyde, P. von Neumann-Cosel, A. Richter, Rev. Mod. Phys. **82**, 2365 (2010)

[3] N. Pietralla, P. von Brentano, A. Lisetskiy, Progress in Particle and Nuclear Physics **60**, 225 (2008)

[4] V.O. Nesterenko, J. Kvasil, P. Vesely, W. Kleinig, P.G. Reinhard, V.Y. Ponomarev, Journal of Physics G: Nuclear and Particle Physics **37**, 064034 (2010)

[5] D.P. Arteaga, P. Ring, Phys. Rev. C **77**, 034317 (2008)

[6] S. Goriely, S. Hilaire, S. Péru, M. Martini, I. Deloncle, F. Lechaftois, Phys. Rev. C **94**, 044306 (2016)

[7] T. Otsuka, T. Suzuki, M. Honma, Y. Utsuno, N. Tsunoda, K. Tsukiyama, M. Hjorth-Jensen, Phys. Rev. Lett. **104**, 012501 (2010)

[8] P. Vesely, J. Kvasil, V.O. Nesterenko, W. Kleinig, P.G. Reinhard, V.Y. Ponomarev, Phys. Rev. C **80**, 031302 (2009)

[9] V. Tselyaev, N. Lyutorovich, J. Speth, P.G. Reinhard, D. Smirnov, Phys. Rev. C **99**, 064329 (2019)

[10] M.B. Chadwick, *et. al.*, Nuclear Data Sheets **112**, 2887 (2011)

[11] J.D. Vergados, H. Ejiri, F. Šimkovic, Reports on Progress in Physics **75**, 106301 (2012)

[12] K. Langanke, G. Martínez-Pinedo, P. von Neumann-Cosel, A. Richter, Phys. Rev. Lett. **93**, 202501 (2004), and references therein.

[13] H.P. Loens, K. Langanke, G. Martinez-Pinedo, K. Sieja, The European Physical Journal A **48**, 34 (2012)

[14] K. Heyde, P. von Neumann-Cosel, A. Richter, Rev. Mod. Phys. **82**, 2365 (2010)

[15] J.P. Ebran, E. Khan, D. Peña Arteaga, D. Vretenar, Phys. Rev. C **83**, 064323 (2011)

[16] A. Tamii, T. Adachi, J. Carter, M. Dozono, H. Fujita, Y. Fujita, K. Hatanaka, H. Hashimoto, T. Kaneda, M. Itoh et al., Nuclear Physics A **788**, 53 (2007), proceedings of the 2nd International Conference on Collective Motion in Nuclei under Extreme Conditions

[17] Y. Fujita, B. Rubio, W. Gelletly, Prog. Part. Nucl. Phys. **66**, 549 (2011)

[18] J. Birkhan, H. Matsubara, P. von Neumann-Cosel, N. Pietralla, V.Y. Ponomarev, A. Richter, A. Tamii, J. Wambach, Phys. Rev. C **93**, 041302 (2016)

[19] R.M. Laszewski, R.J. Holt, H.E. Jackson, Phys. Rev. Lett. **38**, 813 (1977)

[20] J. Speth, P.G. Reinhard, V. Tselyaev, N. Lyutorovich, arXiv:2001.07236 (2020)

[21] G. Kružić, T. Oishi, D. Vale, N. Paar, Phys. Rev. C **102**, 044315 (2020)

[22] T. Oishi, G. Kružić, N. Paar, Journal of Physics G: Nuclear and Particle Physics **47**, 115106 (2020)

[23] G. Kružić, T. Oishi, N. Paar, Phys. Rev. C **103**, 054306 (2021)

[24] T. Oishi, G. Kružić, N. Paar, Eur. Phys. J. A **57**, 180 (2021)

[25] T. Oishi, N. Paar, Phys. Rev. C **100**, 024308 (2019)

[26] S. Bassauer, P. von Neumann-Cosel, P.G. Reinhard, A. Tamii, S. Adachi, C.A. Bertulani, P.Y. Chan, A. D'Alessio, H. Fujioka, H. Fujita et al., Phys. Rev. C **102**, 034327 (2020)

[27] R. Alarcon, R.M. Laszewski, D.S. Dale, Phys. Rev. C **40**, R1097 (1989)

[28] K. Govaert, F. Bauwens, J. Bryssinck, D. De Frenne, E. Jacobs, W. Mondelaers, L. Govor, V.Y. Ponomarev, Phys. Rev. C **57**, 2229 (1998)

[29] P. Vesely, J. Kvasil, V.O. Nesterenko, W. Kleinig, P.G. Reinhard, V.Y. Ponomarev, Phys. Rev. C **80**, 031302(R) (2009)

[30] V.O. Nesterenko, J. Kvasil, P. Vesely, W. Kleinig, P.G. Reinhard, V.Y. Ponomarev, Journal of Physics G: Nuclear and Particle Physics **37**, 064034 (2010)

[31] M. Ichimura, H. Sakai, T. Wakasa, Progress in Particle and Nuclear Physics **56**, 446 (2006)

Shape and electromagnetic properties of the ^{229m}Th isomer

Nikolay Minkov[1,2,*] and *Adriana* Pálffy[2,3]

[1]Institute of Nuclear Research and Nuclear Energy, Bulgarian Academy of Sciences, Tzarigrad Road 72, BG-1784 Sofia, Bulgaria
[2]Max-Planck-Institut für Kernphysik, Saupfercheckweg 1, D-69117 Heidelberg, Germany
[3]Department of Physics, Friedrich-Alexander-Universität Erlangen-Nürnberg, D-91058 Erlangen, Germany

Abstract. We examine the physical conditions, and specifically the role of the quadrupole-octupole deformation, for the emergence of the 8 eV "clock" isomer ^{229m}Th. Our nuclear structure model suggests that such an extremely low-energy state can be the result of a very fine interplay between the shape and single-particle (s.p.) dynamics in the nucleus. We find that the isomer can only appear in a rather limited region of quadrupole-octupole deformation space close to a line along which the ground-state and isomer s.p. orbitals 5/2[633] and 3/2[631], respectively, cross each other providing the isomer-formation quasi-degeneracy condition. The crucial role of the octupole deformation in the formation mechanism is pointed out. Our calculations within the outlined deformation region show a smooth behaviour of the ^{229}Th electromagnetic properties, including the isomer decay rate, allowing for their more precise theoretical determination.

1 Introduction

The ^{229m}Th nuclear isomer attracts much interest related to its exceptionally low energy of approximately 8 eV which appears to be accessible for the contemporary vacuum ultraviolet (VUV) lasers [1] and thus opens the possibility for a number of valuable applications such as the establishing of a new frequency standard, "nuclear clock" [2–4], development of nuclear lasers in the optical range [5] and others [6, 7]. The achieving of these goals, however, requires very precise knowledge of the isomer energy, lifetime and the related nuclear electromagnetic characteristics. Following this motivation several recent experimental studies have been focused on the isomer decay modes and provided estimates about its life time in charged [8] and neutral [9] electronic states. Further, the magnetic dipole moment μ in the isomeric state (IS) was determined for the first time in laser spectroscopy experiments providing the value of $\mu_{\rm IS} = -0.37(6)\mu_N$ [10, 11]. Then, three very recent experiments provided newly updated values for the isomer energy, $E_{IS} = 8.28(17)$ eV [12] from internal conversion electron spectroscopy, $E_{IS} = 8.30(92)$ eV [13] by determining the transition rates and energies from the above level at 29.2 keV and $E_{IS} = 8.10(17)$ eV [14] from a micro-calorimetric determination of absolute γ-ray energy differences.

Yet the direct nuclear decay transition of the ^{229m}Th isomer is not observed experimentally. This imposes the need of detailed theoretical examination and prediction of its nuclear electromagnetic characteristics. The problem is additionally complicated due to the rich deformation properties of the actinide nuclei to which the nucleus ^{229}Th belongs. Trying to address the above aspects recently we initiated a theoretical study [15], exploring a nuclear model approach which incorporates the shape-dynamic properties together with the intrinsic structure characteristics typical for the actinide region where the quadrupole-octupole (QO) deformation manifests. The formalism includes a description of the collective QO vibration-rotation motion coupled to the motion of the single (odd) nucleon within a reflection-asymmetric deformed potential with pairing correlations and fully microscopic treatment of the Coriolis interaction. The approach allows one to determine the energy and radiative decay properties of the ^{229m}Th isomer as a part of the entire low-lying positive- and negative-parity spectrum and transition probabilities observed in the nucleus ^{229}Th. On this basis, we have shown that the extremely small isomer energy can be explained as the consequence of a very fine interplay between the rotation-vibration degrees of freedom and the motion of the unpaired neutron. The model calculations predict for the reduced probability $B(M1)$ for magnetic decay of the isomer a value between 0.006 and 0.008 W.u. which is considerably smaller than earlier deduced values of 0.048 W.u. [16, 17] and 0.014 W.u. [18]. This result provides the currently adopted estimate for the ^{229m}Th nuclear life-time of about 10^4 sec explaining the experimental difficulties to observe the direct radiative decay of the isomer [19, 20] and suggests a new finer accuracy target for further measurements. In a subsequent work [21] we have calculated the magnetic moment of the isomer, $\mu_{\rm IS}$ and of the ground state (GS), $\mu_{\rm GS}$, by taking into account attenuation effects in the spin and collective gyromagnetic factors, without changing the model parameters originally adjusted in Ref. [15]. The result for $\mu_{\rm IS}$ in the range from $\mu_{\rm IS} = -0.25\mu_N$ to $-0.35\mu_N$ is in rather good agreement with the recent experimental values $-0.37(6)$

*e-mail: nminkov@inrne.bas.bg

[10] and (−0.3)–(−0.4) [11]. On the other hand, μ_{GS} was obtained in the range $\mu_{GS} = 0.53 - 0.66\,\mu_N$, overestimating the latest reported and older experimental values of $0.360(7)\,\mu_N$ [22] and $\mu_{GS} = 0.45\mu_N$ [23], respectively, and being in agreement with an earlier theoretical prediction $\mu_{GS} = 0.54\mu_N$ based on the modified Woods-Saxon potential [24].

The aforementioned theoretical studies pointed out the complexity of the ^{229m}Th problem emphasizing on the need of consistent description of all related observables, detailed examination of the physical conditions for the forming of the isomer and further constraining of the arbitrariness in the model description to ensure maximal reliability of the obtained predictions as a tool in the establishing of the new nuclear clock standard. To this end in our very recent work [25] we investigated the shape deformation conditions under which such a low-energy isomer can appear in ^{229m}Th and examined the behaviour of the model solution in the relevant space of quadrupole and octupole deformations clarifying the limits of reliability of our model predictions. In this work we colligate the results of our approach obtained so far trying to outline the frames of the general prescription for the search and investigation of nuclear excitations on the border between nuclear and atomic physics.

In Sec. 2 we briefly present the model formalism. In Sec. 3 we give numerical results and illustrate the main points in the consistent study of ^{229m}Th shape characteristics and their influence on the isomer electromagnetic properties. In Sec. 4 concluding remarks are given.

2 The model approach

The model Hamiltonian is taken in the form [15]

$$H = H_{\text{s.p.}} + H_{\text{pair}} + H_{\text{qo}} + H_{\text{Coriol}}\,, \qquad (1)$$

where $H_{\text{s.p.}}$ is the s.p. Hamiltonian of Deformed Shell Model (DSM) with a Woods-Saxon (WS) potential including axial quadrupole (β_2) and octupole (β_3) deformation parameters [26], which provides the s.p. energies E^K_{sp} with given value of the projection K of the total and s.p. angular momentum operators $\hat{I}$ and $\hat{j}$, respectively on the intrinsic symmetry axis; H_{pair} is the standard Bardeen-Cooper-Schrieffer (BCS) pairing Hamiltonian [27]. This DSM+BCS part provides the quasi-particle (q.p.) spectrum ϵ^K_{qp} as shown in Ref. [28]. H_{qo} represents oscillations of the even–even core with respect to quadrupole and octupole axial deformation variables mixed through a centrifugal (rotation-vibration) interaction [29, 30]; H_{Coriol} involves the Coriolis interaction between the even-even core and the unpaired nucleon (see Eq. (3) in [30]). It is treated as a perturbation with respect to the remaining part of (1) and then incorporated into the QO potential of H_{qo} defined for given angular momentum I, parity π and s.p. band-head projection K_b [15, 31].

The spectrum of Hamiltonian (1) is obtained in the form of coherent (with the same frequency) QO vibrations and rotations built on a q.p. state with $K = K_b$ and parity π^b [15, 31]

$$E^{\text{tot}}_{nk}(I^\pi, K_b) = \epsilon^{K_b}_{\text{qp}} + \hbar\omega\left[2n + 1 + \sqrt{k^2 + b\widetilde{X}(I^\pi, K_b)}\right], \qquad (2)$$

where b is a model inertia parameter; $n = 0, 1, 2, \ldots$ and $k = 1, 2, 3, \ldots$ are radial and angular QO oscillation quantum numbers, respectively, with k odd (even) for the even (odd) parity states of the core [32]; ω is the frequency of the coherent QO mode (CQOM) [29, 30] and $\widetilde{X}(I^\pi, K_b)$ determines the centrifugal term in which the Coriolis mixing is taken into account [25]. The levels of the total QO core plus particle system, determined by a particular n and $k^{(+)}$ ($k^{(-)}$) for the states with given $I^{\pi=+}$ ($I^{\pi=-}$) form a split (quasi) parity doublet [33].

The Coriolis perturbed wave function corresponding to Hamiltonian (1) with the spectrum (2) is obtained in the form

$$\widetilde{\Psi}^{\pi,\pi^b}_{nkIMK_b} = \frac{1}{\widetilde{N}_{I\pi K_b}}\left[\Psi^{\pi,\pi^b}_{nkIMK_b} + A \sum_{\substack{\nu\neq b \\ (K_\nu = K_b\pm 1, \frac{1}{2})}} C^{I\pi}_{K_\nu K_b}\Psi^{\pi,\pi^b}_{nkIMK_\nu}\right], \qquad (3)$$

in which the unperturbed core plus particle wave function $\Psi^{\pi,\pi^b}_{nkIMK_b}$ involving the projection K_b is mixed with the functions based on other s.p. states. In the case of non-zero octupole deformation $\beta = 0$ in the WS potential the s.p. states are characterized by average (mixed) parity and the corresponding s.p. wave functions are taken projected on the experimentally determined band-head parity π^b (for details see Eqs. (20)–(23) in [25]).

Having the wave function (3) we obtain the reduced $B(E1)$, $B(E2)$, $B(E3)$ and $B(M1)$ transition probabilities for the energy spectrum (2) [15] as well as the magnetic moments in the ground and isomeric states, μ_{GS} and μ_{IS}, respectively [21] (for details see Ref. [25]). Because of the Coriolis mixing the states with different K_b values appear connected through electromagnetic transitions which otherwise would be suppressed due to the axial symmetry of the system. This determines the model mechanism allowing for the M1 and E2 radiative decay of the $K^\pi = 3/2^+$ IS to the $K^\pi = 5/2^+$ GS.

3 Transition rates and magnetic moments in ^{229}Th

In Ref. [15] we applied the above DSM-CQOM approach to the low-lying part of the experimental ^{229}Th spectrum [34], as shown in Fig. 1 of [15]. The theoretical spectrum was obtained in the form of an yrast quasi parity-doublet (QPD) built on the 5/2[633] ground-state s.p. orbital and non-yrast QPD built on the 3/2[631] orbital corresponding to the $3/2^+$ isomeric state. The model parameters, β_2 and β_3 entering DSM, the BCS pairing parameters, the collective CQOM parameters and the Coriolis mixing strength were determined so that both states $5/2^+$ GS and the isomeric $3/2^+$ were obtained as a quasi-degenerate pair, showing that the latter can be easily set in the range

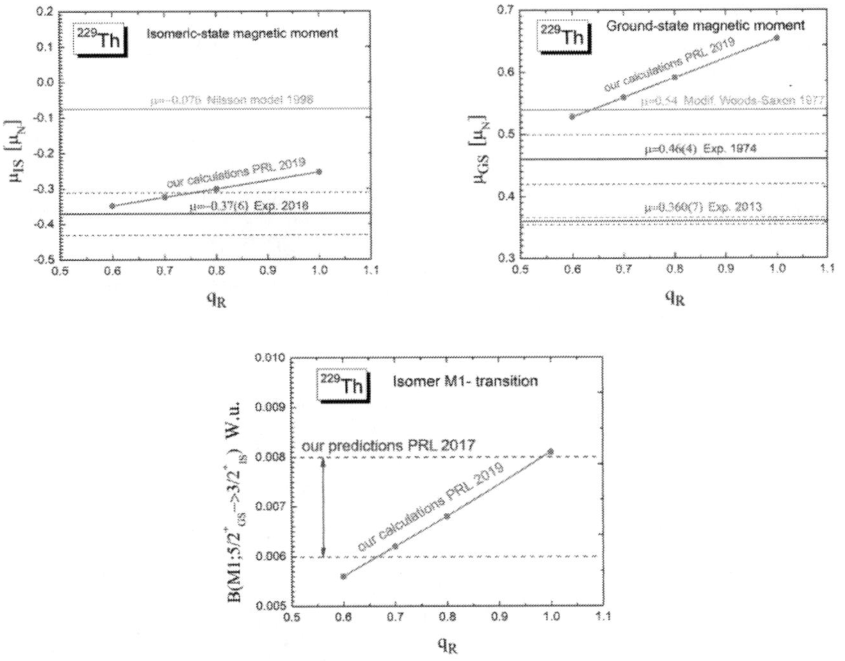

Figure 1. The calculated IS and GS magnetic moments and the isomer B(M1) decay rate in dependence on the quenching factor q_R. The dashed lines mark the experimental error bars for the magnetic moments and theoretical-prediction limits for B(M1). See the text for explanations.

of the experimental one through a very fine tuning of parameters with the excited QPD levels and the available experimental B(E2) and B(M1) values [35] being reproduced reasonably well. On this basis we have made predictions for the M1 and E2 decay probabilities for the $3/2^+$ isomer state given in Table I of [15].

In Ref. [21] we calculated the magnetic moments $\mu_{\rm GS}$ and $\mu_{\rm IS}$ of ^{229}Th corresponding to the parameters of the model description in [15] without further adjustment by only taking the commonly used quenching $q_s = 0.6$ of the spin gyromagnetic factor g_s and by considering several values for the quenching $q_R = 1, 0.8, 0.7, 0.6$ of the collective gyromagnetic factor g_R based on earlier studies of ^{229}Th [36, 37]. Here we illustrate this result in Fig. 1. We see that for $\mu_{\rm IS}$ (upper left plot) the theoretical predictions well approach the experimental value of $-0.37(6)\,\mu_N$ [10] firmly staying inside the uncertainty bars for $q_R = 0.7$ and 0.6. For $\mu_{\rm GS}$ we see (upper right plot) that although the theoretical values obtained with decreasing q_R approach the experimental values $0.360(7)\,\mu_N$ [22] and $\mu_{\rm GS} = 0.45\mu_N$ [23] they still remain outside of the uncertainty bars. At the same time the lower plot of Fig. 1 illustrates that the isomer B(M1) transition rate decreases with q_R. All these results show that more detailed consideration of the magnetic moments together with other observables in ^{229}Th needs to be done in order to assess the limits of reliability of our approach in determining the electromagnetic properties of the 8 eV isomer.

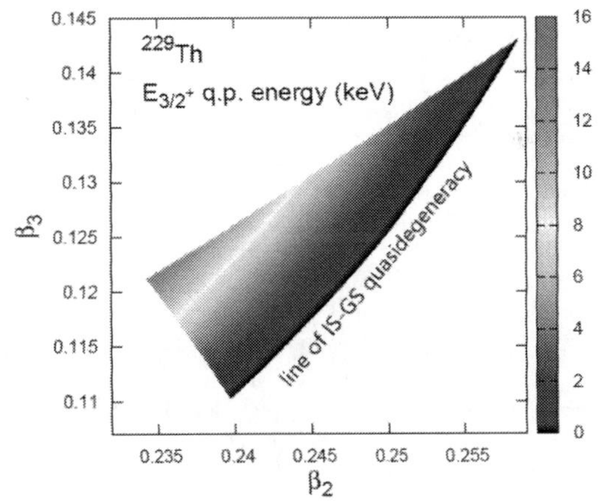

Figure 2. Q.p. energy (in keV) of the $3/2^+$ isomer orbital with respect to the $5/2^+$ GS orbital appearing in DSM within the model-defined space of QO deformations.

Thus, in Ref. [25] we implemented such a complex study of the deformation and electromagnetic properties of the ^{229m}Th isomer. First we implemented DSM calculations over a net in the β_2 and β_3 deformation space outlining the region where the GS and IS s.p. orbitals appear

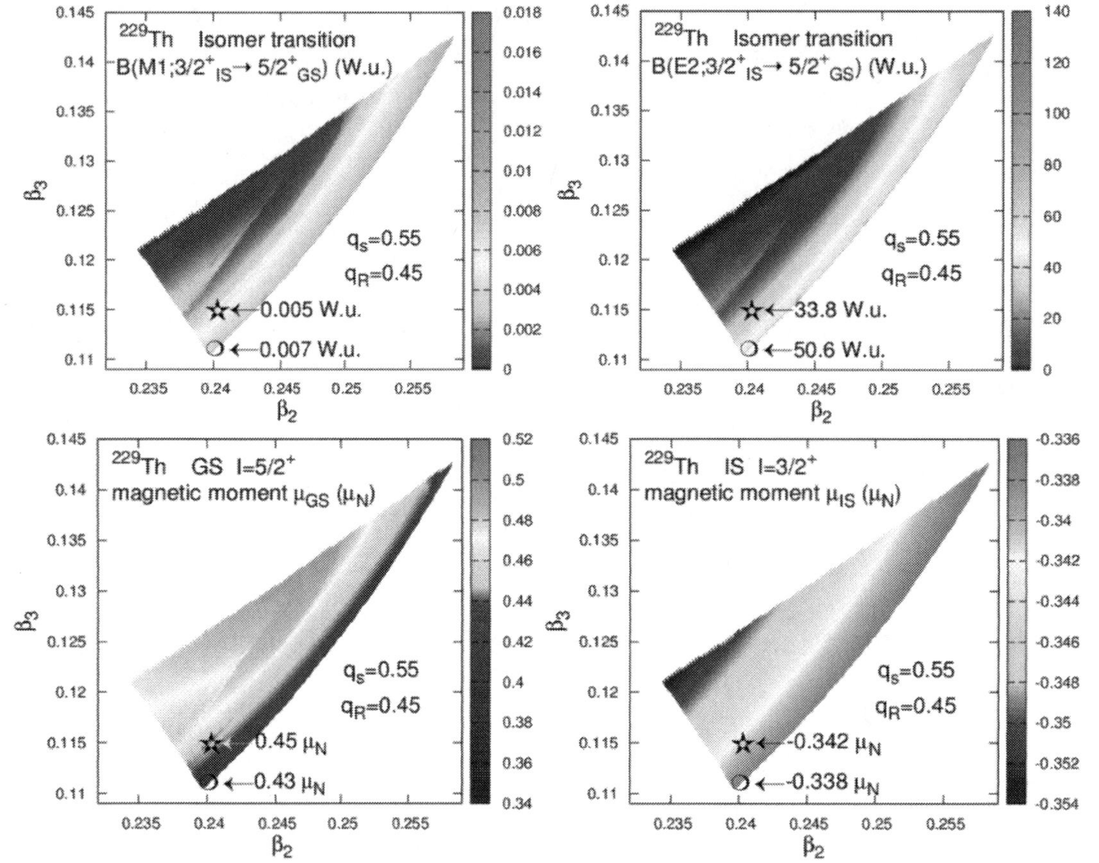

Figure 3. B($M1$), B($E2$) isomer transition rates and $\mu_{\rm GS}$ and $\mu_{\rm IS}$ values obtained by the model fits on a grid within the model-defined QO deformation space at $q_s = 0.55$ and $q_R = 0.45$. The open star indicates the location of the deformations (β_2,β_3)=(0.240, 0.115) adopted in Ref. [15]. The circle indicates the set (β_2,β_3)=(0.240, 0.111) situated closer to the degeneracy line. Adapted from [25].

with the correct K values and average parity. This region is illustrated in Fig. 2 where the q.p. energy corresponding to the $3/2^+$ isomer orbital with respect to the $5/2^+$ GS orbital is given as a function of β_2 and β_3. We can point out two important observations: i) the QO region allowing the correct orbitals configuration essentially includes non-zero octupole deformation; ii) there is a well determined line of IS-GS orbital degeneracy along which one may expect the isomer formation. This result already reveals the crucial role of the deformation for the appearance and the overall properties of the $3/2^+$ isomer. Then, as a second step, we performed massive model DSM-CQOM fits at each point in the deformation grid considering energies, transition rates and magnetic moments all together. This allowed us to examine the behaviour of the obtained model description for all observables of interest within the QO deformation space. In Fig. 3 we illustrate the result obtained for the isomeric B(M1) and B(E2) transition rates and the magnetic moments $\mu_{\rm GS}$ and $\mu_{\rm IS}$ with the quenching factors $q_s = 0.55$ and $q_R = 0.45$. We see that all these quantities exhibit a smooth behaviour in dependence on β_2 and β_3 with the physically interesting theoretical values appearing towards the IS-GS degeneracy line. We also remark that towards this line the model fairly well approaches the experimental isomer energy of about 8 eV. Considering the magnetic moments, now we find a region where $\mu_{\rm GS}$ (lower left plot) already reaches one of the experimental values $\mu_{\rm GS} = 0.45\mu_N$ [23], while for $\mu_{\rm IS}$ (lower right plot) the theoretical predictions always appear close to the experimental value of $-0.37(6)\ \mu_N$ [10]. Also, we see from the upper plots that the behaviour of the isomer transition rates within the QO space suggests a bit smaller values for the B(M1) and somehow larger values for the B(E2) transition compared to those predicted in Ref. [15].

4 Conclusion

Our study and all results presented in this work lead us to the important conclusion that the electromagnetic properties and subsequently the decay and life-time characteristics of the ^{229m}Th isomer are essentially determined by the collective and intrinsic (s.p.) manifestation of nuclear shape deformation. Therefore, its detailed analysis in the aspect of any model description aiming to reach precise determination of the isomer energy and decay properties appears to be of a crucial importance. In a more general aspect the study suggests that the same dynamical mechanism may govern also in other nuclei excitations close to the border of atomic physics energy scale, thus paving the way for the systematic search for similar low-energy isomers in adjacent and remote nuclear regions.

5 Acknowledgments

This work is supported by the Bulgarian National Science Fund (BNSF) under Contract No. KP-06-N48/1. AP gratefully acknowledges funding by the Heisenberg Program of the Deutsche Forschungsgemeinschaft (DFG). The authors also acknowledge the support by Google Inc.

References

[1] B. R. Beck et al., Phys. Rev. Lett. **98**, 142501 (2007); LLNL-PROC-415170 (2009)
[2] C. J. Campbell et al., Phys. Rev. Lett. **108**, 120802 (2012)
[3] E. Peik and C. Tamm, Europhys. Lett. **61**, 181 (2003)
[4] E. Peik and M. Okhapkin, C. R. Phys. **16**, 516 (2015)
[5] E. V. Tkalya, Phys. Rev. Lett. **106**, 162501 (2011)
[6] T. Bürvenich et al., Phys. Rev. Lett. **96**, 142501 (2006)
[7] V. V. Flambaum, Phys. Rev. Lett. **97**, 092502 (2006)
[8] L. von der Wense et al., Nature (London) **533**, 47 (2016)
[9] B. Seiferle, L. von der Wense and P. G. Thirolf, Phys. Rev. Lett. **118**, 042501 (2017)
[10] J. Thielking, M. V. Okhapkin, P. Glowacki, D. M. Meier, L. von der Wense, B. Seiferle, C. E. Düllmann, P. G. Thirolf and E. Peik, Nature (London) **556**, 321 (2018)
[11] R. Müller et al., Phys. Rev. A **98**, 020503(R) (2018)
[12] B. Seiferle et al., Nature (London) **573**, 243 (2019)
[13] A. Yamaguchi et al., Phys. Rev. Lett. **123**, 222501 (2019)
[14] T. Sikorsky et al., Phys. Rev. Lett. **125**, 142503 (2020)
[15] N. Minkov and A. Pálffy, Phys. Rev. Lett. **118**, 212501 (2017)
[16] A. M. Dykhne and E. V. Tkalya, JETP Lett. **67**, 251 (1998)
[17] E. V. Tkalya, C. Schneider, J. Jeet and E. R. Hudson, Phys. Rev. C **92**, 054324 (2015)
[18] E. Ruchowska et al., Phys. Rev. C **73**, 044326 (2006)
[19] J. Jeet et al., Phys. Rev. Lett. **114**, 253001 (2015)
[20] A. Yamaguchi et al., New J. Phys. **17**, 053053 (2015)
[21] N. Minkov and A. Pálffy, Phys. Rev. Lett. **122**, 162502 (2019)
[22] M. S. Safronova, U. I. Safronova, A. G. Radnaev, C. J. Campbell and A. Kuzmich, Phys. Rev. A **88**, 060501(R) (2013)
[23] S. Gerstenkorn et al., J. Phys. (Paris) **35**, 483 (1974)
[24] R. R. Chasman, I. Ahmad, A. M. Friedman and J. R. Erskine, Rev. Mod. Phys. **49**, 833 (1977)
[25] N. Minkov and A. Pálffy, Phys. Rev. C **103**, 014313 (2021)
[26] S. Cwiok, J. Dudek, W. Nazarewicz, J. Skalski and T. Werner, Comp. Phys. Comm. **46**, 379 (1987)
[27] P. Ring and P. Schuck, *The Nuclear Many-Body Problem*, (Springer, Heidelberg, 1980)
[28] P. M. Walker and N. Minkov, Phys. Lett. B **694**, 119 (2010)
[29] N. Minkov, P. Yotov, S. Drenska, W. Scheid, D. Bonatsos, D. Lenis and D. Petrellis, Phys. Rev. C **73**, 044315 (2006)
[30] N. Minkov, S. Drenska, P. Yotov, S. Lalkovski, D. Bonatsos and W. Scheid, Phys. Rev. C **76**, 034324 (2007)
[31] N. Minkov, Physica Scripta **T154**, 014017 (2013)
[32] N. Minkov, S. Drenska, M. Strecker, W. Scheid and H. Lenske, Phys. Rev. C **85**, 034306 (2012)
[33] N. Minkov, S. Drenska, K. Drumev, M. Strecker, H. Lenske and W. Scheid, Phys. Rev. C **88**, 064310 (2013)
[34] http://www.nndc.bnl.gov/ensdf/
[35] http://www.nndc.bnl.gov/nudat2/indx_adopted.jsp
[36] S. G. Nilsson and O. Prior, Kgl. Dan. Vidensk. Selsk. Mat.-Fys. Medd. **32**, No. 16 (1961)
[37] H. Ton, S. Roodbergen, J. Brasz, and J. Blok, Nucl. Phys. A **155**, 245 (1970)

Connecting the proxy-SU(3) symmetry to the shell model

Dennis Bonatsos[1,*], *Andriana* Martinou[1,**], *I.E.* Assimakis[1], *S.K.* Peroulis[1], *S.* Sarantopoulou[1], and *N.* Minkov[2]

[1]Institute of Nuclear and Particle Physics, National Centre of Scientific Research "Demokritos", GR-15310 Aghia Paraskevi, Attiki, Greece
[2]Institute of Nuclear Research and Nuclear Energy, Bulgarian Academy of Sciences, 72 Tzarigrad Road, 1784 Sofia, Bulgaria

Abstract. Proxy-SU(3) symmetry is an approximation scheme extending the Elliott SU(3) algebra of the sd shell to heavier shells. When introduced in 2017, the approximation had been justified by calculations carried out within the Nilsson model. Recently our group managed to map the cartesian basis of the Elliott SU(3) model onto the spherical shell model basis, proving that the proxy-SU(3) approximation corresponds to the replacement of the intruder orbitals by their de Shalit-Goldhaber partners, paving the way for using the proxy-SU(3) approximation in shell model calculations. The connection between the proxy-SU(3) scheme and the spherical shell model has also been worked out in the original framework of the Nilsson model, with identical results.

1 SU(3) symmetry in nuclear structure

The SU(3) symmetry has been used in nuclear structure for a long time [1].

In 1949 the shell model has been introduced [2–5], which is based on a three-dimensional (3D) isotropic harmonic oscillator (HO), with a spin-orbit term added to it. The 3D isotropic HO is known to possess shells labeled by the number of excitation quanta n, characterized by the unitary symmetries U((n+1)(n+2)/2), having SU(3) subalgebras [6–9].

The shell model was considered adequate for describing near-spherical nuclei, thus in 1952 the collective model of Bohr and Mottelson has been introduced [10, 11], in which departure from the spherical shape and from axial symmetry are described by the collective variables β and γ respectively.

A modified version of the shell model, allowing for axial nuclear deformation to be included, has been introduced by Nilsson in 1955 [12, 13], based on a 3D anisotropic HO with cylindrical symmetry [14–19].

In 1958 Elliott proved that deformation within the nuclear sd shell with U(6) symmetry can be described in terms of its SU(3) subalgebra [20–24].

Early attempts since 1972 of extending the SU(3) symmetry to heavy nuclei [25–28] evolved into the Vector Boson Model [29–31], while at the same time the group theoretical structure of the Bohr–Mottelson model, having an overall U(5) symmetry possessing an O(5) subalgebra, has been understood [32].

Beyond the sd nuclear shell the SU(3) symmetry of the 3D isotropic HO is known to be broken by the spin-orbit force, which within each HO shell pushes the orbitals possessing the highest angular momentum j to the shell below. As a consequence, each shell consists by the orbitals left back after this removal, called the normal parity orbitals, plus the orbitals invading from the shell above, having the opposite parity and called the intruder orbitals.

A major step forward has been taken in 1973, with the introduction of the pseudo-SU(3) symmetry [33–39], applicable in heavy nuclei. Within the pseudo-SU(3) framework, the SU(3) symmetry of the 3D isotropic HO is recovered by mapping the incomplete set of normal parity orbitals left in a shell onto the complete set of orbitals of the shell below. This becomes possible by assigning to each orbital a pseudo-orbital angular momentum and a pseudo-spin, while the total angular momentum remains intact. It has been realized later that this mapping is equivalent to a unitary transformation [40–42], and the relativistic mean field origins of the pseudospin symmetry have been understood [43, 44]. Within the pseudo-SU(3) scheme each shell consists of a normal parity part, which possesses a U(n) symmetry and a SU(3) subalgebra, and an intruder part which does not possess any SU(3) structure and has to be treated separately by shell model techniques [37, 38].

In 1974 it was realized that the nuclear quadrupole degree of freedom can be described in terms of a SU(6) algebra [45].

Next year the Interacting Boson Model [46–49] has been introduced, which also has an overall U(6) symmetry built by s-bosons of zero angular momentum and d-bosons of angular momentum two, possessing three limiting symmetries, U(5) for vibrational nuclei, which is equivalent to the Bohr-Mottelson collective model, O(6) for γ-unstable nuclei, and SU(3) for deformed nuclei.

A SU(3) limiting symmetry, appropriate for deformed nuclei, has also been obtained in 1980 within the symplec-

*e-mail: bonat@inp.demokritos.gr
**e-mail: martinou@inp.demokritos.gr

tic model [50, 51], which uses fermion pairs, as well as in 1982 within the Interacting Vector Boson Model [52, 53], in which two vector bosons of angular momentum one are used, and in 1987 within the Fermion Dynamical Symmetry Model [54], in which the total angular momentum of the nucleons is assumed to be split into active and inactive parts instead of orbital angular momentum and spin parts.

In 2017 the proxy-SU(3) symmetry has been introduced [55–57], in which the intruder orbitals in each shell (except the one with the highest projection of the total angular momentum) are replaced by the orbitals which have deserted this shell by sinking into the shell below. As a result of this replacement, each shell regains the relevant U(n) symmetry having a SU(3) subalgebra, with only one orbital (which can accommodate two particles) remaining estranged. However, this orbital is the one lying highest in energy within the shell, thus it should be empty for most of the nuclei living in this shell. Therefore it is expected that its influence on the structure of most nuclei living in the shell should be minimal.

The proxy-SU(3) scheme has been initially justified as a good approximation through calculations [55] carried out within the Nilsson model [12, 13]. In the present work we are going to discuss its justification through its connection to the shell model. However, before doing so, it is instructive to discuss the nature of nucleon pairs related to the development of nuclear deformation.

2 Nucleon pairs favoring deformation

As early as 1953 it has been observed by deShalit and Goldhaber [58] in their studies of β transition probabilities that within the proton–neutron pairs of orbitals (1p3/2, 1d5/2), (1d5/2, 1f7/2), (1f7/2, 1g9/2), (1g9/2, 1h11/2), (1h11/2, 1i13/2) the nucleons of one kind (protons, for example) have a stabilizing effect on pairs of nucleons of the other kind (neutrons in the example), thus favoring the development of nuclear deformation. In the standard shell model notation $|nljm_j\rangle$, in which states are labeled by the number of oscillator quanta n, the orbital angular momentum l, the total angular momentum j, and its z-projection m_j, the orbitals forming a pair differ by $|\Delta n \Delta l \Delta j \Delta m_j\rangle = |0110\rangle$.

A major step forward in our understanding of effective interactions and coupling schemes in nuclei has been taken in 1962 by Talmi [59] through the introduction of seniority [59–62], representing the number of nucleon pairs coupled to non-zero angular momentum, which explained the linear dependence of neutron separation energies on the mass number within series of isotopes.

In 1977 Federman and Pittel [63–65] realized that when adding valence protons and valence neutrons to a nucleus, the proton–neutron pairs (1d5/2, 1d3/2), (1g9/2, 1g7/2), (1h11/2, 1h9/2), and (1i13/2, 1i11/2) are responsible for the onset of deformation, while deformation is then established by the proton–neutron pairs (1d5/2, 1f7/2), (1g9/2, 1h11/2), (1h11/2, 1i13/2), and (1i13/2, 1j15/2), shown in Table 1. In the shell model notation these sets correspond to $|\Delta n \Delta l \Delta j \Delta m_j\rangle = |0010\rangle$ and $|0110\rangle$ respectively, the latter set coinciding with the de Shalit–Goldhaber pairs.

Table 1. Pairs of orbitals playing a leading role in the development of deformation in different mass regions of the nuclear chart according to Federman and Pittel [63–65]. The pairs on the left part of the table contribute in the beginning of the relevant shell, while the pairs on the right become important further within the shell. Adapted from Ref. [83].

	protons	neutrons	protons	neutrons
light	$1d^{5/2}$	$1d^{3/2}$	$1d^{5/2}$	$1f^{7/2}$
intermediate	$1g^{9/2}$	$1g^{7/2}$	$1g^{9/2}$	$1h^{11/2}$
rare earths	$1h^{11/2}$	$1h^{9/2}$	$1h^{11/2}$	$1i^{13/2}$
actinides	$1i^{13/2}$	$1i^{11/2}$	$1i^{13/2}$	$1j^{15/2}$

The decisive role played by proton-neutron pairs has been demonstrated in 1985 through the introduction of the N_pN_n scheme [66, 67] and the P-factor, $P = N_pN_n/(N_p + N_n)$ [68, 69], by showing the systematic dependence of several observables on the competition between the quadrupole deformation, "measured" by the quadrupole-quadrupople interaction through N_pN_n, and the pairing interaction, "measured" through $N_p + N_n$, where N_p (N_n) is the number of valence protons (neutrons).

In 1995 the quasi-SU(3) symmetry [70, 71] has been introduced, based on the proton–neutron pairs (1g9/2, 2d5/2), (1h11/2, 2f7/2), (1i13/2, 2g9/2), expressed as $|\Delta n \Delta l \Delta j \Delta m_j\rangle = |1220\rangle$ in the shell model notation, which lead to enhanced quadrupole collectivity [72].

Following detailed studies of double differences of binding energies [73–77], in 2010 it has been realized [78] that proton-neutron pairs differing by $\Delta K[\Delta N \Delta n_z \Delta \Lambda] = 0[110]$ in the Nilsson notation [12, 13] $K[Nn_z\Lambda]$, where N is the total number of oscillator quanta, n_z is the number of quanta along the z-axis, and Λ, K are the projections along the z-axis of the orbital angular momentum and the total angular momentum respectively, play a major role in the development of nuclear deformation, due to their large spatial overlaps [79]. These pairs correspond to the replacements made within the proxy-SU(3) scheme [55–57]. No relation to the pairs mentioned in the previous paragraphs had been realized at that time.

3 Connecting the cartesian Elliott basis to the spherical shell model basis

In the Elliott model [20–24] the cartesian basis of the 3D isotropic HO is used, $[n_z n_x n_y m_s]$, in which the number of quanta along the z, x, y directions and the z-projection of the spin appear. This basis can be transformed to the spherical basis $[nlm_l m_s]$ in l-s coupling, labeled by the principal quantum number n, the orbital angular momentum l, its z-projection (m_l), and the z-projection of the spin (m_s), through a unitary transformation [80–82]

$$[n_z n_x n_y m_s] = R[nlm_l m_s], \tag{1}$$

the details of which can be found in Ref. [83]. Using Clebsch-Gordan coefficients the spherical basis can be

Table 2. The transformation matrix $R \cdot C$ for $N = 2$. The shell model orbitals appear in the first line, while the Elliott orbitals appear in the first column. These orbitals are used in the harmonic oscillator shell 8-20 (*sd* shell), or in the proxy-SU(3) shell 14-26 after the replacement of the intruder orbitals with their de Shalit–Goldhaber partners. Adapted from Ref. [83].

$\|n,l,j,m_j\rangle$ $\|n_z,n_x,n_y,m_s\rangle$	$\|2s^{1/2}_{-1/2}\rangle$	$\|2s^{1/2}_{1/2}\rangle$	$\|1d^{3/2}_{-3/2}\rangle$	$\|1d^{3/2}_{-1/2}\rangle$	$\|1d^{3/2}_{1/2}\rangle$	$\|1d^{3/2}_{3/2}\rangle$	$\|1d^{5/2}_{-5/2}\rangle$	$\|1d^{5/2}_{-3/2}\rangle$	$\|1d^{5/2}_{-1/2}\rangle$	$\|1d^{5/2}_{1/2}\rangle$	$\|1d^{5/2}_{3/2}\rangle$	$\|1d^{5/2}_{5/2}\rangle$
$\|0,0,2,-\frac{1}{2}\rangle$	$-\frac{1}{\sqrt{3}}$	0	0	$-\frac{1}{\sqrt{15}}$	0	$-\frac{1}{\sqrt{5}}$	$-\frac{1}{2}$	0	$-\frac{1}{\sqrt{10}}$	0	$-\frac{1}{2\sqrt{5}}$	0
$\|0,0,2,\frac{1}{2}\rangle$	0	$-\frac{1}{\sqrt{3}}$	$\frac{1}{\sqrt{5}}$	0	$\frac{1}{\sqrt{15}}$	0	0	$-\frac{1}{2\sqrt{5}}$	0	$-\frac{1}{\sqrt{10}}$	0	$-\frac{1}{2}$
$\|0,1,1,-\frac{1}{2}\rangle$	0	0	0	0	0	$-i\sqrt{\frac{2}{5}}$	$\frac{i}{\sqrt{2}}$	0	0	0	$-\frac{i}{\sqrt{10}}$	0
$\|0,1,1,\frac{1}{2}\rangle$	0	0	$-i\sqrt{\frac{2}{5}}$	0	0	0	0	$\frac{i}{\sqrt{10}}$	0	0	0	$-\frac{i}{\sqrt{2}}$
$\|0,2,0,-\frac{1}{2}\rangle$	$-\frac{1}{\sqrt{3}}$	0	0	$-\frac{1}{\sqrt{15}}$	0	$\frac{1}{\sqrt{5}}$	$\frac{1}{2}$	0	$-\frac{1}{\sqrt{10}}$	0	$\frac{1}{2\sqrt{5}}$	0
$\|0,2,0,\frac{1}{2}\rangle$	0	$-\frac{1}{\sqrt{3}}$	$-\frac{1}{\sqrt{5}}$	0	$\frac{1}{\sqrt{15}}$	0	0	$\frac{1}{2\sqrt{5}}$	0	$-\frac{1}{\sqrt{10}}$	0	$\frac{1}{2}$
$\|1,0,1-\frac{1}{2}\rangle$	0	0	$\frac{i}{\sqrt{10}}$	0	$i\sqrt{\frac{3}{10}}$	0	0	$i\sqrt{\frac{2}{5}}$	0	$\frac{i}{\sqrt{5}}$	0	0
$\|1,0,1,\frac{1}{2}\rangle$	0	0	0	$-i\sqrt{\frac{3}{10}}$	0	$-\frac{i}{\sqrt{10}}$	0	0	$\frac{i}{\sqrt{5}}$	0	$i\sqrt{\frac{2}{5}}$	0
$\|1,1,0,-\frac{1}{2}\rangle$	0	0	$\frac{1}{\sqrt{10}}$	0	$-\sqrt{\frac{3}{10}}$	0	0	$\sqrt{\frac{2}{5}}$	0	$-\frac{1}{\sqrt{5}}$	0	0
$\|1,1,0,\frac{1}{2}\rangle$	0	0	0	$-\sqrt{\frac{3}{10}}$	0	$\frac{1}{\sqrt{10}}$	0	0	$\frac{1}{\sqrt{5}}$	0	$-\sqrt{\frac{2}{5}}$	0
$\|2,0,0,-\frac{1}{2}\rangle$	$-\frac{1}{\sqrt{3}}$	0	0	$\frac{2}{\sqrt{15}}$	0	0	0	0	$\sqrt{\frac{2}{5}}$	0	0	0
$\|2,0,0,\frac{1}{2}\rangle$	0	$-\frac{1}{\sqrt{3}}$	0	0	$-\frac{2}{\sqrt{15}}$	0	0	0	0	$\sqrt{\frac{2}{5}}$	0	0

rewritten as

$$[nlm_lm_s] = C[nljm_j], \tag{2}$$

in which the total angular momentum j and its z-projection appear. Combining these two transformations one obtains

$$[n_zn_xn_ym_s] = RC[nljm_j], \tag{3}$$

i.e., the connection between the cartesian Elliott basis and the shell model basis in j-j coupling. An example of this transformation is shown in Table 2. Details of the calculations and tables for other shells can be found in Ref. [83].

Using the above transformation one sees that the Nilsson 0[110] replacements made within the proxy-SU(3) scheme are "translated" into $|0110\rangle$ replacements within the spherical shell model basis. The resulting correspondence between original shell model orbitals and proxy-SU(3) orbitals is summarized in Table 3. This correspondence paves the way for taking advantage of the proxy-SU(3) symmetry in shell model calculations for heavy nuclei, in a way similar to that of the symmetry-adapted no-core shell model approach [85, 86] used in light nuclei.

The correspondence between Nilsson pairs and shell model pairs has been corroborated by calculations [87] within the Nilsson model, in which the first justification of the proxy-SU(3) scheme has been found [55]. As one can see in Tables 4 and 5, the correspondence used in proxy-SU(3) works only for the Nilsson orbitals which possess the highest total angular momentum j within their shell, which are exactly the orbitals which are replaced within the proxy-SU(3) scheme. In further corroboration of this result, a unitary transformation connecting the orbitals being replaced within the proxy-SU(3) scheme has been found [83] within the shell model basis and is depicted in Fig. 1.

A by-product of the above transformation is that the 0[110] Nilsson pairs identified in Ref. [78] and used within the proxy-SU(3) scheme [55–57] are identical to the de Shalit–Goldhaber pairs [58] and the Federman–Pittel pairs [63–65] within the spherical shell model basis, in which they are expressed as $|0110\rangle$ pairs.

Within the proxy-SU(3) scheme the importance of the highest weight irreducible representations of SU(3) has been demonstrated [88] and used [56, 57] for the successful prediction of the prolate to oblate shape transition at $N = 114$, the dominance of prolate over oblate shapes in the ground states of even-even nuclei, and the prediction of specific islands on the nuclear chart in which shape coexistence can appear [89]. The compatibility of certain predictions of proxy-SU(3) and pseudo-SU(3) has been demonstrated in Refs. [90, 91]. These topics will be discussed in the talks by A. Martinou [92] and S. Sarantopoulou [93].

References

[1] V.K.B. Kota, *SU(3) Symmetry in Atomic Nuclei* (Springer, Singapore, 2020)

[2] M. G. Mayer, Phys. Rev. **74**, 235 (1948)

[3] M. G. Mayer, Phys. Rev. **75**, 1969 (1949)

[4] O. Haxel, J.H.D. Jensen, and H.E. Suess, Phys. Rev. **75**, 1766 (1949)

[5] M.G. Mayer and J.H.D. Jensen, *Elementary Theory of Nuclear Shell Structure* (Wiley, New York, 1955)

[6] B. G. Wybourne, *Classical Groups for Physicists* (Wiley, New York, 1974)

[7] M. Moshinsky and Yu. F. Smirnov, *The Harmonic Oscillator in Modern Physics* (Harwood, Amsterdam, 1996)

https://doi.org/10.1051/epjconf/202125202004

Table 3. The shell model orbitals of the original spin-orbit like shells and of the proxy-SU(3) shells. The magic number 14 is proposed as a sub-shell closure in Ref. [84]. The symmetry of each proxy-SU(3) shell is U(Ω) with $\Omega = \frac{(N+1)(N+2)}{2}$. The orbitals being replaced are denoted by bold letters. Adapted from Ref. [83].

$U(\Omega)$ spin-orbit 3D-HO magic numbers	original orbitals	proxy orbitals
U(3)	$1p^{1/2}_{\pm 1/2}$	$1p^{1/2}_{\pm 1/2}$
6-14	$\mathbf{1d^{5/2}_{\pm 1/2,\pm 3/2}}$	$\mathbf{1p^{3/2}_{\pm 1/2,\pm 3/2}}$
2-8	$\mathbf{1d^{5/2}_{\pm 5/2}}$	-
U(6)	$2s^{1/2}_{\pm 1/2}$	$2s^{1/2}_{\pm 1/2}$
14-28	$1d^{3/2}_{\pm 1/2,\pm 3/2}$	$1d^{3/2}_{\pm 1/2,\pm 3/2}$
8-20	$\mathbf{1f^{7/2}_{\pm 1/2,\pm 3/2,\pm 5/2}}$	$\mathbf{1d^{5/2}_{\pm 1/2,\pm 3/2,\pm 5/2}}$
	$\mathbf{1f^{7/2}_{\pm 7/2}}$	-
U(10)	$2p^{1/2}_{\pm 1/2}$	$2p^{1/2}_{\pm 1/2}$
28-50	$2p^{3/2}_{\pm 1/2,\pm 3/2}$	$2p^{3/2}_{\pm 1/2,\pm 3/2}$
20-40	$1f^{5/2}_{\pm 5/2,\pm 3/2,\pm 1/2}$	$1f^{5/2}_{\pm 5/2,\pm 3/2,\pm 1/2}$
	$\mathbf{1g^{9/2}_{\pm 1/2,\ldots,\pm 7/2}}$	$\mathbf{1f^{7/2}_{\pm 1/2,\ldots,\pm 7/2}}$
	$\mathbf{1g^{9/2}_{\pm 9/2}}$	-
U(15)	$3s^{1/2}_{\pm 1/2}$	$3s^{1/2}_{\pm 1/2}$
50-82	$2d^{3/2}_{\pm 1/2,\pm 3/2}$	$2d^{3/2}_{\pm 1/2,\pm 3/2}$
40-70	$2d^{5/2}_{\pm 1/2,\ldots,\pm 5/2}$	$2d^{5/2}_{\pm 1/2,\ldots,\pm 5/2}$
	$1g^{7/2}_{\pm 1/2,\ldots,\pm 7/2}$	$1g^{7/2}_{\pm 1/2,\ldots,\pm 7/2}$
	$\mathbf{1h^{11/2}_{\pm 1/2,\ldots,\pm 9/2}}$	$\mathbf{1g^{9/2}_{\pm 1/2,\ldots,\pm 9/2}}$
	$\mathbf{1h^{11/2}_{\pm 11/2}}$	-
U(21)	$3p^{1/2}_{\pm 1/2}$	$3p^{1/2}_{\pm 1/2}$
82-126	$3p^{3/2}_{\pm 1/2,\pm 3/2}$	$3p^{3/2}_{\pm 1/2,\pm 3/2}$
70-112	$2f^{5/2}_{\pm 1/2,\ldots,\pm 5/2}$	$2f^{5/2}_{\pm 1/2,\ldots,\pm 5/2}$
	$2f^{7/2}_{\pm 1/2,\ldots,\pm 7/2}$	$2f^{7/2}_{\pm 1/2,\ldots,\pm 7/2}$
	$1h^{9/2}_{\pm 1/2,\ldots,\pm 9/2}$	$1h^{9/2}_{\pm 1/2,\ldots,\pm 9/2}$
	$\mathbf{1i^{13/2}_{\pm 1/2,\ldots,\pm 11/2}}$	$\mathbf{1h^{11/2}_{\pm 1/2,\ldots,\pm 11/2}}$
	$\mathbf{1i^{13/2}_{\pm 13/2}}$	-
U(28)	$4s^{1/2}_{\pm 1/2}$	$4s^{1/2}_{\pm 1/2}$
126-184	$3d^{3/2}_{\pm 1/2,\pm 3/2}$	$3d^{3/2}_{\pm 1/2,\pm 3/2}$
112-168	$3d^{5/2}_{\pm 1/2,\ldots,\pm 5/2}$	$3d^{5/2}_{\pm 1/2,\ldots,\pm 5/2}$
	$2g^{7/2}_{\pm 1/2,\ldots,\pm 7/2}$	$2g^{7/2}_{\pm 1/2,\ldots,\pm 7/2}$
	$2g^{9/2}_{\pm 1/2,\ldots,\pm 9/2}$	$2g^{9/2}_{\pm 1/2,\ldots,\pm 9/2}$
	$1i^{11/2}_{\pm 1/2,\ldots,\pm 11/2}$	$1i^{11/2}_{\pm 1/2,\ldots,\pm 11/2}$
	$\mathbf{1j^{15/2}_{\pm 1/2,\ldots,\pm 13/2}}$	$\mathbf{1i^{13/2}_{\pm 1/2,\ldots,\pm 13/2}}$
	$\mathbf{1j^{15/2}_{\pm 15/2}}$	-

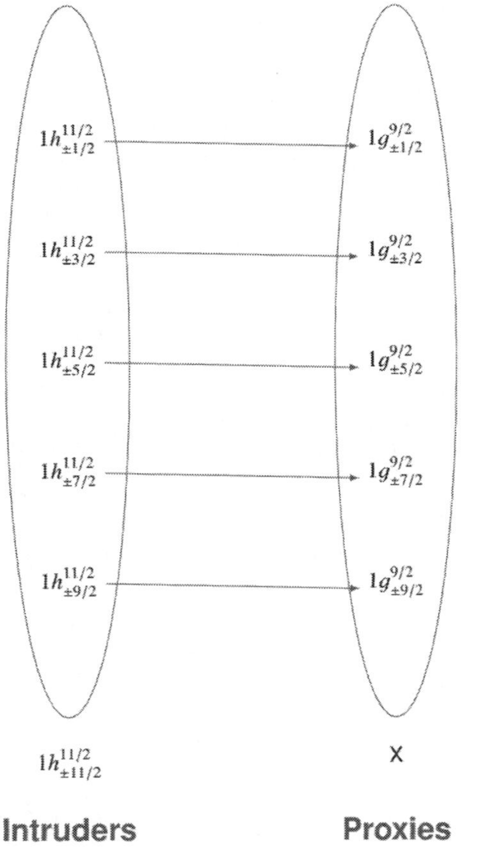

Figure 1. Unitary transformation of the intruder orbitals $1h^{11/2}_{m_j}$ (except for the $1h^{11/2}_{\pm 11/2}$) in the 50-82 shell onto the orbitals $1g^{9/2}_{m_j}$. Adapted from Ref. [83].

[8] F. Iachello, *Lie Algebras and Applications* (Springer, Berlin, 2006)

[9] D. Bonatsos and A. Klein, Ann. Phys. (NY) **169**, 61 (1986)

[10] A. Bohr, Mat. Fys. Medd. K. Dan. Vidensk. Selsk. **26**, no. 14 (1952)

[11] A. Bohr and B. R. Mottelson, *Nuclear Structure Vol. II: Nuclear Deformations* (Benjamin, New York, 1975)

[12] S.G. Nilsson, Mat. Fys. Medd. K. Dan. Vidensk. Selsk. **29**, no. 16 (1955)

[13] S.G. Nilsson and I. Ragnarsson, *Shapes and Shells in Nuclear Structure* (Cambridge University Press, Cambridge, 1995)

[14] Y. Takahashi, Prog. Theor. Phys. **53**, 461 (1975)

[15] R.M. Asherova, Yu.F. Smirnov, V.N. Tolstoy, and A.P. Shustov, Nucl. Phys. A **355**, 25 (1981)

[16] G. Rosensteel and J. P. Draayer, J. Phys. A: Math. Gen. **22**, 1323 (1989)

[17] W. Nazarewicz and J. Dobaczewski, Phys. Rev. Lett. **68**, 154 (1992)

Table 4. Expansions of Nilsson orbitals $K[Nn_z\Lambda]$ in the shell model basis $|Nljm_j\rangle$ for three different values of the deformation ϵ. The Nilsson orbitals shown possess the highest total angular momentum j in their shell. The existence of a leading shell model eigenvector is evident at all deformations. Adapted from Ref. [87].

$\frac{3}{2}[541]$	ϵ	$\|Nljm_j\rangle$	$\|51\frac{3}{2}\frac{3}{2}\rangle$	$\|53\frac{5}{2}\frac{3}{2}\rangle$	$\|53\frac{7}{2}\frac{3}{2}\rangle$	$\|55\frac{9}{2}\frac{3}{2}\rangle$	$\|55\frac{11}{2}\frac{3}{2}\rangle$
	0.05		0.0025	−0.0015	0.0641	−0.0122	0.9979
	0.22		0.0371	−0.0286	0.2565	−0.0640	0.9633
	0.30		0.0601	−0.0506	0.3287	−0.0922	0.9366

$\frac{3}{2}[651]$	ϵ	$\|Nljm_j\rangle$	$\|62\frac{3}{2}\frac{3}{2}\rangle$	$\|62\frac{5}{2}\frac{3}{2}\rangle$	$\|64\frac{7}{2}\frac{3}{2}\rangle$	$\|64\frac{9}{2}\frac{3}{2}\rangle$	$\|66\frac{11}{2}\frac{3}{2}\rangle$	$\|66\frac{13}{2}\frac{3}{2}\rangle$
	0.05		−0.0002	0.0046	−0.0013	0.0821	−0.0086	0.9966
	0.22		−0.0100	0.0711	−0.0278	0.3240	−0.0469	0.9418
	0.30		−0.0207	0.1149	−0.0509	0.4091	−0.0687	0.9010

Table 5. Expansions of Nilsson orbitals $K[Nn_z\Lambda]$ in the shell model basis $|Nljm_j\rangle$ for three different values of the deformation ϵ. The Nilsson orbitals shown do not possess the highest total angular momentum j in their shell. The existence of a leading shell model eigenvector is evident at small deformation, but this is not the case any more at higher deformations, at which several shell model eigenvectors make considerable contributions. Adapted from Ref. [87].

$\frac{1}{2}[431]$	ϵ	$\|Nljm_j\rangle$	$\|40\frac{1}{2}\frac{1}{2}\rangle$	$\|42\frac{3}{2}\frac{1}{2}\rangle$	$\|42\frac{5}{2}\frac{1}{2}\rangle$	$\|44\frac{7}{2}\frac{1}{2}\rangle$	$\|44\frac{9}{2}\frac{1}{2}\rangle$
	0.05		−0.0213	0.1254	−0.0702	0.9893	0.0127
	0.22		−0.2248	0.4393	−0.2791	0.8057	0.1717
	0.30		−0.2630	0.5003	−0.2458	0.7447	0.2559

$\frac{1}{2}[541]$	ϵ	$\|Nljm_j\rangle$	$\|51\frac{1}{2}\frac{1}{2}\rangle$	$\|51\frac{3}{2}\frac{1}{2}\rangle$	$\|53\frac{5}{2}\frac{1}{2}\rangle$	$\|53\frac{7}{2}\frac{1}{2}\rangle$	$\|55\frac{9}{2}\frac{1}{2}\rangle$	$\|55\frac{11}{2}\frac{1}{2}\rangle$
	0.05		−0.0200	0.1770	−0.0295	0.9780	−0.0446	−0.0944
	0.22		−0.2492	0.4619	−0.3768	0.5550	−0.4161	−0.3185
	0.30		−0.3121	0.4331	−0.4829	0.3430	−0.4789	−0.3671

[18] W. Nazarewicz, J. Dobaczewski, and P. Van Isacker, AIP Conf. Proc. **259**, 30 (1992)

[19] D. Bonatsos, C. Daskaloyannis, P. Kolokotronis, and D. Lenis, arXiv: hep-th/9411218

[20] J. P. Elliott, Proc. Roy. Soc. London Ser. A **245**, 128 (1958)

[21] J. P. Elliott, Proc. Roy. Soc. London Ser. A **245**, 562 (1958)

[22] J. P. Elliott and M. Harvey, Proc. Roy. Soc. London Ser. A **272**, 557 (1963)

[23] J. P. Elliott and C. E. Wildson, Proc. Roy. Soc. London Ser. A **302**, 509 (1968)

[24] M. Harvey, in Adv. Nucl. Phys. **1**, ed. M. Baranger and E. Vogt (Plenum, New York, 1968) p. 67

[25] P.P. Raychev, Compt. Rend. Acad. Bulg. Sci. **25**, 1503 (1972)

[26] G.N. Afanas'ev, S.A. Abramov, and P.P. Raychev, Yad. Fiz. **16**, 53 (1972) [Sov. J. Nucl. Phys. **16**, 27 (1973)]

[27] P.P. Raychev, Yad. Fiz. **16**, 1171 (1972) [Sov. J. Nucl. Phys. **16**, 643 (1973)]

[28] P.P. Raychev and R.P. Roussev, Yad. Fiz. **27**, 1501 (1978) [Sov. J. Nucl. Phys. **27**, 792 (1978)]

[29] N. Minkov, S.B. Drenska, P.P. Raychev, R.P. Roussev, and D. Bonatsos, Phys. Rev. C **55**, 2345 (1997)

[30] N. Minkov, S.B. Drenska, P.P. Raychev, R.P. Roussev, and D. Bonatsos, Phys. Rev. C **60**, 034305 (1999)

[31] N. Minkov, S.B. Drenska, P.P. Raychev, R.P. Roussev, and D. Bonatsos, Phys. Rev. C **61**, 064301 (2000)

[32] G.N. Afanas'ev and P.P. Raychev, Fiz. Elem. Chast. At. Yadra **3**, 436 (1972) [Sov. J. Nucl. Phys. **3**, 229 (1972)]

[33] K.T. Hecht and A. Adler, Nucl. Phys. A **137**, 129 (1969)

[34] A. Arima, M. Harvey, and K. Shimizu, Phys. Lett. B **30**, 517 (1969)

[35] R. D. Ratna Raju, J. P. Draayer, and K. T. Hecht, Nucl. Phys. A **202**, 433 (1973)

[36] J. P. Draayer, K. J. Weeks, and K. T. Hecht, Nucl. Phys. A **381**, 1 (1982)

[37] J. P. Draayer and K. J. Weeks, Phys. Rev. Lett. **51**, 1422 (1983)

[38] J. P. Draayer and K. J. Weeks, Ann. Phys. (N.Y.) **156**, 41 (1984)

[39] J.P. Draayer, in *Algebraic Approaches to Nuclear Structure*, ed. R.F. Casten (Harwood, Chur, 1993), p. 423

[40] O. Castaños, M. Moshinsky, and C. Quesne, *Group Theory and Special Symmetries in Nuclear Physics Ann Arbor, 1991*, ed. J. P. Draayer and J. Jänecke (World Scientific, Singapope, 1992) p. 80

[41] O. Castaños, M. Moshinsky, and C. Quesne, Phys. Lett. B **277**, 238 (1992)

[42] O. Castaños, V. Velázquez A., P.O. Hess, and J.G. Hirsch, Phys. Lett. B **321**, 303 (1994)
[43] J.N. Ginocchio, Phys. Rev. Lett. **78**, 436 (1997)
[44] J.N. Ginocchio, J. Phys. G: Nucl. Part. Phys. **25**, 617 (1999)
[45] D. Janssen, R.V. Jolos, and F. Dönau, Nucl. Phys. A **224**, 93 (1974)
[46] A. Arima and F. Iachello, Phys. Rev. Lett **35**, 1069 (1975)
[47] F. Iachello and A. Arima, *The Interacting Boson Model* (Cambridge University Press, Cambridge, 1987)
[48] F. Iachello and P. Van Isacker, *The Interacting Boson-Fermion Model* (Cambridge University Press, Cambridge, 1991)
[49] A. Frank and P. Van Isacker, *Symmetry Methods in Molecules and Nuclei* (S y G editores, México D.F., 2005)
[50] G. Rosensteel and D.J. Rowe, Ann. Phys. (NY) **126**, 343 (1980)
[51] D. J. Rowe and J. L. Wood, *Fundamentals of Nuclear Models: Foundational Models* (World Scientific, Singapore, 2010)
[52] A. Georgieva, P. Raychev, and R. Roussev, J. Phys. G: Nucl. Phys. **8**, 1377 (1982)
[53] A. Georgieva, P. Raychev, and R. Roussev, J. Phys. G: Nucl. Phys. **9**, 521 (1983)
[54] C.-L. Wu, D. H. Feng, X.-G. Chen, J.-Q. Chen, and M. W. Guidry, Phys. Rev. C **36**, 1157 (1987)
[55] D. Bonatsos, I. E. Assimakis, N. Minkov, A. Martinou, R. B. Cakirli, R. F. Casten, and K. Blaum, Phys. Rev. C **95**, 064325 (2017)
[56] D. Bonatsos, I. E. Assimakis, N. Minkov, A. Martinou, S. Sarantopoulou, R. B. Cakirli, R. F. Casten, and K. Blaum, Phys. Rev. C **95**, 064326 (2017)
[57] D. Bonatsos, Eur. Phys. J. A **53**, 148 (2017)
[58] A. de Shalit and M. Goldhaber, Phys. Rev. **92**, 1211 (1953)
[59] I. Talmi, Rev. Mod. Phys. **34**, 704 (1962)
[60] I. Talmi, Nucl. Phys. A **172**, 1 (1971)
[61] I. Talmi, Riv. Nuovo Cim. **3**, 85 (1973)
[62] I. Talmi, *Simple Models of Complex Nuclei* (Harwood, Chur, 1993)
[63] P. Federman and S. Pittel, Phys. Lett. B **69**, 385 (1977)
[64] P. Federman and S. Pittel, Phys. Lett. B **77**, 29 (1978)
[65] P. Federman and S. Pittel, Phys. Rev. C **20**, 820 (1979)
[66] R.F. Casten, Phys. Rev. Lett. **54**, 1991 (1985)
[67] R.F. Casten, Nucl. Phys. A **443**, 1 (1985)
[68] R.F. Casten, D.S. Brenner, and P.E. Haustein, Phys. Rev. Lett. **58**, 658 (1987)
[69] R. F. Casten, *Nuclear Structure from a Simple Perspective* (Oxford University Press, Oxford, 2000)
[70] A. P. Zuker, J. Retamosa, A. Poves, and E. Caurier, Phys. Rev. C **52**, R1741 (1995)
[71] A. P. Zuker, A. Poves, F. Nowacki, and S. M. Lenzi, Phys. Rev. C **92**, 024320 (2015)
[72] K. Kaneko, N. Shimizu, T. Mizusaki, and Y. Sun, Phys. Rev. **103**, L021301 (2021)
[73] R.B. Cakirli, D.S. Brenner, R.F. Casten, and E.A. Millman, Phys. Rev. Lett. **94**, 092501 (2005); erratum Phys. Rev. Lett. **95**, 119903 (2005)
[74] R.B. Cakirli and R.F. Casten, Phys. Rev. Lett. **96**, 132501 (2006)
[75] D.S. Brenner, R.B. Cakirli and R.F. Casten, Phys. Rev. C **73**, 034315 (2006)
[76] M. Stoitsov, R.B. Cakirli, R.F. Casten, W. Nazarewicz, and W. Satula, Phys. Rev. Lett. **98**, 132502 (2007)
[77] R.B. Cakirli, R.F. Casten, R. Winkler, K. Blaum, and M. Kowalska, Phys. Rev. Lett. **102**, 082501 (2009)
[78] R. B. Cakirli, K. Blaum and R. F. Casten, Phys. Rev. C **82**, 061304 (2010)
[79] D. Bonatsos, S. Karampagia, R.B. Cakirli, R.F. Casten, K. Blaum, and L. Amon Susam, Phys. Rev. C **88**, 054309 (2013)
[80] K.T.R. Davies and S.J. Krieger, Can. J. Phys. **69**, 62 (1991)
[81] R.R. Chasman and S. Wahlborn, Nucl. Phys. A **90**, 401 (1967)
[82] E. Chacón and M. de Llano, Rev. Mex. Fís. **12** 57 (1963)
[83] A. Martinou, D. Bonatsos, N. Minkov, I.E. Assimakis, S.K. Peroulis, S. Sarantopoulou, and J. Cseh, Eur. Phys. J. A **56**, 239 (2020)
[84] O. Sorlin and M.-G. Porquet, Prog. Part. Nucl. Phys. **61**, 602 (2008)
[85] K.D. Launey, J.P. Draayer, T. Dytrych, G.-H. Sun, and S.-H. Dong, Int. J. Mod. Phys. E **24**, 1530005 (2015)
[86] K.D. Launey, T. Dytrych, and J.P. Draayer, Prog. Part. Nucl. Phys. **95**, 044312 (2017)
[87] D. Bonatsos, H. Sobhani, and H. Hassanabadi, Eur. Phys. J. Plus **135**, 710 (2020)
[88] A. Martinou, D. Bonatsos, S. Sarantopoulou, I.E. Assimakis, S.K. Peroulis, and N. Minkov, Eur. Phys. J. A **57**, 83 (2021)
[89] A. Martinou, D. Bonatsos, T.J. Mertzimekis, K. Karakatsanis, I.E. Assimakis, S.K. Peroulis, S. Sarantopoulou, and N. Minkov, Eur. Phys. J. A **57**, 84 (2021)
[90] J. Cseh, Phys. Rev. C **101**, 054306 (2020)
[91] D. Bonatsos, A. Martinou, S. Sarantopoulou, I.E. Assimakis, S. Peroulis, and N. Minkov, Eur. Phys. J. ST **229**, 2367 (2020)
[92] A. Martinou *et al.*, these proceedings
[93] S. Sarantopoulou *et al.*, these proceedings

A mechanism for shape coexistence

Andriana Martinou[1]*

[1]Institute of Nuclear and Particle Physics, National Centre of Scientific Research "Demokritos", GR-15310 Aghia Paraskevi, Attiki, Greece

Abstract. The phenomenon of shape coexistence in a nucleus is about the occurence of two different nuclear states with drastically different shapes, lying close in energy. It is commonly seen in the data, that such coexisting states manifest in specific nuclei, which lie within certain islands on the nuclear chart, the islands of shape coexistence. A recently introduced mechanism predicts that these islands derive from the coexistence of two different types of magic numbers: the harmonic oscillator and the spin-orbit like. The algebraic realization of the nuclear Shell Model, the Elliott $SU(3)$ symmetry, along with its extension, the proxy-$SU(3)$ symmetry , are used for the parameter-free theoretical predictions of the islands of shape coexistence.

1 Introduction

The studies about shape coexistence began with the excited 0_2^+ state of ^{16}O [1]. It was then suggested that this excited state of ^{16}O can be explained by a particle-hole (p-h) excitation across the harmonic oscillator shell among the magic numbers 2-8. This suggestion triggered the idea of particle-hole excitations, which became later on the state-of-the-art mechanism for shape coexistence [2].

Several years later, another phenomenon, that of the inversion of states, became known in the region of the Mg isotopes [3–7]. Inversion of states occurs, when the predicted spin and parity of the ground state contradicts with the data. Specifically the predictions are produced with the use of the Shell Model [8, 9] and with the assumption that the particles occupy the valence harmonic oscillator shell, which lies among the magic numbers 8-20 in the Mg isotopes.

The intense study in this region led to the conclusion that the inversion of states begins and ends at a certain neutron number N across an isotopic chain [3]. Thus discussion about the "islands of inversion" began [10, 11]. Both theorists and experimentalists drew the conclusion that the inverted states are due to a partial particle configuration in the valence harmonic oscillator shell and in the slightly excited spin-orbit like shell [12, 13]. The spin-orbit like shells [14] lie among the nucleon magic numbers 6, 14, 28, 50, 82, 126, ..., which are created by a significant spin-orbit force [9]. So the idea that particles are excited from the the harmonic oscillator shell to the neighboring spin-orbit like shell has been established in the light nuclei.

The same idea has been used in Ref. [15] to predict the islands of shape coexistence in the medium mass and heavy nuclei. Again it is supposed that the islands of shape coexistence can be predicted, if one assumes that the particles occupy the valence harmonic oscillator shell along with the valence spin-orbit like shell. But since in medium mass and heavy nuclei, the particles usually already are in the spin-orbit like shells among magic numbers 28, 50, 82, 126, we cannot speak about particle excitations from the valence harmonic oscillator shell to the valence spin-orbit like shell. Instead, the situation is better described by the phrase "shell merging" [12, 13]. Shell merging occurs when the single-particle energy gaps dissolve due to the onset of deformation and thus the valence harmonic oscillator shell merges with the valence spin-orbit like shell. So the key to enter into an island of shape coexistence is the deformation.

Generally the sudden onset of deformation in specific nuclei is attributed to the proton-neutron interaction, when the last protons and neutrons occupy specific Shell Model orbitals, namely the spin-orbit partners or the Federman-Pittel pairs introduced in Ref. [16–18]. Actually the particle-hole excitation mechanism, which is the state-of-the-art mechanism for shape coexistence, is based on the Federman-Pittel pairs (see Ref. [19]). The main idea is that the protons lie near a magic or semi-magic number, above which large single-particle energy gaps extend and that the protons are being excited, so as to create spin-orbit partners with the neutrons. The states created by the spin-orbit partners are highly deformed and thus lower in energy. Obviously the magic or semi-magic numbers play a crucial role in the particle-hole excitations mechanism. Thus, in regions where the magic or semi-magic numbers are not justified, the particle-hole excitation mechanism cannot be used as an explanation for the occurrence of shape coexistence, or inversely, where shape coexistence appears, sub-shell gaps have to be justified. For further discussion see for instance subsection IIIA3 of Ref. [2].

In Ref. [15] an alternative mechanism, namely the dual-shell mechanism, has been introduced in order to explain the phenomenon of shape coexistence. This mechanism has similarities with the particle-hole excitation

*e-mail: martinou@inp.demokritos.gr

mechanism in light nuclei (see section 4 of Ref. [15]) and aligns with the concept of shell merging in medium mass and heavy nuclei (see section 5 of Ref. [15]). The mechanism is realized within the Elliott [20–23] and the proxy-$SU(3)$ [24–28] symmetry. Deformed nuclei are adequately described by the Shell Model $SU(3)$ symmetry, or nowadays called the Elliott $SU(3)$ symmetry [20–23], which is appropriate for the harmonic oscillator shells, along with its extension, the proxy-$SU(3)$ symmetry [24–28], which suits to the spin-orbit like shells. These two algebraic realizations of the nuclear Shell Model use the Shell Model single-particle basis, but in the cartesian coordinate system, in order to give parameter-free predictions about the nuclear deformation.

2 The dual-shell mechanism

The Elliott $SU(3)$ three-dimensional (3D) isotropic harmonic oscillator (HO) Hamiltonian is the sum [29]:

$$H_0 = \sum_{i=1}^{A} h_{0,i}, \tag{1}$$

where A is the mass number and h_{0i} is the Hamiltonian of a single particle i moving into a three dimensional isotropic harmonic oscillator potential. The overall algebraic quadrupole-quadrupole (QQ) interaction in the Elliott $SU(3)$ model is calculated by the formula [30, 31]:

$$QQ = 4C_2 - 3L(L+1), \tag{2}$$

where L is the orbital angular momentum [20, 29] and $\hat{C}_2$ is the second order Casimir operator of the $SU(3)$ algebra obtaining eigenvalues [23, 32]:

$$C_2 = \lambda^2 + \mu^2 + \lambda\mu + 3(\lambda + \mu), \tag{3}$$

with (λ, μ) being the quantum numbers of the Elliott $SU(3)$ symmetry. The C_2 operator is proportional to the square of the nuclear quadrupole deformation parameter β of the Bohr and Mottelson Collective Model [33, 34]:

$$\beta^2 \sim (C_2 + 3). \tag{4}$$

The simplest $SU(3)$ Hamiltonian is [35]:

$$H = H_0 - \frac{\chi}{2}QQ. \tag{5}$$

The parameter χ is the strength of the quadrupole-quadrupole interaction and is given by a global expression in Ref. [35]:

$$\chi = \frac{\hbar\omega}{4N_0}. \tag{6}$$

Alternatively, instead of χ, the parameter κ can be used [35]:

$$\kappa \approx 2\chi. \tag{7}$$

This doubling of the quadrupole strength is in the same spirit with the scaling factor, used in Ref. [25], for the β^2 values. Generally in the $SU(3)$ nuclear models the deformation parameter β^2 and the $B(E2)s$ obtain smaller values than those in the data. So, a scaling is to be used and this is justified by the fact that the $SU(3)$ models use the algebraic quadrupole operator, which is restricted within one nuclear shell, while the data are attributed to the collective quadrupole operator, which applies to to many nuclear shells with the same parity. Nevertheless, once the choice of χ or κ is done, the quadrupole strength will be given by a global expression and it is not treated as a free parameter.

Using the Hamiltonian of Eq. (5) the energy difference among the two bandheads of shape coexistence is the eigenvalue of the difference:

$$H_{HO} - H_{SO} = (H_{0,HO} - H_{0,SO}) + \frac{\chi}{2}(QQ_{SO} - QQ_{HO}), \tag{8}$$

where the subscripts SO and HO stand for the spin-orbit like and the harmonic oscillator valence shells respectively. The difference

$$\Delta(H_0) = H_{0,HO} - H_{0,SO} \leq 0 \tag{9}$$

obtains negative values [15], while

$$\Delta(\frac{\chi}{2}QQ) = (\frac{\chi}{2}QQ)_{SO} - (\frac{\chi}{2}QQ)_{HO} \geq 0 \tag{10}$$

is the difference in the quadrupole terms and obtains positive values within the regions of shape coexistence. Thus the two differences, due to their opposite signs, compete with each other in the Hamiltonian (8) and the energies of the two bandheads of shape coexistence lie close in energy. The islands of shape coexistence appear when the deformation prevails [15]:

$$\Delta(\frac{\chi}{2}QQ) \geq -\Delta(H_0) \geq 0 \tag{11}$$

and till the harmonic oscillator shell closure. The above condition applies for the proton or neutron configurations separately. Thus it implies that the proton (neutron) configuration of the spin-orbit like shell lies lower in energy, than the one deriving from the harmonic oscillator shell. But the situation for the coupled proton-neutron configuration is uncertain and it might result that the configuration of the spin-orbit shell lies higher in energy. Thus, the condition $\Delta(\frac{\chi}{2}QQ) \geq 0$ leads to:

$$C_{2,SO} - C_{2,HO} \geq 0, \tag{12}$$

for the proton or neutron configuration separately and gives the *widest* possible regions of shape coexistence predicted by the dual-shell mechanism.

The values of the C_2 are determined in Eq. (3) by the $SU(3)$ irrep. In the literature, there exist two kinds of Elliott $SU(3)$ irreps as candidates for the description of the low lying in energy nuclear states: the highest weight irrep $(\lambda, \mu)_{hw}$ and the most deformed irrep $(\lambda, \mu)_{def}$. Up to the middle of each valence shell the two irreps coincide $(\lambda, \mu)_{hw} = (\lambda, \mu)_{def}$ and consequently the C_2 operators obtained by them with the use of Eq. (3) are the same, $C_{2,hw} = C_{2,def}$. But after the middle of the shell and before the shell closure the two irreps become different $(\lambda, \mu)_{hw} \neq (\lambda, \mu)_{def}$ with $C_{2,hw} \leq C_{2,def}$. In Ref. [25]

the $(\lambda, \mu)_{def}$ is labeled by C, which means that it is the irrep with the maximum value of the C_2 for a given number of protons or neutrons in a given valence shell. In Refs. [25, 28, 36] it is stressed out that the highest weight irrep has to be used, because this irrep gives the right prolate-oblate transition. Consequently the widest possible islands of shape coexistence determined by the condition (12) are predicted with the use of the highest weight irreps for both spin-orbit like shells and harmonic oscillator shells. The relevant irreps are tabulated in Ref. [15].

3 Results

The condition (12) is satisfied among the proton or neutron numbers:

Z or N : $7-8, 17-20, 34-40, 59-70, 96-112, 145-168$.

These results have been obtained using the highest weight $SU(3)$ irreps [27] of the Elliott and the proxy-$SU(3)$ symmetry [15, 25]. The plots appear in Figs. 1-6. The proton and neutron numbers indicated, identify the possible regions on the nuclear chart, where one may see coexisting structures, with drastically different shapes, lying close in energy.

Examples of shape coexistence appear in every of the above proton or neutron regions. For instance we see parity inversion in neutron numbers $N = 7-8$ in the Be isotopes [37, 38], inversion of states at neutron numbers $N \geq 17$ in the Mg isotopes [3–5], a comeback of the HO magic numbers in the deformation parameter β in Ni isotopes among neutron numbers $34 \leq N \leq 40$ [15], shape coexistence in the Se, Kr, Sr, Zr isotopes at proton numbers $Z = 34, 36, 38, 40$ respectively [2], shape coexistence among $60 \leq N \leq 70$ in the Sn isotopes [39], or after $N = 60$ in Kr isotopes [40], shape coexistence in the rare earths among proton numbers $60 \leq Z \leq 70$ [2], shape coexistence among $96 \leq N \leq 110$ in the Hg isotopes and fission isomers around $N = 145$ in the superheavies [41].

Generally it is not implied that one should see shape coexistence in each and every isotope of the map in Fig. 25 of Ref. [15], but instead it is stated that the regions marked in this map are the widest possible islands of shape coexistence predicted by the dual-shell mechanism.

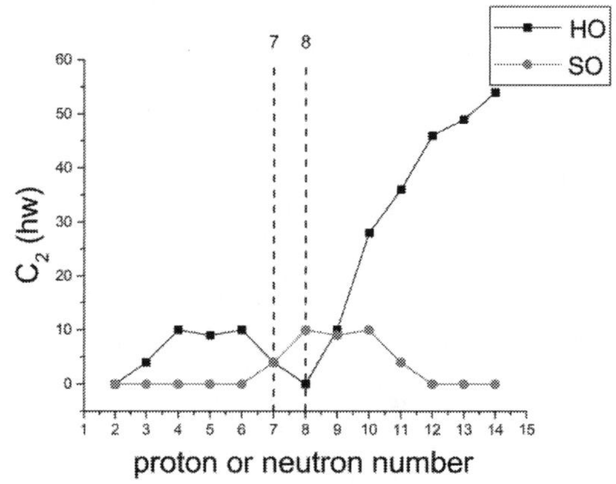

Figure 1. The eigenvalues of the second order Casimir operator of the $SU(3)$ algebra versus the proton (Z) or neutron number (N). Equation (3) has been used along with the irreps (λ, μ) of Table 1 of Ref. [15]. Shape coexistence is predicted among proton or neutron numbers $7-8$.

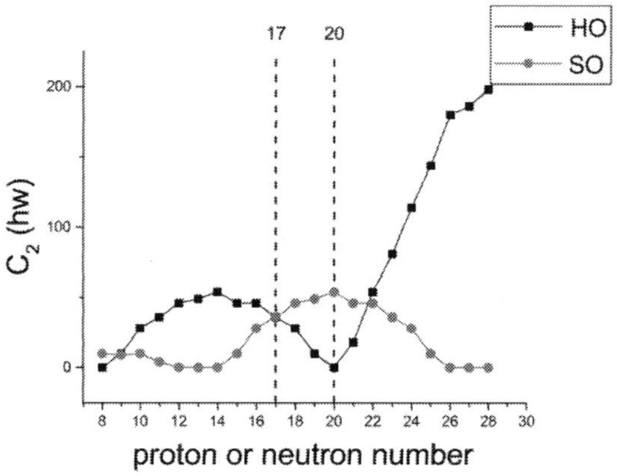

Figure 2. The same as Fig. 1. An island of shape coexistence is predicted within proton or neutron numbers $17-20$.

4 Conclusions

A unified view of the islands of inversion and of shape coexistence has been described. The mechanism supposes that the coexisting structures can occur when the proton or the neutron deformation due the valence spin-orbit like shell is larger than the one of the valence harmonic oscillator shell. This requirement is satisfied in certain islands on the nuclear chart, which coincide with the data about the islands of inversion and the islands of shape coexistence [2].

Nevertheless, although the dual-shell mechanism is promising, since it explains the manifestation of shape coexistence in various well-known paradigm nuclei, it is not at all a complete theory. Thus we do not expect to observe shape coexistence in each an every nucleus of the map of Fig. 25 of Ref. [15]. Much has to be done. The effect of the:

a) proton-neutron interaction,

b) merging of the spin-orbit like shell with the harmonic oscillator shell,

c) realistic interactions in the Hamiltonian (pairing and spin-orbit interaction for instance)

on the predicted islands of shape coexistence, has not been examined yet. The problem is that although studies for the these interactions within the Elliott $SU(3)$ framework began in the late 60's [23], they have never been completed. This dual-shell mechanism appears as a motivation to keep on studying the Elliott $SU(3)$ symmetry, because

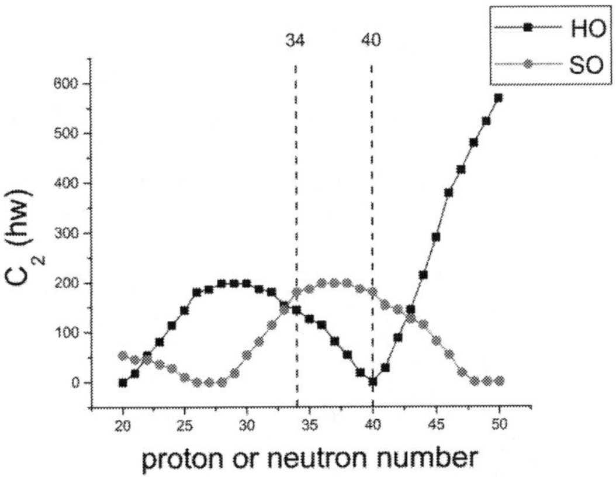

Figure 3. The same as Fig. 1. An island of shape coexistence is predicted within proton or neutron numbers 34 – 40.

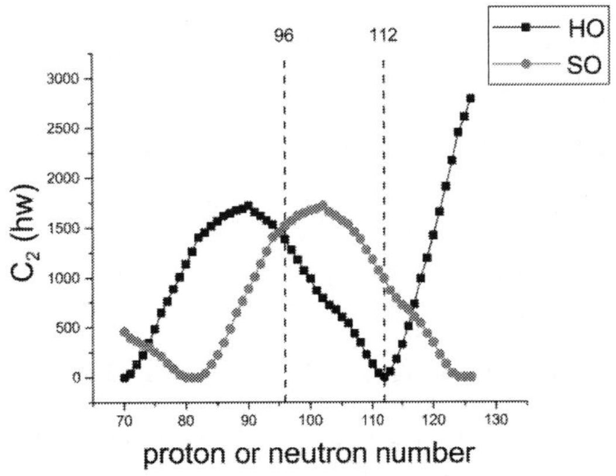

Figure 5. The same as Fig. 1. An island of shape coexistence is predicted within proton or neutron numbers 96 – 112.

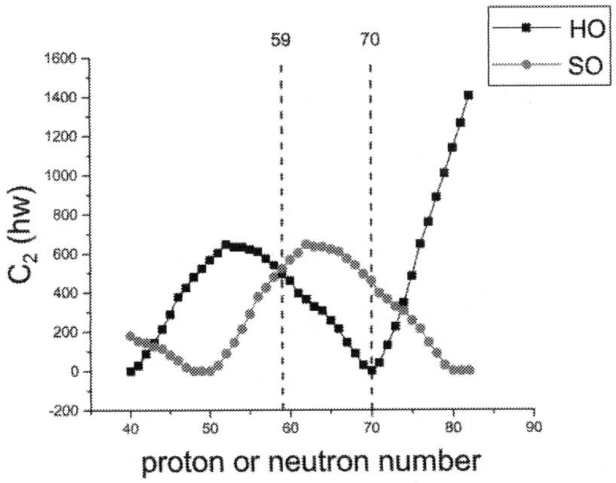

Figure 4. The same as Fig. 1. An island of shape coexistence is predicted within proton or neutron numbers 59 – 70.

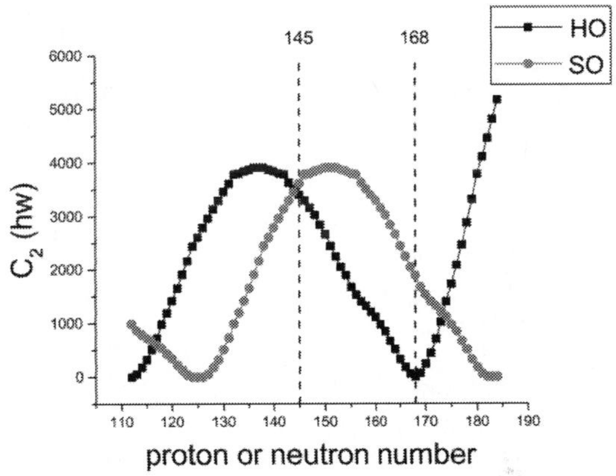

Figure 6. The same as Fig. 1. An island of shape coexistence is predicted within proton or neutron numbers 145 – 168.

it is likely to give elegant solutions into burning problems of nuclear structure.

Acknowledgements

This research is co-financed by Greece and the European Union (European Social Fund- ESF) through the Operational Programme "Human Resources Development, Education and Lifelong Learning" in the context of the project "Reinforcement of Postdoctoral Researchers-2nd Cycle" (MIS-5033021), implemented by the State Scholarships Foundation (IKY).

Ευρωπαϊκή Ένωση
Operational Programme Human Resources Development, Education and Lifelong Learning
Co-financed by Greece and the European Union
ΕΣΠΑ 2014-2020

References

[1] H. Morinaga, Phys. Rev. **101**, 254 (1956)

[2] K. Heyde, J.L. Wood, Rev. Mod. Phys. **83**, 1467 (2011)

[3] P. Himpe, G. Neyens, D. Balabanski, G. Bélier, J. Daugas, F. de Oliveira Santos, M.D. Rydt, K. Flanagan, I. Matea, P. Morel et al., Phys. Lett. B **658**, 203 (2008)

[4] M. Kowalska, D.T. Yordanov, K. Blaum, P. Himpe, P. Lievens, S. Mallion, R. Neugart, G. Neyens, N. Vermeulen, Phys. Rev. C **77**, 034307 (2008)

[5] G. Neyens, M. Kowalska, D. Yordanov, K. Blaum, P. Himpe, P. Lievens, S. Mallion, R. Neugart, N. Vermeulen, Y. Utsuno et al., Phys. Rev. Lett. **94**, 022501 (2005)

[6] W. Schwerdtfeger, P.G. Thirolf, K. Wimmer, D. Habs, H. Mach, T.R. Rodriguez, V. Bildstein, J.L. Egido, L.M. Fraile, R. Gernhäuser et al., Phys. Rev. Lett. **103**, 012501 (2009)
[7] K. Wimmer, T. Kröll, R. Krücken, V. Bildstein, R. Gernhäuser, B. Bastin, N. Bree, J. Diriken, P. Van Duppen, M. Huyse et al., Phys. Rev. Lett. **105**, 252501 (2010)
[8] M.G. Mayer, Phys. Rev. **74**, 235 (1948)
[9] O. Haxel, J.H.D. Jensen, H.E. Suess, Physical Review **75**, 1766 (1949)
[10] A. Poves, *Shape Coexistence and Islands of Inversion Monopole vs Multipole*, in *Proceedings of the Ito International Research Center Symposium "Perspectives of the Physics of Nuclear Structure"* (Journal of the Physical Society of Japan, 2018)
[11] B.A. Brown, Physics **3**, 104 (2010)
[12] F. Nowacki, A. Poves, J. Phys.: Conference Series **966**, 012023 (2018)
[13] E. Caurier, F. Nowacki, A. Poves, Phys. Rev. C **90**, 014302 (2014)
[14] O. Sorlin, M.G. Porquet, Progr. Part. Nucl. Phys. **61**, 602 (2008)
[15] A. Martinou, D. Bonatsos, T.J. Mertzimekis, K.E. Karakatsanis, I.E. Assimakis, S.K. Peroulis, S. Sarantopoulou, N. Minkov, European Physical Journal A **57**, 84 (2021)
[16] P. Federman, S. Pittel, Phys. Lett. B **69**, 385 (1977)
[17] P. Federman, S. Pittel, Physics Letters B **77**, 29 (1978)
[18] P. Federman, S. Pittel, Phys. Rev. C **20**, 820 (1979)
[19] K. Heyde, P.V. Isacker, R. Casten, J. Wood, Phys. Lett. B **155**, 303 (1985)
[20] J.P. Elliott, Proceedings of the Royal Society of London. Series A. Mathematical and Physical Sciences **245**, 128 (1958)
[21] J.P. Elliott, Proceedings of the Royal Society of London. Series A. Mathematical and Physical Sciences **245**, 562 (1958)
[22] J.P. Elliott, M. Harvey, Proceedings of the Royal Society of London. Series A. Mathematical and Physical Sciences **272**, 557 (1963)
[23] J.P. Elliott, C.E. Wilsdon, Proceedings of the Royal Society of London. Series A. Mathematical and Physical Sciences **302**, 509 (1968)
[24] D. Bonatsos, I.E. Assimakis, N. Minkov, A. Martinou, R.B. Cakirli, R.F. Casten, K. Blaum, Phys. Rev. C **95**, 064325 (2017)
[25] D. Bonatsos, I.E. Assimakis, N. Minkov, A. Martinou, S. Sarantopoulou, R.B. Cakirli, R.F. Casten, K. Blaum, Phys. Rev. C **95**, 064326 (2017)
[26] D. Bonatsos, European Physical Journal A **53**, 148 (2017)
[27] A. Martinou, D. Bonatsos, N. Minkov, I.E. Assimakis, S.K. Peroulis, S. Sarantopoulou, J. Cseh, European Physical Journal A **56**, 239 (2020)
[28] A. Martinou, D. Bonatsos, K.E. Karakatsanis, S. Sarantopoulou, I.E. Assimakis, S.K. Peroulis, N. Minkov, European Physical Journal A **57**, 83 (2021)
[29] H.J. Lipkin, *Lie groups for pedestrians* (Dover, New York, 2002)
[30] J. Draayer, K.J. Weeks, Ann. Phys. **156**, 41 (1984)
[31] J.P. Draayer, *Algebraic approaches to nuclear structure: Interacting boson and fermion models* (Harwood Academic Publishers, New York, United States, 1993), Vol. 6 of *Contemporary Concepts in Physics*, chap. 7, pp. 423–546, ISBN 3-7186-0538-4
[32] P.O. Lipas, *Algebraic approaches to nuclear structure : interacting boson and fermion models* (Harwood Academic Publishers, 1993), Vol. 6 of *Contemporary Concepts in Physics*, chap. 2, pp. 47–12
[33] A. Bohr, B.R. Mottelson, *Nuclear Structure*, Vol. II (World Scientific Publishing Company, 1998)
[34] O. Castaños, J.P. Draayer, Y. Leschber, Zeitschrift für Physik A Atomic Nuclei **329**, 33 (1988)
[35] D.J. Rowe, G. Thiamova, J.L. Wood, Phys. Rev. Lett. **97**, 202501 (2006)
[36] D. Bonatsos, A. Martinou, S. Sarantopoulou, I.E. Assimakis, S.K. Peroulis, N. Minkov, Eur. Phys. J. Spec. Top. **229**, 2367 (2020)
[37] Y. Hirayama, T. Shimoda, H. Izumi, H. Yano, M. Yagi, A. Hatakeyama, C. Levy, K. Jackson, H. Miyatake, Nucl. Phys. A **746**, 71 (2004)
[38] F. Cappuzzello, A. Cunsolo, S. Fortier, A. Foti, M. Khaled, H. Laurent, H. Lenske, J. Maison, A. Melita, C. Nociforo et al., Phys. Lett. B **516**, 21 (2001)
[39] J.L. Wood, K. Heyde, W. Nazarewicz, M. Huyse, P.V. Duppen, Phys. Rep. **215**, 101 (1992)
[40] F. Flavigny, P. Doornenbal, A. Obertelli, J.P. Delaroche, M. Girod, J. Libert, T. Rodriguez, G. Authelet, H. Baba, D. Calvet et al., Physical Review Letters **118** (2017)
[41] P. Thirolf, D. Habs, Progr. Part. Nucl. Phys. **49**, 325 (2002)

Highest weight irreducible representations favored by nuclear forces within SU(3)-symmetric fermionic systems

Andriana Martinou[1], *S.* Sarantopoulou[1], *K.E.* Karakatsanis[1,2], and *Dennis* Bonatsos[1,*]

[1]Institute of Nuclear and Particle Physics, National Centre of Scientific Research "Demokritos", GR-15310 Aghia Paraskevi, Attiki, Greece

[2]Department of Physics, University of Zagreb, HR-10000 Zagreb, Croatia

Abstract. The consequences of the attractive, short-range nucleon-nucleon (NN) interaction on the wave functions of nuclear models bearing the SU(3) symmetry are reviewed. The NN interaction favors the most symmetric spatial SU(3) irreducible representation (irrep), which corresponds to the maximal spatial overlap among the fermions. The consideration of the highest weight (hw) irreps in nuclei and in alkali metal clusters, leads to the prediction of a prolate to oblate shape transition beyond the mid–shell region. Subsequently, the consequences of the use of the hw irreps on the binding energies and two-neutron separation energies in the rare earth region are discussed within the proxy-SU(3) scheme, by considering a very simple Hamiltonian, containing only the three dimensional (3D) isotropic harmonic oscillator (HO) term and the quadrupole-quadrupole interaction. This Hamiltonian conserves the SU(3) symmetry and treats the nucleus as a rigid rotator.

1 Most important SU(3) irreducible representations in atomic nuclei

The SU(3 symmetry has a long history of being used in nuclear structure [1]. Basic features of several nuclear observables are determined, often in a parameter-free way, by the irreducible representations (irreps) [2, 3] in which the various energy bands live. In the simplest possible terms, the irreps represent the internal structure of the states bearing specific symmetry characteristics imposed by SU(3). The SU(3) irreps in nuclear physics are usually described in the Elliott notation [4–9] by (λ, μ), with $\lambda = f_1 - f_2$ and $\mu = f_2$, where f_1, f_2 are the number of boxes in the first and the second line of the relevant Young diagram [2, 3], exhibiting pictorially the symmetrization and antisymmetrization contents of the irrep, since boxes on the same horizontal line correspond to symmetrization, while boxes in the same vertical column represent antisymmetrization.

One can distinguish two large families of algebraic collective models, based on their building blocks, which can be either fermions (as initially expected from the microscopic point of view, since nucleons are fermions) or bosons (under specific assumptions and approximations, allowing a system of correlated fermions to be described by bosons, as boson mapping techniques [10] indicate).

Initially introduced by Elliott in 1958 [4, 5] in the sd shell, the fermionic SU(3) symmetry is based on the SU(3) symmetry of the three-dimensional (3D) isotropic harmonic oscillator (HO) [2, 3, 11]. The 3D isotropic HO SU(3) symmetry is known to be destroyed in higher shells by the spin-orbit interaction, which is the essential ingredient leading to the interpretation of the nuclear magic numbers 2, 8, 20, 28, 50, 82, 126, ... through the shell model [12–15]. Possible extensions of the SU(3) symmetry to the shells above the sd shell have been proposed within the pseudo-SU(3) model (1973) [16–27], the quasi-SU(3) model (1995) [28–30], and the proxy-SU(3) model (2017) [31–37]. In addition, the symplectic model (1980) [38, 39] and the Fermion Dynamical Symmetry Model (FDSM, 1987) [40] have been introduced. Within these models the ground state band (gsb) and possibly additional collective bands (like the γ_1 band, the first $K = 4$ band, etc) are contained within a (λ, μ) irrep with $\mu > 0$.

SU(3) symmetry is also present within bosonic models, like the Vector Boson Model (VBM, 1972) [41–47], the Interacting Boson Model (IBM, 1975) [48–51], and the Interacting Vector Boson Model (IVBM, 1982) [52, 53]. Within the popular IBM, the gsb sits alone within the SU(3) irrep $(2N, 0)$, where N is the number of bosons corresponding to the relevant nucleus, while the β_1 and γ_1 collective bands are accommodated within the $(2N - 4, 2)$ irrep.

The drastic difference between fermionic models and the IBM as far as the irrep accommodating the gsb is concerned, is a consequence of the short range nature [54, 55] of the nucleon-nucleon interaction. In a fermionic system, the overall wave function has to be antisymmetric. The short range nature of the nucleon-nucleon interaction requires the irrep for the spatial part of the wave function to be as symmetric as possible, the spin-isospin part of the wave function characterized by the irrep conjugate to the spatial one. Looking into the mathematical details [36] one sees that this "balance" between the spatial and the

*e-mail: bonat@inp.demokritos.gr

spin-isospin part of the wave function leads in most cases to $\mu > 0$. In the case of bosons of one type, as it is the case in the simplest version of the IBM (called IBM-1) [49], there are no restrictions imposed by the spin-isospin part of the wave function, since on one hand the difference between protons and neutrons is ignored, thus isospin plays no role, while on the other hand the Pauli principle is absent, since we are dealing with bosons. Therefore the spatial part of the wave function can become totally symmetric without meeting any spin-isospin obstacle, thus acquiring $\mu = 0$.

Even within fermionic models, the following question concerning the irrep in which the gsb sits arises. Should this be the highest weight (hw) irrep or the irrep with the highest eigenvalue of the second order Casimir operator of SU(3), which is known to correspond to the highest eigenvalue of the operator representing the quadrupole-quadrupole interaction, being responsible for nuclear deformation? Detailed arguments presented in Ref. [36] prove that the hw irrep is the one which should be used. In particular, the percentage of the symmetric components out of the total in a SU(3) wave function is introduced, through which it is found, that no SU(3) irrep is more symmetric than the hw irrep for a given number of valence particles in a 3D isotropic HO shell. This choice makes no difference in the lower part of each shell, but it shows up in the upper half of each shell, having as a consequence the occurrence of a transition from prolate to oblate shapes, seen experimentally both in rare earth nuclei and in metallic atomic clusters. It is remarkable that the hw irrep provides similar predictions for the nuclear shapes within both the pseudo-SU(3) and proxy-SU(3) models, despite the significant differences in their assumptions [56]. A by-product of the use of the hw irrep is the theoretical justification of the dominance of prolate over oblate shapes in the ground states of even-even nuclei [32, 36], a puzzle which has stayed unresolved for a long time [57, 58].

In the remainder of this work we are going to examine the consequences of choosing the hw irrep on the calculation of binding energies and two-neutron separation energies within the proxy-SU(3) scheme.

2 Binding energies within the proxy-SU(3) scheme

The simplest Hamiltonian of the Elliott SU(3) Model [4–6] is used (see Chapter 4 of Ref. [59]):

$$H = H_0 - \frac{\chi}{2} QQ, \tag{1}$$

where H_0 is the 3D isotropic HO Hamiltonian for the many-particle system:

$$H_0 = \sum_{i=1}^{A} \left(\frac{\mathbf{p}_i^2}{2m} + \frac{1}{2} m\omega^2 \mathbf{r}_i^2 \right) \tag{2}$$

with $\mathbf{p}_i, \mathbf{r}_i, m$ being the momentum, the position and the mass of the particle, while ω is the oscillation frequency and A is the mass number. If one sets $\hbar\omega = 1$ in Eq. (2), then the eigenvalue of the H_0 is labeled N_0 and is calculated by the expression [60]:

$$N_0 = \sum_{i=1}^{A} \left(N_i + \frac{3}{2} \right) \tag{3}$$

with N_i being the number of the harmonic oscillator quanta of the orbit of the ith particle.

The parameter $\chi/2$ is the strength of the quadrupole-quadrupole (QQ) interaction, given by [61]:

$$\frac{\chi}{2} = \frac{\hbar\omega}{8N_0}, \tag{4}$$

where N_0 is calculated by Eq. (3), while [62]:

$$\hbar\omega = \frac{41}{A^{1/3}} \mathrm{MeV}. \tag{5}$$

In the Elliott SU(3) symmetry the many-quanta wave function is identified by the quantum numbers (λ, μ). For the nuclear ground state band, the (λ, μ) refer to the highest weight irreducible representations (irreps). The second order Casimir operator of the SU(3) symmetry is given by [7, 22]:

$$C_2 = \lambda^2 + \mu^2 + \lambda\mu + 3(\lambda + \mu). \tag{6}$$

If L is the orbital angular momentum of the nuclear state, then the value of the quadrupole-quadrupole interaction in a SU(3) wave function is [22]:

$$QQ = 4C_2 - 3L(L+1). \tag{7}$$

Supposing that the depth of the 3D isotropic HO potential is V_0 the binding energy for a nucleus with Z protons and N neutrons is given by:

$$BE(Z, N) = AV_0 - \left(H_0 - \frac{\chi}{2} QQ \right). \tag{8}$$

The above represents the minimum energy amount one has to offer, in order to extract the particles from the 3D-HO potential. The value of the depth of the 3D-HO potential is given approximately by [62]:

$$V_0 \approx 50 \left(1 - \frac{N - Z}{A} \right) \mathrm{MeV}, \tag{9}$$

but in the calculation of the binding energies with the proxy-SU(3) symmetry we have treated V_0 as a free parameter. The values of V_0 have been found using a polynomial fit $V_0 = c_1 + c_2 N + c_3 N^2$ on the data. In the rare earths this fitting procedure has given $V_0 \approx 44$ MeV, which is rational in comparison with the results of Eq. (9).

The two proton (S_{2p}) or two neutron (S_{2n}) separation energies are given by:

$$S_{2p} = BE(Z, N) - BE(Z-2, N), \tag{10}$$

$$S_{2n} = BE(Z, N) - BE(Z, N-2). \tag{11}$$

2.1 Calculation of N_0 in the harmonic oscillator shells

The Elliott SU(3) symmetry is valid for the 3D isotropic HO shells among the magic numbers 2, 8, 20, 40, 70, 112,

168... Each 3D-HO shell has degeneracy $(\mathcal{N}+1)(\mathcal{N}+2)$. So the first shell with $\mathcal{N}=0$ has up to 2 identical nucleons, the $\mathcal{N}=1$ shell has up to 6 identical nucleons etc.

Suppose for instance a nucleus with 6 protons and 8 neutrons, *i.e.* , ^{14}C. The first 2 protons are in an orbital with $\mathcal{N}=0$, while the rest 4 are in orbitals with $\mathcal{N}=1$. For the proton configuration:

$$N_{0,Z}=2\left(0+\frac{3}{2}\right)+4\left(1+\frac{3}{2}\right)=13. \qquad (12)$$

Similarly the 8 neutrons are distributed as follows: the first 2 have $\mathcal{N}=0$ and the last 6 are in $\mathcal{N}=1$ orbitals. So for the neutron configuration:

$$N_{0,N}=2\left(0+\frac{3}{2}\right)+6\left(1+\frac{3}{2}\right)=18. \qquad (13)$$

Finally the overall N_0 is given by the summation:

$$N_0=N_{0,Z}+N_{0,N}, \qquad (14)$$

which means that ^{14}C is characterized by $N_0=31$.

In a similar fashion one may reproduce the values of N_0 given in Table 1 for the 3D-HO shells.

Table 1. N_0 values for the 3D isotropic HO shells.

p	N_0	$\mathcal{N}$	p	N_0	$\mathcal{N}$	p	N_0	$\mathcal{N}$
2	3	0	42	161	4	86	419	5
			44	172		88	432	
4	8	1	46	183		90	445	
6	13		48	194		92	458	
8	18		50	205		94	471	
			52	216		96	484	
10	29	2	54	227		98	497	
12	32		56	238		100	510	
14	38		58	249		102	523	
16	46		60	260		104	536	
18	53		62	271		106	549	
20	60		64	282		108	562	
			66	293		110	575	
22	69	3	68	304		112	588	
24	78		70	315				
26	87					114	603	6
28	96		72	328	5	116	618	
30	105		74	341		118	633	
32	114		76	354		120	648	
34	123		78	367		122	663	
36	132		80	380		124	678	
38	141		82	393		126	693	
40	150		84	406				

2.2 Calculation of N_0 in the proxy, spin-orbit like shells

The spin-orbit like shells extend among the spin-orbit like magic numbers 6, 14, 28, 50, 82, 126... Such shells consist by the normal parity orbitals with $\mathcal{N}$ number of quanta and by the intruder orbitals with $\mathcal{N}+1$ quanta. Within the proxy-SU(3) symmetry [31], a unitary transformation [34] can be applied in the intruder orbitals. This transformation reduces the number of quanta by 1 unit, which means that $\mathcal{N}+1\rightarrow\mathcal{N}$, while the last orbital, which lies highest in energy and therefore is empty for most nuclei, is ignored. Consequently the proxy, spin-orbit like magic numbers become 6-12, 14-26, 28-48, 50-80, 82-124... and each proxy, spin-orbit like shell consists by orbitals with common $\mathcal{N}_{proxy}$=1, 2, 3, 4, 5, 6, ... quanta respectively.

As a result if a nucleus has some valence neutrons in the spin-orbit like 6-14 shell, then the value of N_0 is the one of the closed core $N_{0,core}$ with 6 neutrons plus the $N_{0,val}$ due to the valence neutrons in the 6-12 proxy shell with $\mathcal{N}_{proxy}=1$:

$$N_{0,SO}=N_{0,core}+N_{0,val}. \qquad (15)$$

In the example of ^{14}C, if the 8 neutrons follow the 6-14 spin-orbit like shell then:

$$N_{0N,SO}=13+(8-6)\left(1+\frac{3}{2}\right), \qquad (16)$$

where the value 13 has been taken from Table 1 for the core with 6 particles, while (8-6) gives the number of valence neutrons in the 6-12 proxy shell, which in turn consists by orbitals with $\mathcal{N}_{proxy}=1$ quanta. In general:

$$N_{0,SO}=N_{0,core}+\sum_{i=1}^{A_{val}}\left(\mathcal{N}_{i,proxy}+\frac{3}{2}\right). \qquad (17)$$

2.3 Results

Using Eq. (8) we calculated the binding energies for the rare earth nuclei within the proxy-SU(3) symmetry. In order to make clear the difference between the highest weight irreps and the irreps possessing the highest eigenvalue of the second order Casimir operator, we have made two calculations, labeled respectively by "hw" and "Casimir" in the relevant figures. The necessary irreps, obtained from Tables I and II of Ref. [32], are shown in Table 2, along with the corresponding eigenvalues of the second order Casimir operator. Only the upper part of the neutron shell is shown, since in the lower part of the shell the "hw" and "C" irreps are identical and can be found in Table II of Ref. [32]. Results for the Nd isotopes are shown in Figs. 1 and 2. In the lower half of the neutron shell, in which data exist, both the "hw" and "C" claculations yield identical results, as expected. In the upper half of the shell the "hw" and "C" irreps coming out from Table I of Ref. [32] are considerably different, as seen in Table 2, but their influence on the binding energies is minimal, as seen in Figs. 1 and 2. The value of V_0 is the only free parameter used in this calculation. As seen in Fig. 3, $V_0\approx$ 44MeV for the Nd isotopes.

S_{2n} have been calculated using Eq. (11), the results for the Nd isotopes being shown in Fig. 4. The "hw" proxy-SU(3) predictions for the S_{2n} exhibit a staircase behavior, which is similar to the one exhibited by the data, while the "C" predictions in the upper half of the neutron shell are considerably different, especially immediately above the

Table 2. Highest weight SU(3) irreps, labeled by "hw", and SU(3) irreps possessing the highest eigenvalue of the second order Casimir operator of Eq. (6), labeled by "C", are listed for the upper half of the neutron 82-126 shell, together with the corresponding eigenvalues of the second order Casimir operator of SU(3), given by Eq. (6). See subsection 2.3 for further discussion.

N	irrep hw	irrep C	C_2 hw	C_2 C
104	54,12	20,44	3906	3408
106	50,16	26,40	3754	3514
108	48,16	28,38	3520	3490
110	48,12	26,38	3204	3300
112	50,4	20,40	2878	2980
114	40,14	24,34	2518	2722
116	32,20	24,30	1900	2358
118	26,22	20,28	1876	1888
120	22,20	22,20	1450	1450
122	20,14	20,14	978	978
124	20,4	20,4	568	568

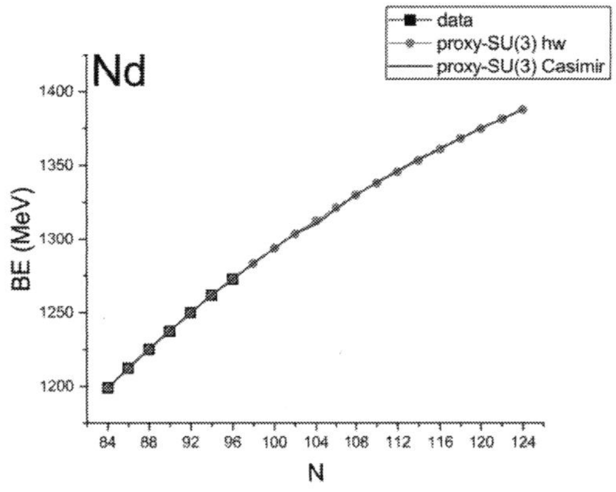

Figure 1. The predictions for the binding energies of the Nd isotopes within the proxy-SU(3) symmetry, calculated with the highest weight SU(3) irrep, labeled by "hw", or with the SU(3) irrep possessing the highest eigenvalue of the second order Casimir operator given by Eq. (6), labeled by "Casimir". The data have been taken from the Atomic Mass Evaluation 2016 [63]. See subsection 2.3 for further discussion.

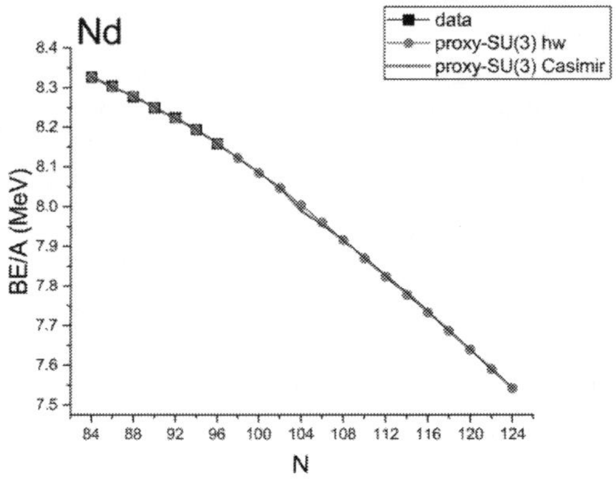

Figure 2. The predictions for the binding energies *BE* over the mass number *A* for the Nd isotopes within the proxy-SU(3) symmetry, calculated with the highest weight SU(3) irrep, labeled by "hw", or with the SU(3) irrep possessing the highest eigenvalue of the second order Casimir operator given by Eq. (6), labeled by "Casimir". The data have been taken from the Atomic Mass Evaluation 2016 [63]. See subsection 2.3 for further discussion.

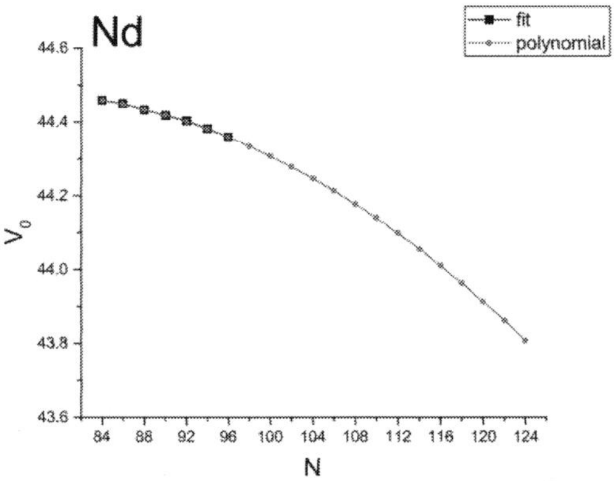

Figure 3. The fitted values of V_0 for the Nd isotopes. A polynomial fit $V_0 = c_1 + c_2 N + c_3 N^2$ has been used. The values of the constants $c_1 = 42.86274$, $c_2 = 0.04289$, $c_3 = -2.84426 \cdot 10^{-4}$ have been calculated using the *Origin* code.

midshell, where the Casimir eigenvalues of the "hw" and "C" cases differ most, as seen in Table 2. It is encouraging that good coincidence with the data is achieved in the "hw" case with the use of the simplest SU(3) Hamiltonian of Eq. (1). The inclusion of the spin-orbit interaction [7] will mix the highest weight SU(3) irrep with other SU(3) irreps (see Eq. 4.13b of Ref. [9]), thus we expect that the spin-orbit interaction will soften this staircase behavior. It would be interesting to extend the present calculations to higher Z, in which data exist in the upper half of the neutron shell, where the differences between the "hw" and "C" irreps are large, in order to see which of the two approaches leads to better agreement to the data. Work in this direction is in progress.

Acknowledgements

This research is co-financed by Greece and the European Union (European Social Fund- ESF) through the Operational Programme "Human Resources Development, Education and Lifelong Learning 2014-2020" in the context of the project "Nucleon Separation Energies" (MIS

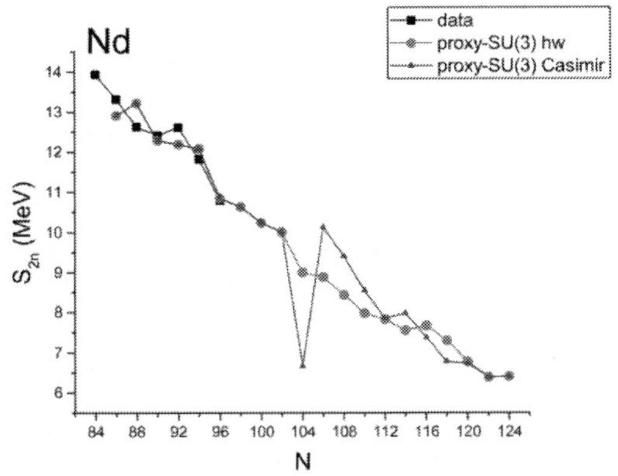

Figure 4. The two neutron separation energies for the Nd isotopes. Eq. (11) has been used for the calculation. Only the depth V_0 has been used as a free parameter, which has been fitted to data. Whenever no data exist, a linear fit has been used (see Fig. 3). Calculations have been carried out both with the highest weight SU(3) irrep, labeled by "hw", and with the SU(3) irrep possessing the highest eigenvalue of the second order Casimir operator given by Eq. (6), labeled by "Casimir". See subsection 2.3 for further discussion.

5047793).

Operational Programme Human Resources Development, Education and Lifelong Learning
Co-financed by Greece and the European Union
ΕΣΠΑ 2014-2020

References

[1] V.K.B. Kota, *SU*(3) *Symmetry in Atomic Nuclei* (Springer, Singapore, 2020)
[2] B. G. Wybourne, *Classical Groups for Physicists* (Wiley, New York, 1974)
[3] F. Iachello, *Lie Algebras and Applications* (Springer, Berlin, 2006)
[4] J. P. Elliott, Proc. Roy. Soc. London Ser. A **245**, 128 (1958)
[5] J. P. Elliott, Proc. Roy. Soc. London Ser. A **245**, 562 (1958)
[6] J. P. Elliott and M. Harvey, Proc. Roy. Soc. London Ser. A **272**, 557 (1963)
[7] J. P. Elliott and C. E. Wilsdon, Proc. Roy. Soc. London Ser. A **302**, 509 (1968)
[8] M. Harvey, in Adv. Nucl. Phys. **1**, ed. M. Baranger and E. Vogt (Plenum, New York, 1968) p. 67
[9] C. E. Wilsdon, PhD Thesis, University of Sussex (1965), https://ethos.bl.uk/Home.do, Ethos ID: 477554
[10] A. Klein and E. R. Marshalek, Rev. Mod. Phys. **63**, 375 (1991)
[11] M. Moshinsky and Yu. F. Smirnov, *The Harmonic Oscillator in Modern Physics* (Harwood, Amsterdam, 1996)
[12] M. G. Mayer, Phys. Rev. **74**, 235 (1948)
[13] M. G. Mayer, Phys. Rev. **75**, 1969 (1949)
[14] O. Haxel, J.H.D. Jensen, and H.E. Suess, Phys. Rev. **75**, 1766 (1949)
[15] M.G. Mayer and J.H.D. Jensen, *Elementary Theory of Nuclear Shell Structure* (Wiley, New York, 1955)
[16] K.T. Hecht and A. Adler, Nucl. Phys. A **137**, 129 (1969)
[17] A. Arima, M. Harvey, and K. Shimizu, Phys. Lett. B **30**, 517 (1969)
[18] R. D. Ratna Raju, J. P. Draayer, and K. T. Hecht, Nucl. Phys. A **202**, 433 (1973)
[19] J. P. Draayer, K. J. Weeks, and K. T. Hecht, Nucl. Phys. A **381**, 1 (1982)
[20] J. P. Draayer and K. J. Weeks, Phys. Rev. Lett. **51**, 1422 (1983)
[21] J. P. Draayer and K. J. Weeks, Ann. Phys. (N.Y.) **156**, 41 (1984)
[22] J.P. Draayer, in *Algebraic Approaches to Nuclear Structure*, ed. R.F. Casten (Harwood, Chur, 1993), p. 423
[23] O. Castaños, M. Moshinsky, and C. Quesne, *Group Theory and Special Symmetries in Nuclear Physics Ann Arbor, 1991*, ed. J. P. Draayer and J. Jänecke (World Scientific, Singapope, 1992) p. 80
[24] O. Castaños, M. Moshinsky, and C. Quesne, Phys. Lett. B **277**, 238 (1992)
[25] O. Castaños, V. Velázquez A., P.O. Hess, and J.G. Hirsch, Phys. Lett. B **321**, 303 (1994)
[26] J.N. Ginocchio, Phys. Rev. Lett. **78**, 436 (1997)
[27] J.N. Ginocchio, J. Phys. G: Nucl. Part. Phys. **25**, 617 (1999)
[28] A. P. Zuker, J. Retamosa, A. Poves, and E. Caurier, Phys. Rev. C **52**, R1741 (1995)
[29] A. P. Zuker, A. Poves, F. Nowacki, and S. M. Lenzi, Phys. Rev. C **92**, 024320 (2015)
[30] K. Kaneko, N. Shimizu, T. Mizusaki, and Y. Sun, Phys. Rev. **103**, L021301 (2021)
[31] D. Bonatsos, I. E. Assimakis, N. Minkov, A. Martinou, R. B. Cakirli, R. F. Casten, and K. Blaum, Phys. Rev. C **95**, 064325 (2017)
[32] D. Bonatsos, I. E. Assimakis, N. Minkov, A. Martinou, S. Sarantopoulou, R. B. Cakirli, R. F. Casten, and K. Blaum, Phys. Rev. C **95**, 064326 (2017)
[33] D. Bonatsos, Eur. Phys. J. A **53**, 148 (2017)
[34] A. Martinou, D. Bonatsos, N. Minkov, I.E. Assimakis, S.K. Peroulis, S. Sarantopoulou, and J. Cseh, Eur. Phys. J. A **56**, 239 (2020)
[35] D. Bonatsos, H. Sobhani, and H. Hassanabadi, Eur. Phys. J. Plus **135**, 710 (2020)
[36] A. Martinou, D. Bonatsos, S. Sarantopoulou, I.E. Assimakis, S.K. Peroulis, and N. Minkov, Eur. Phys. J. A **57**, 83 (2021)
[37] A. Martinou, D. Bonatsos, T.J. Mertzimekis, K. Karakatsanis, I.E. Assimakis, S.K. Peroulis, S. Sarantopoulou, and N. Minkov, Eur. Phys. J. A **57**, 84 (2021)
[38] G. Rosensteel and D.J. Rowe, Ann. Phys. (NY) **126**, 343 (1980)

[39] D. J. Rowe and J. L. Wood, *Fundamentals of Nuclear Models: Foundational Models* (World Scientific, Singapore, 2010)

[40] C.-L. Wu, D. H. Feng, X.-G. Chen, J.-Q. Chen, and M. W. Guidry, Phys. Rev. C **36**, 1157 (1987)

[41] P.P. Raychev, Compt. Rend. Acad. Bulg. Sci. **25**, 1503 (1972)

[42] G.N. Afanas'ev, S.A. Abramov, and P.P. Raychev, Yad. Fiz. **16**, 53 (1972) [Sov. J. Nucl. Phys. **16**, 27 (1973)]

[43] P.P. Raychev, Yad. Fiz. **16**, 1171 (1972) [Sov. J. Nucl. Phys. **16**, 643 (1973)]

[44] P.P. Raychev and R.P. Roussev, Yad. Fiz. **27**, 1501 (1978) [Sov. J. Nucl. Phys. **27**, 792 (1978)]

[45] N. Minkov, S.B. Drenska, P.P. Raychev, R.P. Roussev, and D. Bonatsos, Phys. Rev. C **55**, 2345 (1997)

[46] N. Minkov, S.B. Drenska, P.P. Raychev, R.P. Roussev, and D. Bonatsos, Phys. Rev. C **60**, 034305 (1999)

[47] N. Minkov, S.B. Drenska, P.P. Raychev, R.P. Roussev, and D. Bonatsos, Phys. Rev. C **61**, 064301 (2000)

[48] A. Arima and F. Iachello, Phys. Rev. Lett **35**, 1069 (1975)

[49] F. Iachello and A. Arima, *The Interacting Boson Model* (Cambridge University Press, Cambridge, 1987)

[50] F. Iachello and P. Van Isacker, *The Interacting Boson-Fermion Model* (Cambridge University Press, Cambridge, 1991)

[51] A. Frank and P. Van Isacker, *Symmetry Methods in Molecules and Nuclei* (S y G editores, México D.F., 2005)

[52] A. Georgieva, P. Raychev, and R. Roussev, J. Phys. G: Nucl. Phys. **8**, 1377 (1982)

[53] A. Georgieva, P. Raychev, and R. Roussev, J. Phys. G: Nucl. Phys. **9**, 521 (1983)

[54] P. Ring and P. Schuck, *The Nuclear Many-Body Problem* (Springer, Berlin, 1980)

[55] R. F. Casten, *Nuclear Structure from a Simple Perspective* (Oxford University Press, Oxford, 2000)

[56] D. Bonatsos, A. Martinou, S. Sarantopoulou, I.E. Assimakis, S. Peroulis, and N. Minkov, Eur. Phys. J. ST **229**, 2367 (2020)

[57] I. Hamamoto and B.R. Mottelson, Phys. Rev. C **79**, 034317 (2009)

[58] I. Hamamoto and B. Mottelson, Scholarpedia **7**, 10693 (2012)

[59] Lipkin H., *Lie groups for pedestrians*, Dover Publications, New York (2014)

[60] D. J. Rowe, J. Phys. G: Nucl. Part. Phys. **43**, 024011 (2016)

[61] D. J. Rowe, G. Thiamova, and J. L. Wood, PRL **97**, 202501 (2006).

[62] S.G. Nilsson and I. Ragnarsson, *Shapes and Shells in Nuclear Structure* (Cambridge University Press, Cambridge, 1995)

[63] Meng Wang, G. Audi, F.G. Kondev, W.J. Huang, S. Naimi, Xing Xu, Chinese Physics C **41**, No. 3, 030003 (2017).

Structure of $^{190-200}$Hg within the covariant density functional theory

Vaia Prassa[1,*], *Konstantinos E.* Karakatsanis[2,3], and *George A.* Lalazissis[3]

[1]Department of Computer Science and Telecommunications, University of Thessaly, Lamia GR-35100, Greece
[2]Physics Department, Faculty of Science, University of Zagreb, 10000 Zagreb, Croatia
[3]Physics Department, Aristotle University of Thessaloniki, Thessaloniki GR-54124, Greece

Abstract. We study in detail the chain of even - even mercury isotopes $^{190-200}$Hg using the relativistic point coupling model. A five-dimensional collective Hamiltonian (5DCH) model, with parameters determined by constrained self-consistent mean-field (SCMF) calculations based on the relativistic density-dependent point-coupling (DD-PC1) energy density functional, and a finite-range pairing interaction is used to calculate the low-energy excitation spectrum and the B(E2) transitions rates of even-even nuclei. The calculations suggest coexisting configurations in ^{190}Hg, increased collectivity in the isotopes $^{192-198}$Hg and a more spherical structure in ^{200}Hg.

1 Introduction

One of the most intriguing subjects of research in nuclear structure studies is the shape evolution in atomic nuclei. The variation of nuclear shapes far from stability depends on the residual interaction of the valence nucleons outside closed shells that can polarize the nuclear core leading to axially deformed ground state configurations or triaxial structures. In a number of cases different shape configurations can coexist.

The phenomena of phase transitions and phase coexistence in even-even nuclei near shell closures have been extensively investigated both theoretically and experimentally (see Refs.[1–6] for reviews). In the region of $Z = 82$ near the neutron midshell $N = 104$ the phenomena of phase coexistence [6] and phase transitions [7] were first observed in studies of hyperfine structure [8]. Later spectroscopic studies [9–18] revealed that the structure of those isotopes was defined by intruder prolate deformed configurations coexisting with less deformed oblate ground states. The low-lying excited states of the intruder band exhibit a parabola shape as a function of neutron number, starting from ^{188}Hg down to the midshell $N = 104$, with a minimum observed at ^{182}Hg and going up to ^{180}Hg and ^{178}Hg [19, 20]. On the other hand, in the heavier transitional isotopes with $190 < A < 200$, the observed energy levels of the yrast band remain almost constant. Although, the isotopes between the stable ^{200}Hg and the beginning of the midshell in ^{190}Hg have been investigated by different experiments [21–32], there are still crucial observables that remain to be measured.

Theoretical studies based on the Gogny [33–35], the relativistic mean field (RMF) interactions [36, 37], and the Nilsson-Strutinsky method [38] have generally confirmed these experimental findings. A systematic study of the low-lying states in the lead region has been performed within the number and angular-momentum projected generator coordinate method with axial symmetry, employing the Skyrme energy density functional (EDF) [39]. Excitation energies, electromagnetic transition rates, deformation properties, and ground-state properties relevant to the shape coexistence in Hg isotopes, have been investigated using the interacting boson model (IBM) [40–42]. A recent study within the Elliott and the proxy-SU(3) models [43] suggests that the evolution of shape coexistence in the neutron deficient Hg isotopes is accompanied by a merging of the spin-orbit (SO) - like shell with the open harmonic oscillator (HO) shell [43].

In this contribution we present contrained SCMF calculations for even-even $^{190-200}$Hg isotopes within the relativistic Hartree-Bogoliubov [44] method with the density-dependent point-coupling (DD-PC1) [45] energy density functional in the particle-hole channel and a separable pairing force [46] in the particle-particle channel. The DD-PC1 density functional has been successfully applied to various studies of nuclear structure phenomena related to quantum phase transitions [47–50], shape coexistence [51] and the effect of collective correlations on the ground state and fission properties of superheavy nuclei [52, 53].

A five-dimensional collective Hamiltonian (5DCH) with quadrupole deformations as dynamical collective coordinates [54, 55] is used to calculate the low-energy excitation spectrum and the B(E2) transitions rates of $^{190-200}$Hg isotopes. The microscopic self-consistent solutions of deformation-constrained triaxial relativistic Hartree-Bogolyubov (RHB) calculations, the single-particle wave functions, occupation probabilities, and quasiparticle energies, are used to calculate the Hamiltonian parameters. The moments of inertia are calculated with the Inglis-Belyaev formula [56, 57] and the mass parameters with the cranking approximation [58]. The col-

*e-mail: vprassa@uth.gr

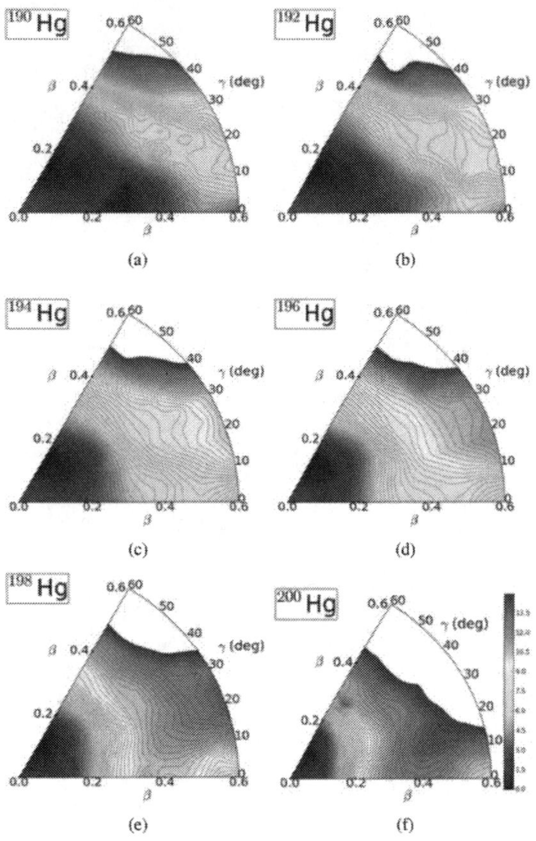

Figure 1: Self-consistent RHB triaxial quadrupole energy surfaces of even-even $^{190-200}$Hg isotopes in the $\beta-\gamma$ plane ($0° < \gamma < 60°$). All energies are normalized with respect to the binding energy of the corresponding ground state.

lective potential is obtained by subtracting the zero-point energy corrections[58] from the total energy that corresponds to the solution of constrained triaxial SCMF calculations. The resulting collective potential and inertia parameters as functions of the collective coordinates determine the dynamics of the 5DCH. Calculations shown here have been partially presented in [59].

2 Potential energy surfaces

To illustrate the rapid change of equilibrium shapes in Fig. 1 we present the potential energy surfaces of even-even $^{190-200}$Hg within the SCMF framework with the DD-PC1 functional and a separable pairing force. Starting with the lighter isotope ^{190}Hg the energy surface is γ-soft with two minima within an energy difference of 500keV, which indicates a case of shape coexistence of the two different configurations. The more pronounced minimum is oblate deformed at $\beta \approx 0.15$ and the second one is prolate at $\beta \approx 0.25$. In ^{192}Hg the energy surface is still rather flat in the γ-direction with the equilibrium configuration on the oblate side at $0.1 < \beta < 0.2$. The prolate minimum diminishes and only the oblate one is seen in $^{194-198}$Hg. The single oblate minimum becomes less deformed and approaches $\beta = 0$ for ^{200}Hg, which implies a structural change from weakly oblate deformed to nearly spherical states.

The present calculations, based on the relativistic DD-PC1 functional, are consistent with other theoretical efforts in this region (using the interacting boson model based on the Gogny-D1M EDF [40], the D1 [33] and D1S [34, 60] parametrizations of the Gogny-EDF, the Skyrme-SLy4 EDF [35],other Skyrme[39],the relativistic NL3 parametrization [36], and the relativistic PC-PK1 functional [61]).

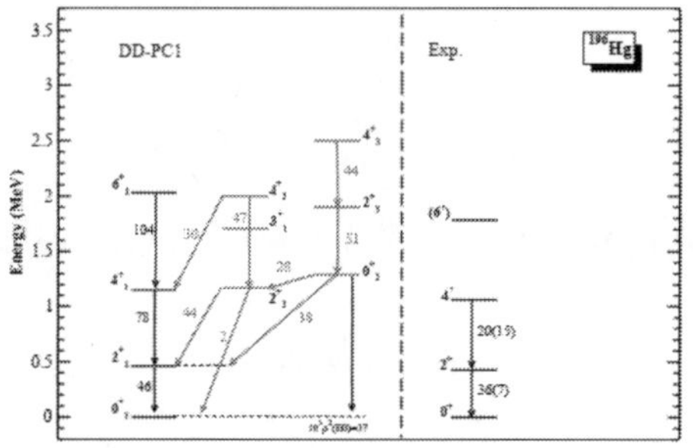

Figure 2: Low-lying level scheme of the even-even ^{196}Hg nucleus. The excitation energies, the B(E2) values (in Weisskopf units) and the $\rho^2(E0; 0_2^+ \longrightarrow 0_1^+)$ obtained with the 5DCH based on the DD-PC1 functional are shown. The experimental data are taken from Ref. [21].

3 Spectroscopic properties

The constrained self-consistent solutions of the relativistic Hartree-Bogoliubov (RHB) equations at each point on the energy surface determine the mass parameters the three moments of inertia and the zero-point energy corrections as functions of the deformation parameters β and γ. The diagonalization of the Hamiltonian yields the excitation spectra and collective wave functions that are used in the calculation of various observables, e.g., electromagnetic transition probabilities B(E2) and electric monopole transition strengths $\rho(E0)$. Physical observables are calculated in the full configuration space and there are no effective charges in the model.

As an example in Fig. 2 we display the low-lying collective spectrum of ^{196}Hg, in comparison to available data for the excitation energies and reduced electric quadrupole transition probabilities B(E2) in Weisskopf units (W.u.) taken from Refs. [21]. In addition to the yrast ground-state band, in deformed and transitional nuclei excited states are also assigned to (quasi-) β and γ-bands. The comparison with the few existing experimental data shows a

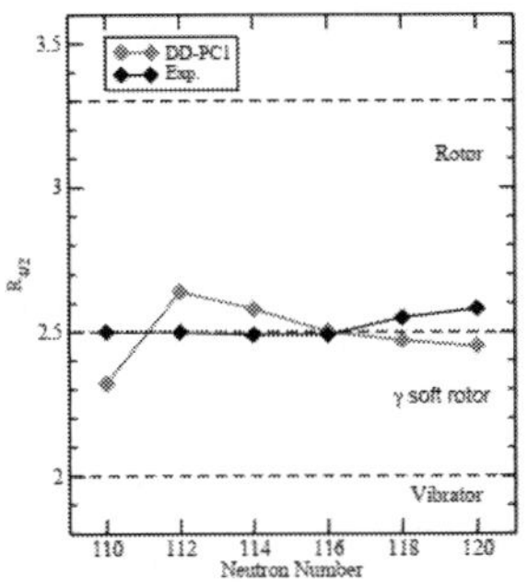

Figure 3: The ratio $R_{4/2}$ of excitation energies of the yrast states 4_1^+ and 2_1^+ as a function of the neutron number in even-even $^{190-200}$Hg isotopes calculated using the 5DCH model based on the DD-PC1 functional. The experimental data are taken from Ref. [21]

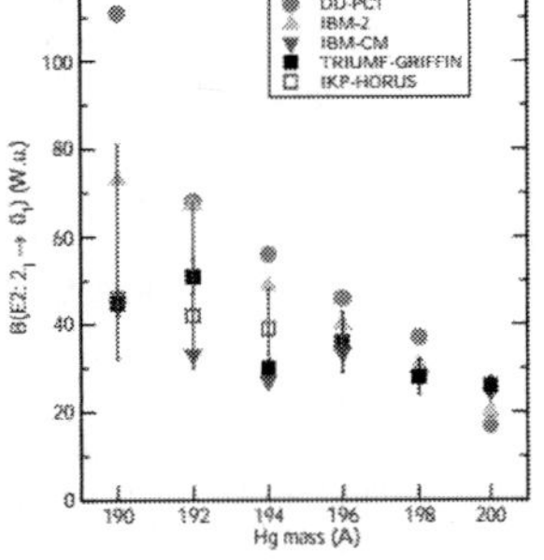

Figure 4: $B(E2; 2_1^+ \longrightarrow 0_1^+)$ values (in Weisskopf units) calculated using the 5DCH model with the DD-PC1 functional. Theoretical calculations with the IBM-2 [40], and IBM-CM [41] are given for comparison. Experimental values marked as TRIUMF-GRIFFIN are from Ref. [21] and those marked as IKP-HORUS are from Ref. [22].

rather reasonable agreement of the excitation energy levels in the yrast band for $J^\pi < 6^+$. The theoretical reduced electric-quadrupole transition probabilities BE(2) (in W.u.) are generally larger than the data. Although a reasonable agreement within the experimental errors is observed for the first excited state 2_1^+, the calculated value for the $B(E2; 4_1^+ \longrightarrow 2_1^+)$ overestimates the experimental value considerably. This indicates that there is probably more mixing between the theoretical states than what can be inferred from the data.

One of the distinct characteristics of shape transitions between axially symmetric rotors, γ-soft rotors and spherical vibrators is the evolution of the ratio $R_{4/2}$ of excitation energies of the yrast states 4_1^+ and 2_1^+. For an axially symmetric rotor $R_{4/2} = 3.33$, values between 2.5 and 2 are typical for a nucleus characterized by a γ -unstable potential and, finally, $R_{4/2} = 2.0$ for a spherical vibrator. In Fig. 3 we plot the theoretical values of $R_{4/2}$ as function of the neutron number of even-even $^{190-200}$Hg isotopes in comparison to data taken from Ref. [21]. The calculated $R_{4/2}$ ratio starts at 2.32 in ^{190}Hg increases rapidly to 2.64 in ^{192}Hg and then decreases gradually to 2.45 in ^{200}Hg. The experimental values in this region vary slighty around 2.5 as the neutron number increases. The crossing between the 2_1^+ and 4_1^+ normal and intruder states at $N = 110$ in Hg is probably the reason for the drop of the ratio $R_{4/2}$ in ^{190}Hg, as reported in Ref. [42]. This effect is less pronounced in the Pt isotopes where the ratio is around 3.3 for $N = 102$ up to $N = 106$ and then decreases gradually with neutron number to approximately 2.5 for $N = 110 - 118$ (cf. Ref. [61]). The results disclose the γ-softness of the potential energy surfaces in $^{190-200}$Hg. The strong configuration mixing predicted by the collective Hamiltonian model is probably the cause of the theoretical deviations from data.

In Fig. 4 the calculated $B(E2; 2_1^+ \longrightarrow 0_1^+)$ values (in Weisskopf units) within the 5DCH framework based on the DD-PC1 functional are shown. Theoretical results with the interacting boson model, IBM-2 [40] and the interacting boson model with configuration mixing (CM), IBM-CM [41] are presented for comparison. Our calculations reproduce the general decreasing trend with neutron number, however for all isotopes $^{190-198}$Hg an increased collectivity in the 2_1^+ yrast states is observed compared to data. For the isotopes $^{192-200}$Hg the accuracy of the calculations compared to the experimental values is of the same quality as the one reached with the IBM-2 model in Ref. [40].

The most noticeable discrepancies between the theoretical calculations and the measured values are consistently in the lighter isotope ^{190}Hg. The potential energy surface of ^{190}Hg exhibits two minima, a dominant oblate configuration and a prolate one at larger defomation that are degenarate in energy with a rather flat path connecting them going through the triaxial region. The inclusion of dynamical correlations yields an oblate deformed but γ-soft 0_1^+ state and a 0_2^+ state, within an energy gap of 400keV, that is predominantly on the prolate side but with oblate admixtures. The relatively large overlap between the 0_1^+ and 0_2^+ wavefunctions, the large electric monopole transition strength $\rho(E0)$ from the 0_2^+ to the 0_1^+ state ($\rho^2(E0; 0_2^+ \longrightarrow 0_1^+) \times 10^3 = 150$) and the large B(E2) values of the interband transitions, suggest a strong mixing between the two configurations and support the hypothesis of shape-coexistence at $N = 110$ in Hg.

4 Overview and conclusions

Deformation constrained SCMF calculations have been performed with the relativistic Hartree-Bogoliubov method based on the universal energy density functional DD-PC1 and a separable pairing interaction. The triaxial

$(\beta - \gamma)$ deformation energy surfaces obtained from the SCMF calculations for the even-even $^{190-200}$Hg isotopes predict a very interesting nuclear structure evolution: coexistence of prolate and oblate deformed minima in ^{190}Hg, γ-soft oblate deformed potential in $^{192-198}$Hg, and an almost spherical deformation in ^{200}Hg.

A quadrupole collective Hamiltonian, with parameters determined by self-consistent constrained triaxial RHB calculations, has been used to calculate the low-energy spectra, B(E2) transitions rates and electric monopole transition strengths $\rho(E0)$ of Hg isotopes at neutron number $N = 110 - 120$. The calculated excitations energies of the low-lying yrast band in ^{196}Hg reproduce the experimental values, however the B(E2) reduced transition probabilities for the 2_1^+ and 4_1^+ yrast states overestimate the data. The two low-lying bands based on the lowest excited vibrational state that appear in the energy spectrum support the hypothesis of increased collectivity in the theoretical calculations than what can be deduced from the data. The evolution with neutron number of the ratio $R_{4/2}$ validate the above assumptions. In ^{190}Hg, the triaxial SCMF calculations of the energy surface and the results of the quadrupole collective Hamiltonian model suggest shape-coexistence of a dominant oblate configuration and a more prolate deformed intruder state.

Theoretical and experimental efforts in the region of even-even neutron deficient $^{190-200}$Hg isotopes predict shape transitions from nearly spherical configurations in ^{200}Hg to γ-softness in $^{192-198}$Hg and shape coexistence in ^{190}Hg. The results presented in this work verify these findings, demonstrating the potential of the semi-empirical relativistic EDFs including the explicit treatment of collective correlations using a microscopic collective Hamiltonian.

5 Acknowledgments

V.P.'s research was conducted within the "Call Fellowship for Postdoctoral research" implemented by the University of Thessaly and funded by the Stavros Niarchos Foundation. The work of K.E.K. was financed within the Tenure Track Pilot Programme of the Croatian Science Foundation and the École Polytechnique Fédérale de Lausanne, and the Project No. TTP-2018-07-3554 Exotic Nuclear Structure and Dynamics with funds of the Croatian-Swiss Research Programme. Some of the results presented in this work have been produced using the AUTH Compute Infrastructure and Resources.

References

[1] K. Heyde, P. Van Isacker, M. Waroquier, J. L. Wood, and R. A. Meyer, Phys. Rep. **102**, 291 (1983).

[2] J. L. Wood, K. Heyde, W. Nazarewicz, M. Huyse, and P. van Duppen, Phys. Rep. **215**, 101 (1992).

[3] A. N. Andreyev, M. Huyse, P. Van Duppen, L. Weissman, D. Ackermann, J. Gerl, F. P. Hessberger, S. Hofmann, A. Kleinbohl, G. Munzenberg et al., Nature (London) **405**, 430 (2000).

[4] R. Julin, K. Helariutta, and M. Muikku, J. Phys. G **27**, R109 (2001).

[5] T. Grahn, A. Dewald, O. Möller, R. Julin, C. W. Beausang, S. Christen, I. G. Darby, S. Eeckhaudt, P. T. Greenlees, A. Görgen et al., Nucl. Phys. A **801**, 83 (2008).

[6] K. Heyde and J. L. Wood, Rev. Mod. Phys. **83**, 1467 (2011).

[7] P. Cejnar, J. Jolie, and R. F. Casten, Rev. Mod. Phys. **82**, 2155 (2010).

[8] Bonn, J., G. Huber, H.-J. Kluge, L. Kugler, and E.W. Otten, 1972, Phys. Lett. B **38**, 308 (1972).

[9] Rud, N., D. Ward, H. R. Andrews, R. L. Graham, and J. S. Geiger, Phys. Rev. Lett. **31**, 1421 (1973).

[10] Proetel, D., R. M. Diamond, and F. S. Stephens, Phys. Lett. B **48**, 102 (1974).

[11] Hamilton, J. H., *et al.*, Phys. Rev. Lett. **35**, 562 (1975).

[12] Cole, J. D., *et al.*, Phys. Rev. Lett. **37**, 1185 (1976).

[13] G. D. Dracoulis, Phys. Scr. **2000**, 54 (2000).

[14] N. Bree, K. Wrzosek-Lipska, A. Petts, A. Andreyev, B. Bastin, M. Bender, A. Blazhev, B. Bruyneel, P. A. Butler, *et al*, Phys. Rev. Lett. **112**, 162701 (2014).

[15] L. P. Gaffney, M. Hackstein, R. D. Page, T. Grahn, M. Scheck, P. A. Butler, P. F. Bertone, N. Bree, R. J. Carroll,M. P. Carpenter et al., Phys. Rev. C **89**, 024307 (2014).

[16] R. Julin, T. Grahn, J. Pakarinen, and P. Rahkila, J. Phys. G: Nucl. Part. Phys. **43**, 024004 (2016).

[17] B. A. Marsh, T. Day Goodacre *et al.*, Nat. Phys. **14**, 1163 (2018).

[18] S. Sels, T. Day Goodacre, B. A. Marsh *et al.*, Phys. Rev. C **99**, 044306 (2019).

[19] C. Müller-Gatermann *et al.*, Phys. Rev. C **99**, 054325 (2019).

[20] J. Elseviers, A. N. Andreyev, S. Antalic, A. Barzakh, N. Bree, T. E. Cocolios, V. F. Comas, J. Diriken, D. Fedorov, V. N. Fedosseyev *et al.*, Phys. Rev. C **84**, 034307 (2011).

[21] B. Olaizola *et al.*, Phys. Rev. C **100**, 024301 (2019).

[22] A. Esmaylzadeh, L. M. Gerhard, V. Karayonchev, J.-M. Régis, J. Jolie, M. Bast, A. Blazhev, T. Braunroth, M. Dannhoff, F. Dunkel, C. Fransen, G. Häfner, L. Knafla, M. Ley, C. Müller-Gatermann, K. Schomacker, N. Warr, and K.-O. Zell, Phys. Rev. C **98**, 014313 (2018).

[23] National Nuclear Data Center (NNDC), Nuclear Levels and Gamma Search.

[24] M. O. Kortelahti, E. F. Zganjar, J. L. Wood, C. R. Bingham, H. K. Carter, K. S. Toth, J. H. Hamilton, J. Kormicki, L. Chaturvedi, and W. B. Newbolt, Phys. Rev. C **43**, 484 (1991).

[25] A. Bockisch, K. Bharuth-Ram, A. M. Kleinfeld, and K. P. Lieb, Z. Phys. A **289**, 231 (1979).

[26] Th. J. De Boer, E.W. Ten Napel, and J. Block, Physica **29**, 1013 (1963).

[27] M. Guttormsen,A.Von Grumbkow,Y.K.Agarwal,K. P. Blume, K. H. Hübel, J. Recht, P. Schuler, H.Kluge,

https://doi.org/10.1051/epjconf/202125202007

K.H.Maier, A.Maj, and N. Roy, Nucl. Phys. A **398**, 119 (1983).

[28] R. Kroth, S.K. Bhattacherjee, C. Günther, M. Guttormsen, K. Hardt, H.Hübel, and A.Kleinrahm, Phys. Lett.B **97**, 197 (1980).

[29] V. Kölschbach, P. Schüler, K. Hardt, D. Rosendaal, C. Günther, K. Euler, R. Tölle, M. Marten-Tölle, and P. Zeyen, Nucl. Phys. A **439**, 189 (1985).

[30] S. A. Hjorth, I. Y. Lee, J. R. Beene, C. Roulet, D. R. Haenni, N. R. Johnson, F. E. Obenshain, and G. R. Young, Phys. Rev. Lett. **45**, 878 (1980).

[31] R. M. Lieder, H. Beuscher, W. F. Davidson, A. Neskakis, and C. Mayer-Böricke, Nucl. Phys. A **248**, 317 (1975).

[32] R. Kroth, K. Hardt, M. Guttormsen, G. Mikus, J. Recht, W. Vilter, H. Hübel, and C. Günther, Phys. Lett. B **99**, 209 (1981).

[33] M. Girod and P. G. Reinhard, Phys. Lett. B **117**, 1 (1982).

[34] J. P. Delaroche, M. Girod, G. Bastin, I. Deloncle, F. Hannachi, J. Libert, M. G. Porquet, C. Bourgeois, D. Hojman, P. Kilcher *et al.*, Phys. Rev. C **50**, 2332 (1994).

[35] O. Moreno, P. Sarriguren, R. Álvarez-Rodríguez, and E. Moya de Guerra, Phys. Rev. C **73**, 054302 (2006).

[36] T. Nikšić, D. Vretenar, P. Ring, and G. A. Lalazissis, Phys. Rev. C **65**, 054320 (2002).

[37] R. Fossion, Dennis Bonatsos, and G. A. Lalazissis, Phys. Rev. C **73**, 044310 (2006).

[38] R. Bengtsson and W. Nazarewicz, Z. Phys. A **334**, 269 (1989).

[39] J. M. Yao, M. Bender, and P.-H. Heenen, Phys. Rev. C **87**, 034322 (2013).

[40] K. Nomura, R. Rodríguez-Guzmán, and L. M. Robledo, Phys. Rev. C **87**, 064313, (2013).

[41] J. E. García-Ramos and K. Heyde, Phys. Rev. C **89**, 014306, (2014).

[42] J. E. García-Ramos and K. Heyde, EPJ Web of Conferences **178**, 05005 (2018).

[43] A. Martinou, D. Bonatsos, T.J. Mertzimekis, K. Karakatsanis, I.E. Assimakis, S.K. Peroulis, S. Sarantopoulou, and N. Minkov, Eur. Phys. J. A **57**, 84 (2021).

[44] D. Vretenar, A. Afanasjev, G. Lalazissis, and P. Ring, Phys. Rep. **409**, 101 (2005).

[45] T. Nikšić, D. Vretenar, and P. Ring, Phys. Rev. C **78**, 034318 (2008).

[46] Y. Tian, Z. Y. Ma, and P. Ring, Phys. Lett. B **676**, 44 (2009).

[47] K. Nomura, T. Nikšić, and D. Vretenar, Phys. Rev. C **94**,064310 (2016).

[48] K. Nomura, T. Nikšić, and D. Vretenar, Phys. Rev. C **96**, 014304 (2017).

[49] K. Nomura, D. Vretenar, T. Nikšić, and B.-N. Lu, Phys. Rev. C **89**, 024312 (2014).

[50] V. Prassa, T. Nikšić, and D. Vretenar, Phys. Rev. C **88**, 044324 (2013).

[51] Z. P. Li, T. Nikšić, and D. Vretenar, J. Phys. G: Nucl. Part. Phys. **43**, 024005 (2016).

[52] V. Prassa, T. Nikšić, G. A. Lalazissis, and D. Vretenar, Phys. Rev. C **86**, 024317 (2012).

[53] Z. Shi, A. V. Afanasjev, Z. P. Li, and J. Meng, Phys. Rev. C **99**, 064316 (2019).

[54] Z. P. Li, T. Nikšić, D. Vretenar, P. Ring, and J.Meng, Phys. Rev. C **81**, 064321 (2010).

[55] T. Nikšić, D. Vretenar, and P. Ring, Prog. Part. Nucl. Phys. **66**, 519 (2011).

[56] D. Inglis, Phys. Rev. **103**, 1786 (1956).

[57] S. Beliaev, Nucl. Phys. **24**, 322 (1961).

[58] M. Girod and B. Grammaticos, Nucl. Phys. A **330**, 40 (1979).

[59] V. Prassa and K.E. Karakatsanis, International Journal of Modern Physics E (accepted for Publication) (2021).

[60] http://www-phynu.cea.fr/science_en_ligne/carte_potentiels_microscopiques/carte_potentiel_nucleaire_eng.htm

[61] X. Q. Yang, L. J. Wang, J. Xiang, X. Y. Wu, and Z. P. Li, Phys. Rev. C **103**, 054321 (2021).

EPJ Web of Conferences **252**, 03001 (2021) https://doi.org/10.1051/epjconf/202125203001
HINPw6

Weak Interaction Physics at *TwinSol*

J. J. Kolata[1],*, *Maxime* Brodeur, *Tan* Ahn, *Dan* Bardayan, *Patrick* O'Malley, and *J. S.* Randhawa

[1]Department of Physics, University of Notre Dame, Notre Dame, Indiana 46556 (USA).

Abstract. A program to investigate the unitarity of the Cabibbo-Kobayashi-Maskawa (CKM) quark-mixing matrix by studying super-allowed mixed mirror β decays has been initiated at the *TwinSol* facility at Notre Dame. These mixed Fermi/Gamow-Teller (F-GT) decays, occurring between T=1/2 isospin doublets in mirror nuclei, provide a complimentary check on the data from super-allowed pure Fermi decays from $0^+$to 0^+ states. The first part of the program, involving the measurement of the lifetimes of the relevant nuclei to the required accuracy of one part in 10^3 or better, has nearly been completed. However, the additional complication introduced by F-GT mixing requires the use of an ion trap to measure the mixing ratio ρ with similar accuracy. The lifetime measurements, as well as progress in installing an ion trap at *TwinSol*, will be discussed. In addition, since the ion trap will require a dedicated beam line for its operation, an opportunity presented itself to greatly improve the performance of *TwinSol* for reaction studies with exotic nuclei. This took the form of an added dipole switching magnet coupled to a third solenoid to form the new *TriSol* facility currently under construction. The expected properties of *TriSol*, and its application to reaction studies of interest for nuclear astrophysics, will also be discussed.

1 Precision Tests of the Standard Model

One important test of the Standard Model (SM) of high-energy physics is via the unitarity of the CKM matrix:

$$\text{CKM} = \begin{bmatrix} V_{ud} & V_{us} & V_{ub} \\ V_{cd} & V_{cs} & V_{cb} \\ V_{td} & V_{ts} & V_{tb} \end{bmatrix} ; |V_{ud}|^2 + |V_{us}|^2 + |V_{ub}|^2 = 1.$$

Here, (u d s c b t) indicate the quarks of the SM. Non-unitarity of this matrix would imply the possibility of physics beyond the SM. The largest element in the summation, $|V_{ud}|^2$, is accessible through nuclear β decay, and its most precise measurement is via super-allowed pure Fermi decays from $0^+$to 0^+ states (Figure 1). However, the inferred value must also be accurate so it's important to check it using other techniques. One possibility is via mixed-mirror β decay. The required parameters for such a measurement include the half-life, Q-value, branching ratios, and F-GT mixing ratio ρ for the transition. All of these, along with some nuclear correction factors, must be determined with a precision of better than one part in 10^3. A program to determine the relevant half-lives using radioactive beams from *TwinSol* will be described below, together with progress on construction of an ion trap to measure ρ.

1.1 Half-life measurements

Half-life measurements have been made for eight T = 1/2 mixed-mirror transitions; five have been published (Refs. [2–6]) and three more are in preparation [7–9].

*e-mail: jkolata@nd.edu

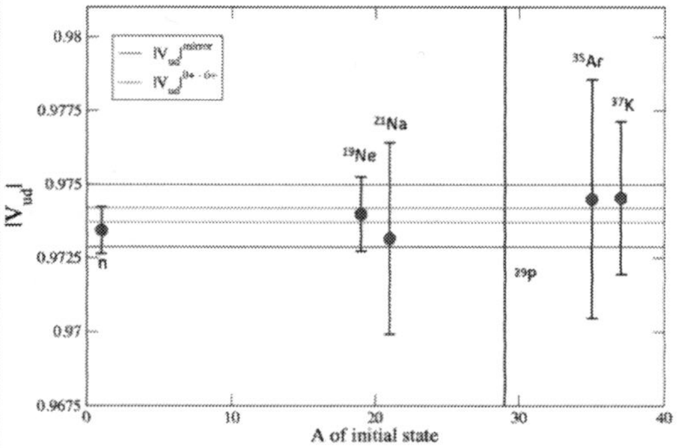

Figure 1. $|V_{ud}|$ from neutron lifetime measurements, $0^+$to 0^+ Fermi decay (red error bar), and mixed-mirror decay (blue error bar). Data from Ref. [1], including latest theoretical corrections. The mixed-mirror results agree with those from pure Fermi decay but are a factor of 4.2 less precise.

^{20}F and ^{28}Al, part of a T=1 triplet, have also been studied [10, 11]. These nuclei can provide information about possible second-class currents [12, 13] and also to test the strong form of the conserved vector current (CVC) hypothesis [14].

The experimental apparatus includes the following elements: (1) an Au stopper for the secondary beam was used to eliminate radioactive contaminants formed by the beam incident on impurities in Ta stoppers, (2) the stopper was rotated from the implantation location to the counting location, (3) the primary beam was removed during

https://doi.org/10.1051/epjconf/202125203001

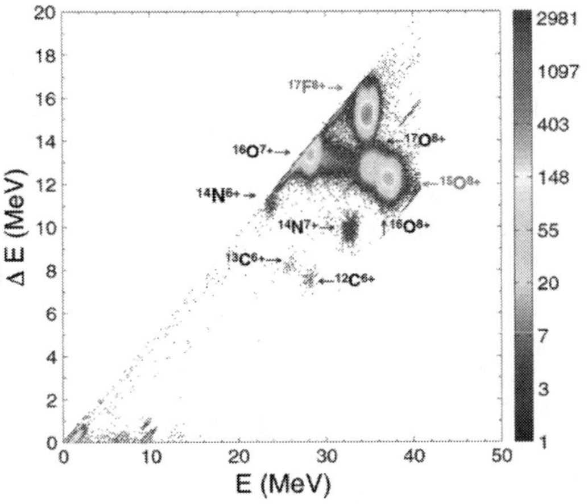

Figure 2. ΔE-E plot for ^{17}F production, showing a 0.24% contamination of ^{15}O.

the counting period using a pre-tandem electrostatic deflector, (4) the detector was a 1 mm thick plastic scintillator to reduce γ-ray background, and (5) a quartz-window photomultiplier detector, which reduced the background by a factor of ten from borosilicate glass windows, was used.

TwinSol typically produces a "cocktail" beam, but it is usually possible to adjust the energy of the primary beam so that only stable ions are produced along with the desired radioactive beam. Nevertheless, small radioactive contaminants, of the order of less than 1%, are sometimes observed (Figure 2). Even such a small contamination can significantly alter the measured half-life when uncertainties of less than 0.1% are required. Since small contaminant levels cannot always be observed using a Si detector telescope, we also generate a so-called *chop plot* in each case by removing successively larger amounts of early-time data and fitting the resulting data set. If there is no contamination, the measured half-life should remain the same until the remaining amount of data is too small to produce statistically significant fits. Such a plot is shown in Figure 3. The contaminant is considered by using a two-component fit including its half-life and intensity.

The data were also tested for the presence of very long-lived contaminants by fitting with a three-component fit where one of the components had a half-life much greater than that of the nucleus being investigated. It should be noted that all the fits were performed using the so-called "summed fit" procedure described in detail in Ref. [15], which avoids the bias introduced when least-squares fitting is used. Corrections for dead-time were applied, and various systematic errors were estimated. For more details on the analysis, refer to [2–6]. An illustration of the results obtained by these methods in shown in Figure 4. This result, for ^{11}C, is the most precise of all superallowed mixed-mirror decays. As the lightest of such systems, ^{11}C provides a laboratory for testing the CVC hypothesis. See Ref. [4] for further discussion.

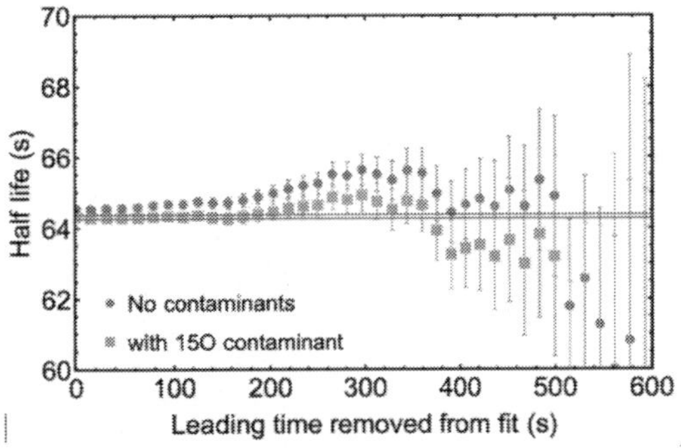

Figure 3. Chop plot for ^{17}F production showing the improvement obtained by including a 0.2% contamination of ^{15}O.

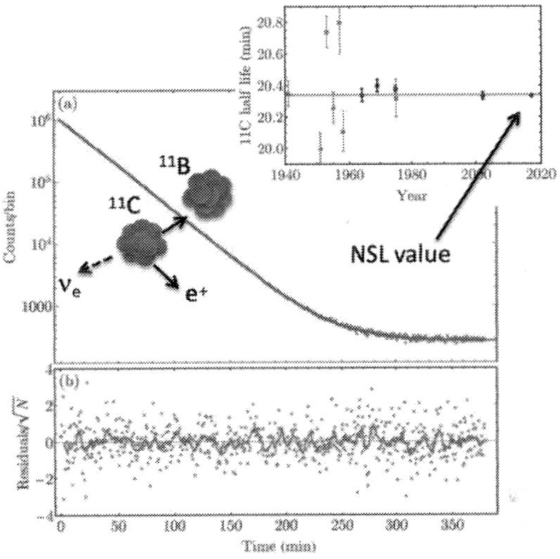

Figure 4. Decay-time plot and fit for ^{11}C. The lower portion shows the residuals divided by the square root of the number of counts in the particular bin together with an associated five-point running average. The inset shows the measured half-life in various experiments as a function of the year of the measurement.

The fractional uncertainties in f_V values deduced for the relevant mixed-mirror transitions are shown in Figure 5. The eight isotopes measured at Notre Dame are indicated. In all cases except for ^{41}Sc (preliminary analysis) and ^{29}P, half-life precision better than one part in 10^3 was achieved.

These are world-average values computed according to criteria from the Particle Data Group [16]. In the case of ^{29}P, one existing high-precision measurement could not be excluded, which inflated the error bar on the world average (see Ref. [6] for more details).

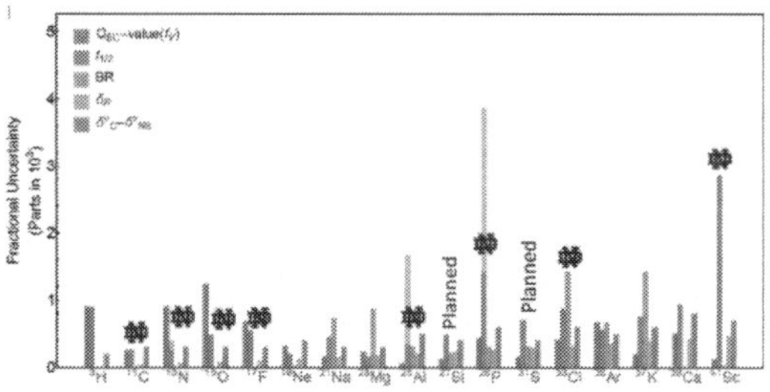

Figure 5. Fractional errors for f_V values of mixed-mirror transitions.

1.2 Progress on ion-trap commissioning

As mentioned above, the f_V values alone for mixed-mirror transitions are insufficient to compute $|V_{ud}|$; the F-GT mixing ratio ρ must also be determined with comparable accuracy and precision. There are three methods to determine ρ, by measuring: (a) the β asymmetry parameter A_β, (b) the neutrino asymmetry parameter B_ν, or (c) the β-ν angular correlation $a_{\beta\nu}$ [17]. At present, ρ values have only been obtained with sufficient precision for 5 isotopes (see Figure 1) . There is also data for ^{17}F, but this experiment [18] measured A_β which was found to be too insensitive to determine a value for $|V_{ud}|$. In order to add to this data set, the ion trap "St. Benedict" is currently under construction at Notre Dame. A diagram of this system is illustrated in Figure 6 below.

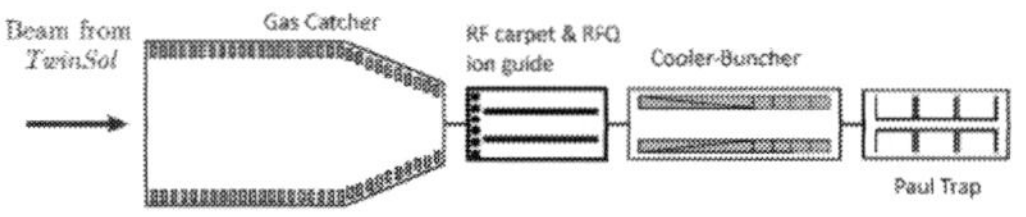

Figure 6. Schematic dagram of the ion trap "St. Benedict".

The gas catcher and cooler-buncher have been completed and tested. Commissioning of the RF carpet and RFQ is currently underway. The first trap experiment will likely be a measurement of $a_{\beta\nu}$ for ^{17}F, which is one of the most intense beams ($\approx 2\times10^6$ particles per second) from *TwinSol*, and also the most sensitive to possible tensor currents in mirror transitions. Later experiments will focus on ^{15}O and ^{13}N. These light systems are of particular interest since they can also be used to probe for scalar or tensor components in β decay.

2 *TriSol*

Since the St. Benedict ion trap requires a dedicated beam line, it was necessary to modify the system in order to preserve the ability to carry out other radioactive beam experiments. This took the form of adding a dipole switching magnet and a third solenoid to form *TriSol*. The original *TwinSol* (Figure 7) had its secondary focal point at a distance of about 5.5 m from the primary target. Under these conditions, the beam spot size was typically 5 mm full width at half maximum (FWHM). However, in order to enable γ-ray and neutron experiments, the focal point was shifted to 12 m so that the beam passed through a 1.25 m thick high-density concrete shielding wall which basically eliminated background coming from the primary target (see Figure 8). This had two undesirable effects: the diameter of the beam spot was increased to about 25 mm FWHM and the transmission to the secondary target was reduced by about a factor of two. The addition of a third solenoid allowed for the adjustment of the beam emittance in order to greatly improve this situation. The optics were now more complicated, so it was necessary to use the LISE++ program [19] from Michigan State University to optimize the settings of the various elements. A sample beam envelope for a 50 MeV ^{20}Ne beam is shown in Figure 9. The beam spot is now 10 mm FWHM and its angular divergence is $\pm$ 1.75°. For lower-Z beams such as ^{6}He or ^{8}B, the spot size is projected to be 5-6 mm FWHM. In addition, there is room behind the third solenoid to move the focal point further upstream to trade spot size for divergence for those experiments that require this.

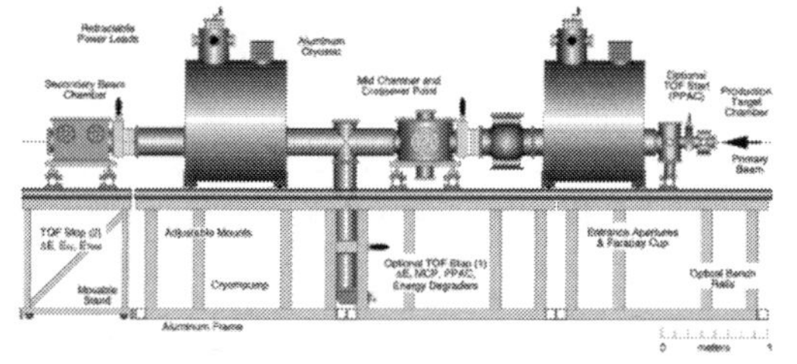

Figure 7. The original *TwinSol*.

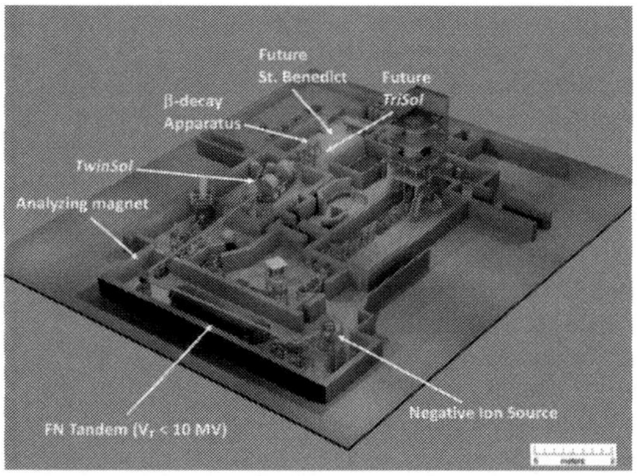

Figure 8. Laboratory diagram showing the location of various elements.

3 Astrophysical applications of *TriSol*

An example of the astrophysically relevant experiments that can be carried out at *TriSol* is the study of ^{20}Ne+^{20}Ne fusion cross sections at low energy in order to constrain

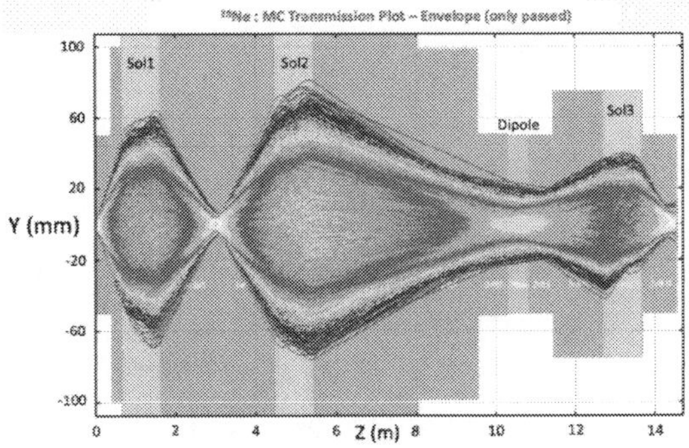

Figure 9. Calculated profile for a 50 MeV ^{20}Ne beam.

pynconuclear burning in accreting neutron stars. A proposal has been submitted to FRIB PAC-1 to measure the ^{26}Ne+^{20}Ne fusion cross sections using the Active Target Time Projection Chamber (AT-TPC) [20]. A comparison with lighter systems, such as ^{20}Ne+^{20}Ne and ^{22}Ne+^{22}Ne, is necessary to validate the anticipated results from this experiment. The former of these can be studied at *TriSol* using the ^{20}Ne beam discussed in the previous section.

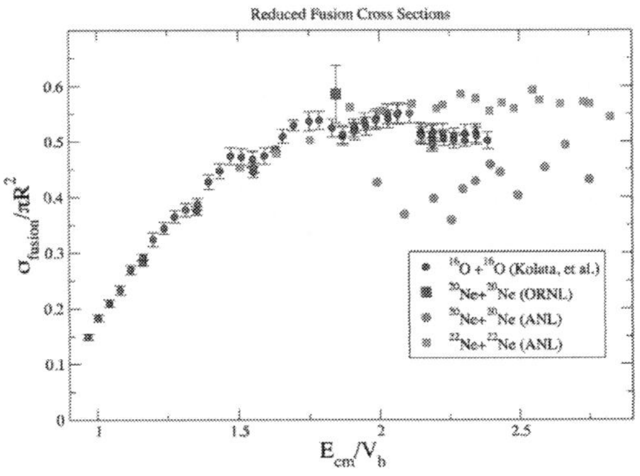

Figure 10. Reduced fusion cross sections for Ne+Ne systems. The "ANL" data are from [21] and the "ORNL" data from [22]. A comparison with ^{16}O+^{16}O fusion [23] is also illustrated.

Existing data on Ne+Ne fusion in the region of the Coulomb barrier is illustrated in Figure 10. These are "reduced" fusion cross sections where the effect of nuclear radius R and barrier V_b are removed. Here, $R=1.44(A_1^{1/3}+A_2^{1/3})$ and $V_b=1.44(Z_1Z_2/R)$ where R is in fm, V_b is in MeV, and σ_{fusion} is in fm^2. It can be seen that the ^{20}Ne+^{20}Ne reduced fusion cross section is considerably smaller than that for the other systems, except possibly for the one data point from Oak Ridge National Laboratory (ORNL).

A more interesting observation is the oscillations that appear in this data, which are similar to the oscillations observed for ^{16}O+^{16}O occurring due to anti-symmetrization in this identical-particle system. This eliminates odd partial waves; however, it's also necessary for the system to have a "surface-transparent" potential in order to reduce the width of the even-partial-wave "resonances" so they can be observed [23]. The existence of similar oscillations in the ^{20}Ne+^{20}Ne system, and the hint of similar structure for ^{22}Ne+^{22}Ne, will give important information on the optical-model potentials for these systems if these results can be confirmed.

The role of these data in testing predictions for the astrophysical S-factor for Ne+Ne systems is illustrated in Figure 11. The "ORNL" set here includes higher energy data from Ref. [22] and the blue curve is a coupled-channels calculation using the H13 potential from that reference. The red curve is the prediction from Ref. [24] using the Sao Paulo potential and a polynomial approximation, which becomes increasingly unrealistic above E_{cm} = 30 MeV.

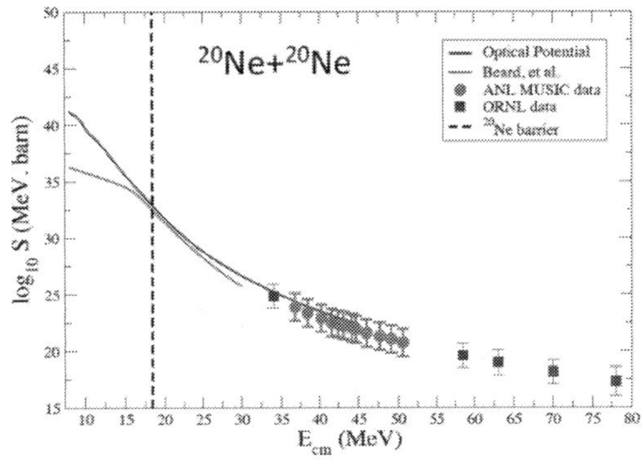

Figure 11. Comparison of the astrophysical S-factor prediction from Ref. [24] with experimental data.

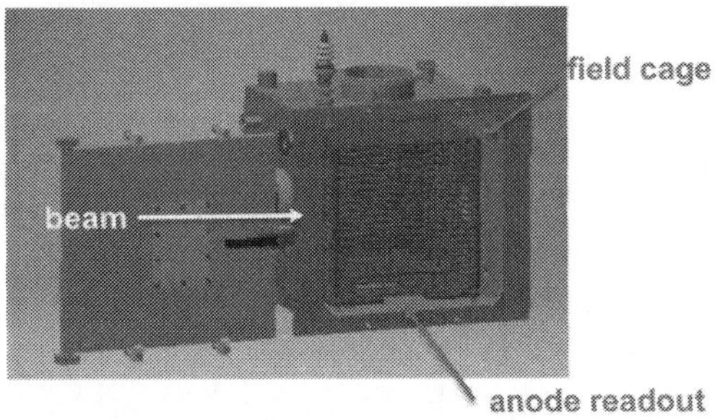

Figure 12. Schematic diagram of the ND-cube time projection chamber

A planned experiment using the 50 MeV ^{20}Ne beam from *TriSol* should easily reach the region below E_{cm}=16 MeV where the disagreement between the Beard, *et al.* [24] and coupled-channels calculations increases. It will be carried out using the ND-Cube time projection chamber (TPC) [25], illustrated in Fig. 12, which has the advantage of very high efficiency and the ability to measure an excitation function over a wide energy range as the incident beam slows down in the target gas which is also the detection medium. The ND-Cube has a rectangular geometry in a (40 cm)^3 chamber. The field cage defines a 20 cm wide by 30 cm long active area. The drift direction is perpendicular to the pad plane, which is a Micro Pattern Gas Detector (MPGD) having 1000 hexagonal pads. In addition, it incorporates a Micromegas [26] with thick GEM [27] amplification (see [25] for more detail). This detector has an advantage over Multisampling Ionization Chamber (MUSIC) detectors [28] for fusion studies near the barrier due to its large segmentation, which allows for high position resolution near the end of the ion track where the particle energy changes rapidly with the penetration distance. The results of an experiment to measure ^{7}Li+^{40}Ar fusion cross sections near the barrier is shown in Figure 13. The horizontal error bars indicate the averaging interval used and the vertical error bars are purely statistical. The data are compared with a DWBA prediction using an optical model potential derived from a fit to Li-ion data on a wide range of targets at energies from 13-153 MeV [29]. The result agrees rather well with experiment. This bodes well for future studies of near- and sub-barrier fusion with with the ND-Cube detector.

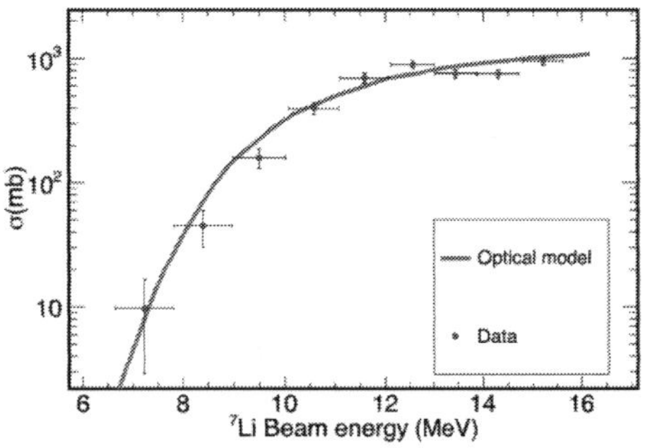

Figure 13. ^{7}Li+^{40}Ar fusion cross sections near the barrier measured with the ND-Cube TPC. Data taken for 3 hrs. at ≈ 1400 particles per second, with a P10 gas target.

4 Acknowledgements

The weak-interactions collaborators include Drs. Maxime Brodeur, Dan Bardayan, Frederick Becchetti (University of Michigan), James J. Kolata, and Patrick O'Malley together with graduate students Dan Burdette, Biying Liu, Jacob Long, and Adrian Valverde. The *TriSol* project collaborators include Drs. Tan Ahn, Dan Bardayan, James J. Kolata, Patrick O'Malley and Jaspreet Randhawa together with undergrad student Sydney Coil. Thanks are also due to a number of Notre Dame Nuclear Science Laboratory graduate students who operated the accelerator during the various experiments. This work was partially supported by the U.S. National Science Foundation under Grant Nos. PHY-2011890 and PHY-1725711 (MRI), and by the University of Notre Dame.

References

[1] Leendert Hayen, Phys. Rev. D 103, 113001 (2021).
[2] M. Brodeur *et al.*, Phys. Rev. C93, 025503 (2016).
[3] J. Long *et al.*, Phys. Rev. C96, 015502 (2017).
[4] A. Valverde *et al.*, Phys. Rev. C97, 035503 (2018).
[5] D. Burdette *et al.*, Phys. Rev. C101, 055504 (2020).
[6] J. Long *et al.*, Phys. Rev. C101, 015501 (2020).
[7] J. Long *et al.*, Phys. Rev. C, submitted (^{13}N).
[8] P. O'Malley *et al.*, Phys. Rev. C, in preparation (^{33}Cl).
[9] B. Liu *et al.*, Phys. Rev. C, in preparation (^{41}Sc).
[10] D. Burdette *et al.*, Phys. Rev. C99, 015501 (2019).
[11] B. Liu *et al.*, Phys. Rev. C, in preparation (^{28}Al).
[12] D.H. Wilkinson and D.E. Alburger, Phys. Rev. Lett. 24, 1134 (1970).
[13] N.A. Smirnova and C. Volpe, Nucl. Phys. A 714, 441 (2003).
[14] Nathal Severijns, Marcus Beck, and Oscar Naviliat-Cuncic, Rev. Mod. Phys. 78, 991 (2006).
[15] V.T. Koslowsky *et al.*, Nucl. Instrum. Methods Phys. Res., Sect. A 401, 289 (1997).
[16] https://pdg.lbl.gov
[17] Nathal Severijns and Oscar Naviliat-Cuncic, Ann. Rev. Nucl. Part. Sci. 61, 23 (2011).
[18] N. Severijns *et al.*, Phys. Rev. Lett. 63, 1050 (1989).
[19] http://lise.nscl.msu.edu/lise.html
[20] Jaspreet S. Randhawa and Rahul Jain, Proposal to FRIB PAC-1, Feb. 12, 2012.
[21] Preliminary data from Argonne National Laboratory, shown with permission from Dr. Melina Avila-Coronado.
[22] D. Shapira *et al.*, Phys. Rev. C28, 1148 (1983).
[23] J.J. Kolata *et al.*, Phys. Rev. C16, 891 (1977).
[24] M. Beard *et al.*, At. Data Nucl. Data Tables 96, 541 (2010).
[25] T. Ahn *et al.*, Nucl. Instrum. Methods Phys. Res., Sect. A (in preparation).
[26] G. Charpak *et al.*, Nucl. Instrum. Methods Phys. Res. Sect. A 478, 26 (2002).
[27] F. Sauli, Nucl. Instrum. Methods Phys. Res. Sect. A 386, 531 (1997).
[28] P. Carnelli *et al.*, Nucl. Instrum. Methods Phys. Res. Sect. A 799, 197 (2015).
[29] J. Cook, Nucl. Phys. A 388, 153 (1982).

Advances in ^{211}At production at Texas A&M University

S.J. Yennello[1,2,*], *L.A.* McIntosh[1], *J.D.* Burns[3], *E.E.* Tereshatov[1], *G.* Tabacaru[1], *L.* McCann[1,2], *S.* Schultz[1,2], *K.* Lofton[1,2], *A.* Abbott[1], *G.* Avila[1], *M.* Berko[1,4], *E.* Engelthaler[1,5], *K.* Hagel[1], *A.* Hannaman[1], *B.* Harvey[1], *A.* Hood[1], *M.* McCarthy[1], *A.B.* McIntosh[1], *M.* Sorensen[1], *Z* Tobin[1], and *A.* Vonder Haar[1]

[1]Cyclotron Institute, Texas A&M University, College Station, TX, USA
[2]Department of Chemistry, Texas A&M University, College Station, TX, USA
[3]Nuclear Engineering and Science Center, Texas A&M University, College Station, TX, USA
[4]New Jersey Institute of Technology, Newark, NJ, USA
[5]University of Dallas, Dallas, TX, USA

Abstract. Alpha emitting radionuclides with medically relevant half-lives are interesting for treatment of tumors and other diseases because they deposit large amounts of energy close to the location of the radioisotope. Researchers at the Cyclotron Institute at Texas A&M University are developing a program to produce ^{211}At, an alpha emitter with a medically relevant half-life. The properties of ^{211}At make it a great candidate for targeted alpha therapy for cancer due to its short half-life (7.2 h). Astatine-211 has now been produced multiple times and reliability of this process is being improved.

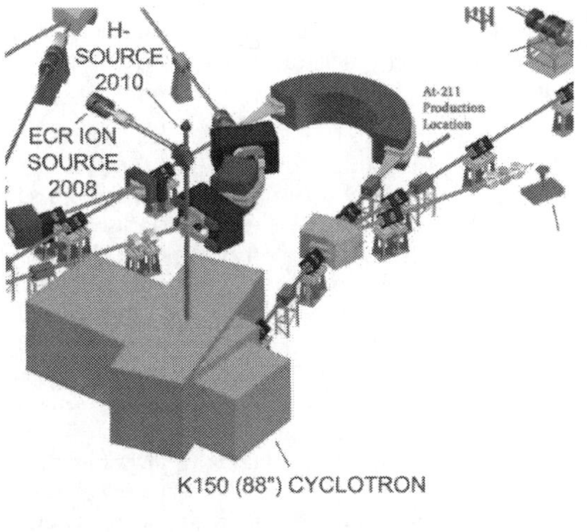

Figure 1. Isotope production line from K150 cyclotron.

1 Introduction

While radioisotopes have long been used to treat cancer, the ability to specifically target cancerous cells has only been available in special cases. With the advent of targeting agents such as monoclonal antibodies, the ability to use radioisotopes which deposit a large amount of energy and do not move far from the targeting agent has become an area of considerable interest. Targeted alpha therapy is taking advantage of these advances to use alpha emitters with biologically relevant half lives. Astatine (At) has emerged as an element of interest for these purposes. There are no known stable isotopes of astatine. The small amounts of At present in the Earth's crust are due to the decay of other elements. Natural presence of astatine is very fleeting as no astatine isotope has a half-life greater than 8 hours.

Targeted alpha therapy drugs [1–11] have gained a large amount of interest following the success of Xofigo®, [12] based on the α emitting $^{223}RaCl_2$, in treating metastatic castration-resistant prostate cancer. Astatine-211 shows promise for cancer treatments when connected to a targeting agent, especially for non-localized cancers [7]. The major impediment to its use in clinical trials remains the availability of production and reliable separation chemistry [8, 13]. However, because of its short half-life, a rapid system with high yield is necessary for recovering the ^{211}At.

At the Cyclotron Institute (CI) at Texas A&M University we have developed the ability to produce ^{211}At by bombardment of a natural Bi target with an alpha beam from the K150 cyclotron and developed a purification system based on a newly discovered chemical interaction of At with a class of chemicals known as ketones.

2 At-211 production

Astatine-211 is produced via a nuclear reaction resulting from a metal plate of bismuth (Bi) being bombarded by α-particles [13–17] accelerated by the Texas A&M K150 cyclotron shown in Figure 1. The ^{211}At is trapped in the Bi target and needs to be extracted through a chemical process. Generally completely dissolving the target followed

*e-mail: yennello@tamu.edu

Table 1. At-211 production. The plus signs denote irradiations done using the 2^+ charge state of the alpha beam, activities were calculated using a different method for the asterisks (*).

Irradiations	Max Beam (pμA)	Ave Beam (pμA)	Irr time (h)	Activity at EoB (mCi)
December 2019$^+$	4.4	2.7	8	24 ± 2
March 2020$^+$	3.5	3.2	9	41 ± 3
June 2020	4.5	2.1 (Unstable)	9.4	8.0 ± 1.3
August 2020	2.6	2.4	9.6	21 ± 2
September 2020	7.4	5.1	7.3	22 ± 2*
October 2020	5	4.0	7.9	12 ± 1
November 2020	7.2	4.2	9.7	24 ± 2
December 2020	6.8	4.8	13.6	47 ± 5*
April 2021	8.2	5.61	15.6	17 ± 2
May 2021	5	3.36	15.5	11 ± 1
June 2021a	8.4	6.5	15.3	24 ± 3
June 2021b	11	8	18.75	38 ± 4

Figure 2. A bismuth target produced by melting bismuth pellets on a hot plate at 300°C and spreading the bismuth using an ultrasonic soldering iron. The surface is mostly smooth with some ridges and valleys with an average thickness of 66 microns.

by chemical processing results in a more consistent ^{211}At yield relative to removal of the ^{211}At by dry distillation. Recent improvements have been made to the ability to reliably produce ^{211}At at the Cyclotron Institute using the (α,2n) reaction twelve times over the past 18 months. A summary of the amount of beam available, length of irradiation, and total amount of ^{211}At produced during each of the runs is shown in Table 1. The amount of ^{211}At produced is determined by dissolving the target and counting the dissolved solution on an HPGe detector. The asterisk for the amount produced in September 2020 and December 2020 is because the target melted during this irradiation, and the amount of activity had to be estimated by a different means than the other runs. The first two irradiation were done with an alpha beam in the +2 charge state (as denoted by $^+$ in Table 1. The targets were transferred to Radiochemical Lab 118 within the CI, where the target was dissolved in $HNO_3$and radiochemical experiments were carried out.

Producing a robust bismuth target remains one of the greatest challenges towards astatine production. Bismuth has a relatively low melting point - only 270° C - resulting in the potential for melting of the target during irradiation [8]. This melting process alters the geometry of the irradiated bismuth (reducing total astatine production) and can cause the astatine to become volatile. To prevent melting of bismuth targets, we have made significant advancements in the cooling of the bismuth target during irradiation by removing interfaces between the chilled water and the target. The current design uses a single jet, chilled water system flowing directly on the back of the target frame. Additionally, efforts are being made to produce the optimal targets for irradiation and subsequently characterize them for integrity and impurities.

To increase the surface interaction and conductivity between the bismuth and aluminum interface, an ultrasonic soldering iron (S-Bonder SB-9210, S-Bond Technologies, LLC, Hatfield, PA 19440) was used. This method resulted in a stronger adhesion of the bismuth to the aluminum surface and allowed thinner targets, on the order of 50-100 microns, to be produced (shown in Figure 2). Since alpha particles only have enough energy to produce astatine in the first 70 μm of the bismuth, targets thicker than this are absorbing the alpha energy as heat, further contributing to melting. The thinner targets produced by the ultrasonic soldering iron show promise for maximum ^{211}At production as well as decreased probability of melting due to the strong bismuth-aluminum interface and 70 μm target depth.

To investigate the uniformity of the beam at the target position, a Segmented Faraday Cup-Viewer (SFCV) was assembled. The body is made of 3D-printed plastic, with 37 brass screws arranged at regular intervals. Each of these screws acts as a small Faraday Cup, from which beam current can be read via a LabView program, as shown on the left in Figure 3. The body of the viewer was coated with adhesive and CdS, which caused it to fluoresce in the presence of beam, as shown on the right of Figure 3. This gives a much more detailed view of where the beam is situated on the target. A more uniform beam on the target lessens the heat load in any one spot leading to a lower probability of melting the target.

 https://doi.org/10.1051/epjconf/202125203002

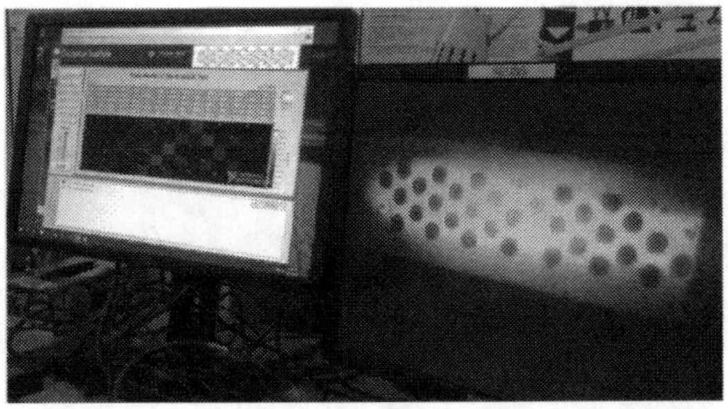

Figure 3. Photograph of Segmented Faraday Cup-Viewer (SFCV) being used to tune the alpha beam. The SFCV is 3D printed, with 37 screws which are wired to read the beam current at each disparate position across the target position. The surface has material adhered to it which phosphoresces in the presence of the beam. On the right is a video feed of the SFCV with alpha beam on the target position. On the left is the LabView program reading in the beam current off of each of the disparate screws.

Figure 4. Air Monitoring System. Left panel shows a cutaway of the system, where air from the isotope production chamber enters from the top and flows around the pink cylinder, pulled by a pump attached to the gold nozzle. The yellow cylinder indicates the presence of a silicon PIPS detector which detects alpha particles emitted from isotopes caught by the fiberglass filter (not depicted here). Right panel shows a photograph of the actual device. The white card is the fiberglass filter.

Another innovation was electrically isolating the target mounting structure from the rest of the beamline, so that the beam current can be read directly off the target throughout the entire overnight irradiation. This means other Faraday Cups do not need to be dropped in to know how much beam is being delivered to the bismuth target during the irradiation.

Another new aspect of the isotope production process was focused on the safety of those who remove the target from the beamline post-irradiation. Astatine-211 is known to be able to volatilize, particularly if the bismuth target melts during the irradiation. To determine whether this has occurred with any given target, the Air Handling Monitoring System depicted in Figure 4 was developed. The top

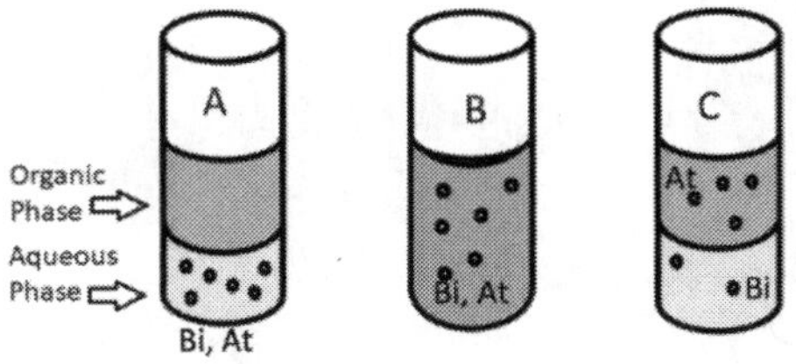

Figure 5. Schematic of liquid-liquid extraction process. The organic phase is added to the aqueous phase, containing Bi and At (A). The vial is shaken (B). When the organic and aqueous layers again separate (C), At is in the organic phase, and Bi remains in the aqueous.

of the system attaches to the target chamber. A pump is attached to the bottom of the system. A filter to catch ^{211}At can be inserted and removed below the silicon detector, as shown. If there is astatine in the air, the silicon detector will measure a characteristic alpha energy, the rate of which will alert the team that it is unsafe to open the chamber at this time. As a double-check, the filter can be removed and counted by a radiation probe, to see whether or not it has alpha contamination.

3 At-211 Purification

The near quantitative separation, purification, and recovery of ^{211}At has been demonstrated at the CI through an extraction chromatography process that uses liquid-liquid extraction as depicted in Figure 5 via porous beads impregnated with 3-octanone, resulting in an elution yield of 92-95%[18]. Moreover, the rapid nature of this process, less than 20 min, was achieved after dissolution of a Bi metal target in HNO_3 following retrieval from the beamline after α particle bombardment. The solution was directly loaded onto the column (Figure 6) with no volume or acidity adjustment. The column was washed with HNO_3 and H_2O, and ^{211}At was eluted with ethanol, collecting roughly 87 - 93% in 1 mL. This process of recovering high purity ^{211}At, in near quantitative yields, represents a significant advance in At separations and purification (see Figure 7).

The development of the efficient extraction chromatography process was based on gaining a fundamental understanding of how the At behaves in nitric acid systems. Through a collaboration with Professor M.B. Hall in the Department of Chemistry experimental data was coupled with theoretic computational chemistry to develop new insights into At chemistry. These studies showed that At exists in the trivalent oxidation state as the AtO^+ species in aqueous nitric acid system. The heavy-element nature of At, and the induced spin-orbit coupling, has a signifi-

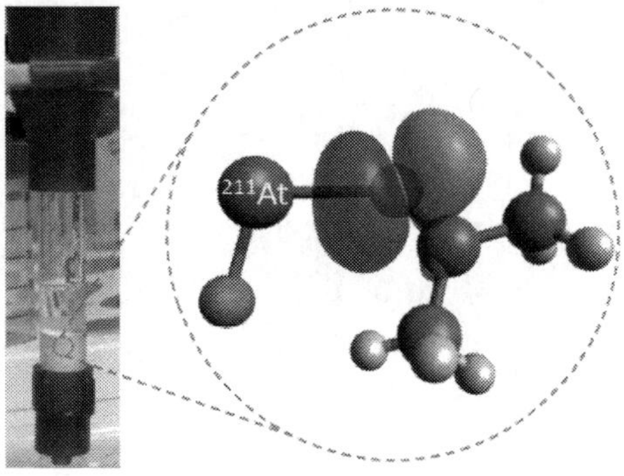

Figure 6. Left: Photo of ^{211}At purification column. Right: Depiction of DFT calculation demonstrating how At-211 interacts with a ketone.

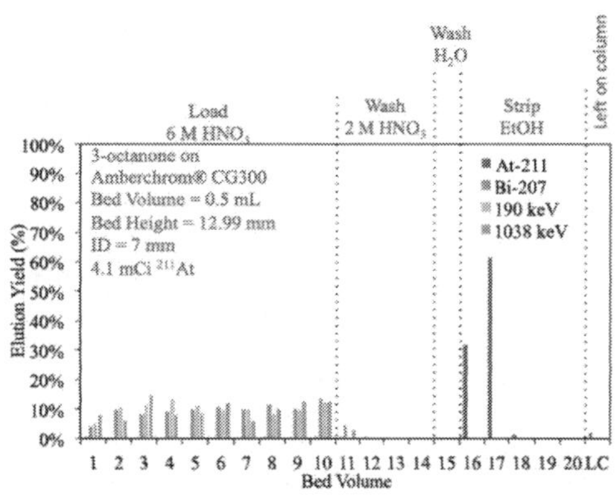

Figure 7. Chromatogram of an aliquot of HNO_3 containing the dissolved bombarded target solution with a 3-octanone impregnated Amberchrom®CG300M resin bed.

cant effect on the electronic structure of the AtO^+ molecular cation, which assumes a pseudo-closed-shell electronic configuration. This closed-shell configuration leads to an enhanced interaction with ketone molecules through a slightly covalent interaction between the sp_2 hybrid lone pair of the carbonyl oxygen and the lowest unoccupied molecular orbital (LUMO), a non-degenerate π^* orbital, of the AtO^+, as shown on the right hand side of Figure 6[17].

4 Characterization

A CdTe detector is utilized to measure the x-rays from the ^{211}At that has been produced before it is removed from the irradiation chamber. Figure 8 is an example of the spectrum from the Bi target at the end of the irradfiation. The 4 peaks at 76.8 keV, 79.2 keV, 89.5 keV, and 92.4 keV are clearly visible indicating the production of astatine.

Two High Purity Germanium (HPGe) detectors are used to measure the hundreds of liquid samples that are

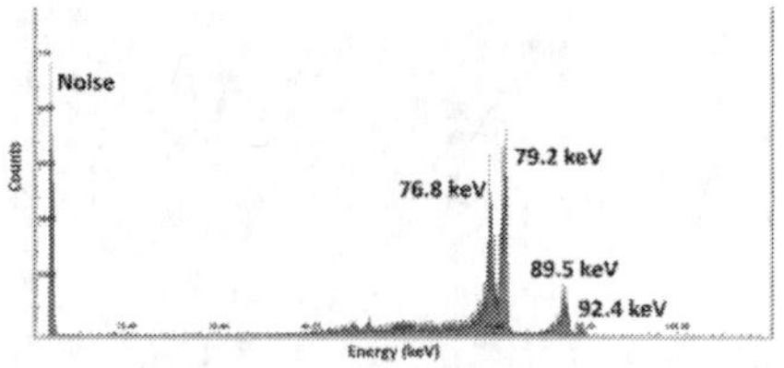

Figure 8. CdTe spectrum showing the 4 astatine peaks are 76.8 keV, 79.2 keV, 89.5 keV, and 92.4 keV, respectively.

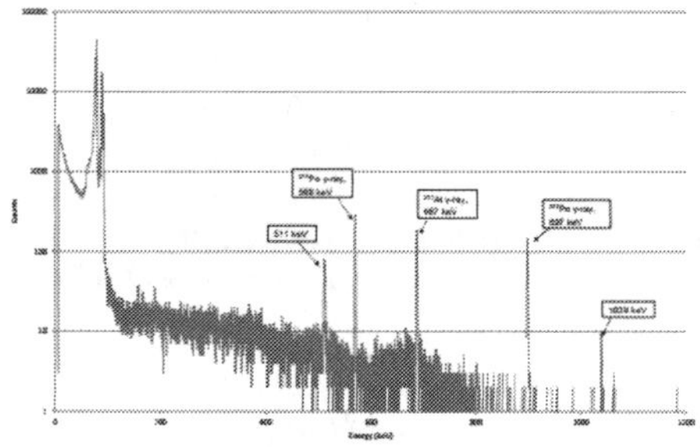

Figure 9. Typical HPGe Spectrum showing the characteristic γ-ray of ^{211}At at 687 keV, as well as peaks belonging to ^{211}Po and ^{66}Ga.

produced during the separation chemistry. A spectrum of one of our samples containing ^{211}At is shown in Figure 9. Below 100 keV, there are many x-rays. The characteristic γ-ray of ^{211}At is pointed out at 687 keV. The two γ-rays of ^{211}Po are also labeled at 569 and 897 keV. The other two γ-rays belong to ^{66}Ga according to our decay curve studies[18], which is separated out by the separation chemistry.

5 Summary

Great strides are being made at the CI to improve every aspect of ^{211}At production. The stability and amount of alpha beam on target, the ability of the targets to handle more beam, and the ability to reliably dissolve the target and separate out the ^{211}At have all been improved. These advancements are allowing the more reliable production of ^{211}At for chemistry experiments and may transfer to the ability to produce other alpha emitting isotopes in the future.

6 Acknowledgements

The authors are indebted to the operations staff at the Texas A&M University Cyclotron Institute, the Radiological Safety program and M.B. Hall for their contributions to this work. This work was supported by the U.S. Department of Energy Isotope Program, managed by the Office of Science for Isotope R&D and Production under Award No. DE-SC0020958; U.S. Department of Energy under Awards No. DE-NA0003841 and DE-FG02-93ER40773;

Texas A&M University through the Bright Chair in Nuclear Science and the Nuclear Solutions Institute; Texas A&M University System National Laboratories Office and Los Alamos National Laboratory through the joint collaborative research program. LM acknowledges support from the National Science Foundation.

References

[1] Makvandi, M.; Dupis, E.; Engle, J. W.; Nortier, F. M.; Fassbender, M. E.; Simon, S.; Birnbaum, E. R.; Atcher, R. W.; John, K. D.; Rixe, O.; Norenberg, J. P. Target. Oncol. 2018, 13 (2), 189.

[2] Li, Y.; Hamlin, D. K.; Chyan, M.-K.; Wong, R.; Dorman, E. F.; Emery, R. C.; Woodle, D. R.; Manger, R. L.; Nartea, M.; Kenoyer, A. L.; Orozco, J. J.; Green, D. J.; Press, O. W.; Storb, R.; Sandmaier, B. M.; Wilbur, D. S. PLoS One 2018, 13 (10), e0205135.

[3] Lambrecht, R. M.; Mirzadeh, S. Int. J. Appl. Radiat. Isot. 1985, 36 (6), 443.

[4] Johnson, E. L. L.; Turkington, T. G. G.; Jaszczak, R. J. J.; Gilland, D. R. R.; Vaidyanathan, G.; Greer, K. L. L.; Coleman, R. E. E.; Zalutsky, M. R. R. Nucl. Med. Biol. 1995, 22 (1), 45.

[5] Dekempeneer, Y.; Keyaerts, M.; Krasniqi, A.; Puttemans, J.; Muyldermans, S.; Lahoutte, T.; D?huyvetter, M.; Devoogdt, N. Expert Opin. Biol. Ther. 2016, 16 (8), 1035.

[6] Kraeber-Bodere, F.; Rousseau, C.; Bodet-Milin, C.; Mathieu, C.; Guerard, F.; Frampas, E.; Carlier, T.; Chouin, N.; Haddad, F.; Chatal, J.-F.; Faivre-Chauvet, A.; Cherel, M.; Barbet, J. Int. J. Mol. Sci. 2015, 16 (2), 3932.

[7] Elgqvist, J.; Frost, S.; Pouget, J.-P.; Albertsson, P. Front. Oncol. 2014, 3, 324.

[8] Zalutsky, M. R.; Pruszynski, M. Curr. Radiopharm. 2011, 4 (3), 177.

[9] Wilbur, D. S. Curr. Radiopharm. 2011, 4 (3), 214.

[10] Andersson, H.; Cederkrantz, E.; Back, T.; Divgi, C.; Elgqvist, J.; Himmelman, J.; Horvath, G.; Jacobsson, L.; Jensen, H.; Lindegren, S.; Palm, S.; Hultborn, R. J. Nucl. Med. 2009, 50 (7), 1153.

[11] McDevitt, M. R.; Sgouros, G.; Finn, R. D.; Humm, J. L.; Jurcic, J. G.; Larson, S. M.; Scheinberg, D. A. Eur. J. Nucl. Med. Mol. Imaging 1998, 25 (9), 1341.

[12] Kluetz, P. G.; Pierce, W.; Maher, V. E.; Zhang, H.; Tang, S.; Song, P.; Liu, Q.; Haber, M. T.; Leutzinger, E. E.; Al-Hakim, A.; Chen, W.; Palmby, T.; Alebachew, E.; Sridhara, R.; Ibrahim, A.; Justice, R.; Pazdur, R. Clin. Cancer Res. 2014, 20 (1), 9.

[13] Zalutsky, M. R.; Reardon, D. A.; Pozzi, O. R.; Vaidyanathan, G.; Bigner, D. D. Nucl. Med. Biol. 2007, 34 (7), 779.

[14] Stickler, J. D.; Hofstetter, K. J. Phys. Rev. C 1974, 9 (3), 1064.

[15] Wilbur, D. Curr. Radiopharm. 2008, 1 (3), 144.

[16] Lucignani, G. Eur. J. Nucl. Med. Mol. Imaging 2008, 35 (9), 1729.

[17] Burns, J. D.; Tereshatov, E. E.; McCarthy, M. A.; McIntosh, L. A.; Tabacaru, G. C.; Yang, X.; Hall, M. B.; Yennello, S. J. Chem. Commun. 2020, 56 (63), 9004.

[18] Burns, J. D.; Tereshatov, E. E.; Avila, G.; Glennon, K. J.; Hannaman, A.; Lofton, K. N.; McCann, L. A.; McCarthy, M. A.; McIntosh, L. A.; Schultz, S. J.; Tabacaru, G. C.; Vonder Haar, A. L.; Yennello, S. J. Sep. Purif. Technol. 2021, 256, 117794.

Upgrade of the MAGNEX spectrometer toward the high-intensity phase of NUMEN

Manuela Cavallaro[1], *Clementina* Agodi[1], *Giuseppe A.* Brischetto[1,2], *Salvatore* Calabrese[1], *Daniela* Calvo[3], *Vittoria* Capirossi[3,4], *Francesco* Cappuzzello[1,2], *Diana* Carbone[1], *Irene* Ciraldo[1,2], *Franck* Delaunay[1,2,5], *Paolo* Finocchiaro[1], *Maria* Fisichella[1], *Felice* Iazzi[3,4], *Federico* Pinna[3,4], *Diego* Sartirana[3], *Alessandro* Spatafora[1,2], *Onoufrios* Sgouros[1], *Vasilis* Soukeras[1,2], *Domenico* Torresi[1], *Salvatore* Tudisco[1]

For the NUMEN collaboration

[1]INFN – Laboratori Nazionali del Sud, Catania, Italy
[2]Dipartimento di Fisica e Astronomia "Ettore Majorana", University of Catania, Catania, Italy
[3]DISAT, Politecnico di Torino, Torino, Italy
[4]INFN – Sezione di Torino, Torino, Italy
[5]LPC Caen, Normandie Univ, ENSICAEN, UNICAEN, CNRS/IN2P3, Caen, France

Abstract. The NUMEN experimental activity with accelerated beams is performed at INFN–Laboratori Nazionali del Sud (LNS) in Catania using the Superconducting Cyclotron and the MAGNEX magnetic spectrometer. The scientific motivation of NUMEN is to extract experiment-driven information on the nuclear matrix elements entering in the expression of the $0\nu\beta\beta$ decay half-life. The reaction cross sections involved, especially for the double charge exchange process, are very low, thus limiting the present exploration to a few selected isotopes of interest in the context of typically low-yield experimental runs. In order to make feasible a systematic study of all the candidate nuclei, a major upgrade of the LNS facility is foreseen to increase the experimental yield by more than two orders of magnitude. To this purpose, frontier technologies are being developed for the accelerator and the detection systems. An updated description of the choices derived from the recent R&D activity on the target system and MAGNEX focal plane detector is given.

1 Introduction

MAGNEX [1] is a large acceptance magnetic spectrometer installed at INFN-Laboratori Nazionali del Sud in Catania (Italy). It analyses in momentum and detects the products of different kinds of nuclear reactions induced by ion beams from both the Tandem Van der Graaff and Superconductive Cyclotron (CS) accelerators. Since 2014 it has been the main facility used for the NUMEN (NUclear Matrix Elements for Neutrinoless double beta decay) project [2].

NUMEN aims at providing experiment-driven information on the Nuclear Matrix Elements (NME) involved in the expression of the decay half-life of neutrinoless double beta decay ($0\nu\beta\beta$), by measuring the cross-sections of nuclear Double Charge Exchange (DCE) reactions [3], [4]. However, the DCE cross sections are very small (a few nb), thus requiring a high sensitivity [5]. For this reason, as discussed in [2], the present limits of the facility in terms of beam power delivered by the CS accelerator and the acceptable rate of a few kHz for the MAGNEX focal plane detector (FPD) have limited so far the exploration of DCE to a small number of cases (e.g., ^{12}C, ^{40}Ca, ^{48}Ti, ^{76}Ge, ^{76}Se, ^{116}Cd, ^{116}Sn, ^{130}Te) with beam power of a few W. The systematic exploration of all the nuclei of interest to $0\nu\beta\beta$ decay, as foreseen in NUMEN, requires an upgraded set-up able to work with kW beam power [6]. Thus, NUMEN has triggered the current upgrade of the INFN-LNS CS and associated infrastructures to develop such high-intensity ion beams.

An intense R&D activity in beam transport, target technology, mechanical integration, detectors and electronics has developed in the last years and is ongoing within the NUMEN project.

In the present paper, an overview of the different activities undertaken for the upgrade of the MAGNEX spectrometer in preparation of the high-intensity phase of NUMEN is given. A picture of the MAGNEX spectrometer in its present, pre-upgrade state is shown in Fig. 1.

[1] Corresponding author: manuela.cavallaro@lns.infn.it

Fig. 1. View of the MAGNEX spectrometer at INFN-LNS

2 The upgrades

2.1 Development of targets for high-intensity beams

The goal of the target development activity is to design and produce a number of thin solid targets of specific isotopes, that will be irradiated with intense ion beams during the NUMEN experiments.

The main issues are related to the thickness of the target, which affects the energy resolution of the reaction products, and the dissipation of the heat produced in the target by the beam energy loss. The proposed solution is to prepare each target by depositing a film of the isotope of interest, with a thickness of a few hundred $\mu g/cm^2$, on a 2-μm thick HOPG (Highly Oriented Pyrolithic Graphite) backing, which will withstand the target layer, fully strip the reaction products, and quickly transfer the heat to a target cooling system [7].

During the R&D phase, the evaporation techniques have been explored and their parameters (temperature, duration, use of a buffer, buffer material, etc.) have been optimized for a few target materials, with the aim of minimizing the thickness non-uniformity and rugosity which can degrade the energy resolution [8]. Moreover, the target cooling technique has been studied and optimized, evaluating, for each target, the maximum allowed ion beam intensity to avoid melting [9].

Fig. 2 shows a sketch of the target holder [10] to be mounted in the MAGNEX scattering chamber during the experimental run of NUMEN phase 4. Its shape has been designed to facilitate the heat flow toward the cold finger located at the bottom. In particular, the base is wide to ensure a large contact surface with the cold finger of the cryocooler. The top part is thin to guarantee a wide solid angle aperture for the ancillary detectors surrounding the target (e.g. γ-ray calorimeter). The isotopically enriched target will be located at the bottom position. Three more target positions are present to allocate a pure HOPG target for background evaluation, an empty hole for the beam transport procedures and an alumina layer needed for the beam optics optimization.

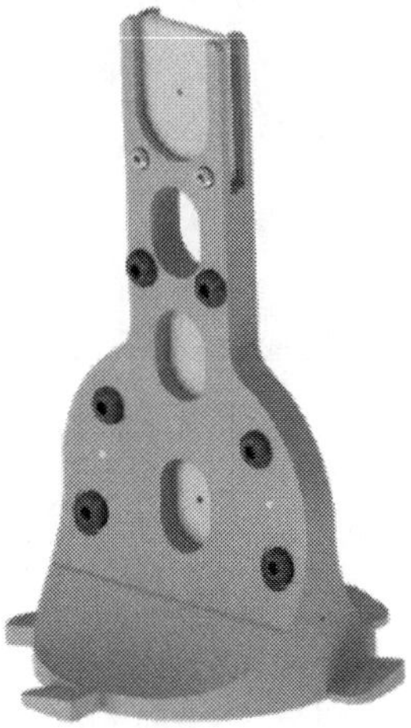

Fig. 2. Sketch of the NUMEN target holder

The target characterization methods have also been the subject of R&D studies [8]. The determination of the thickness of the target and HOPG substrate and of their uniformity is of crucial importance. Several analysis techniques have been adopted. Rutherford Backscattering Spectroscopy (RBS) and Alpha Particle Transmission (APT) have been exploited to measure the thickness and the uniformity of the target layer and substrate. Both techniques can provide accurate evaluations of the average thickness of the sample, but only with APT can the thickness uniformity be directly quantified. The thickness obtained by APT can be verified with RBS, while the roughness of the isotopic deposition can be evaluated with Field Emission Scanning Electron Microscopy, which provides images of the sample surface topography.

A picture of the vacuum chamber used for APT measurements at INFN-LNS is shown in Fig. 3. A rotating plate divided in 4 sections, which allocate several target frames is visible, together an alpha-particle source and a silicon detector used to measure the residual energy of the alpha-particles going through the target.

Fig. 3. Alpha-particle transmission setup at INFN-LNS

2.2 The gas tracker system

A gas tracker for the focal plane detector of the MAGNEX spectrometer is under development within the NUMEN project [11], [12]. The role of the tracker is to perform a three-dimensional measurement of the reaction ejectile trajectories crossing the focal plane. Such an operation is needed for particle identification purposes and to allow the high order reconstruction of the trajectories. Two main characteristics are required. The first is a high-resolution measurement of the phase space parameters of the ion tracks at the focal plane: X_{foc}, Y_{foc}, θ_{foc}, φ_{foc}, where X_{foc} is the horizontal coordinate (dispersive direction along the focal plane), θ_{foc} the horizontal angle, Y_{foc} the vertical coordinate and φ_{foc} the vertical angle. The required resolution is better than 0.6 mm for X_{foc} and Y_{foc} and lower than 5 mrad for θ_{foc} and φ_{foc}. Such a precise and accurate particle ray reconstruction is mandatory for the determination of the momentum vector (scattering angle and kinetic energy) of the ejectiles at the target position [13]. The second requirement is the ability to withstand the expected high rate of impinging particles (about 50 kHz/cm).

The tracker has an active volume of 1200 x 150 x 108 mm^3, like the present one [14], [15]. A reduced size prototype, shown in Fig. 4, has been developed and built during the R&D phase of NUMEN. This prototype, with mechanical and electrical features very similar to those of the final tracker, except for a shorter length which allowed easier tests in a small gas chamber. The applied voltages, the gas pressure and flowing system, the multiplication technology, and the front-end electronics are among the main features tested with the reduced-size prototype.

The prototype, as well as the full-size detector, works as a time projection chamber and consists of three main sections:

- a drift region, which is the active volume of the detector, crossed by the ejectiles of interest,
- an electron multiplication stage, based on multiple Thick Gas Electron Multipliers (m-THGEM),
- a segmented read-out electrode.

The incident charged particle crosses a thin mylar window (typical thickness between 1.5 and 6 μm) and leaves a track of ionized atoms and electrons in the low-pressure gas of the drift region between the cathode and the electron multiplication stage. Under the uniform electric field, the electrons drift with constant velocity toward the m-THGEM. As they reach the multiplication element, electrons are guided into the holes by the strong electric field. The produced electron avalanches reach the segmented read-out electrode, where the projection of the particle trajectory on the horizontal plane can be reconstructed. By the measurement of the drift time of the electrons in the gas, the projection of the trajectory on the vertical plane can also be reconstructed.

Detailed results from the prototype characterization in terms of measured currents and of different segmentation geometries for the anodic read-out plate are given in [11], [12].

Fig. 4. View of the reduced-size tracker prototype.

2.3 The particle-identification wall

The new gas tracker of the MAGNEX spectrometer is not designed to provide accurate information on ion energy loss. Consequently, particle identification (PID) must be demanded from a dedicated wall of telescope detectors downstream of the tracker. Fig. 5 shows the schematic layout of the PID wall system.

The most important requirements for the design of a detection system capable of unambiguously identifying ions in the region of interest for NUMEN ejectiles (typically O, F and Ne) are:

- radiation hardness (the heavy-ion fluence rate will be of the order of 10^{11} ions/cm^2·yr),
- energy resolution (better than 2%, to maintain the present performance in terms of atomic number and mass identification [16] and sensitivity in the cross-section measurements [5]),
- time resolution (~5 ns, to guarantee an accurate measurement of the drift time of the primary electrons in the gas tracker),
- degree of segmentation (to keep the pile-up probability below 3% in the whole FPD),
- geometrical efficiency (which should be high enough to obtain an accurate measurement of the absolute cross section and to reduce the background from events with partial charge collection),
- detector thickness (which must stop the ejectiles of interest in a wide dynamical range of incident energies, typically 15 to 60 MeV/u),
- working in a low-pressure gas environment (typical pressure 10-100 mbar).

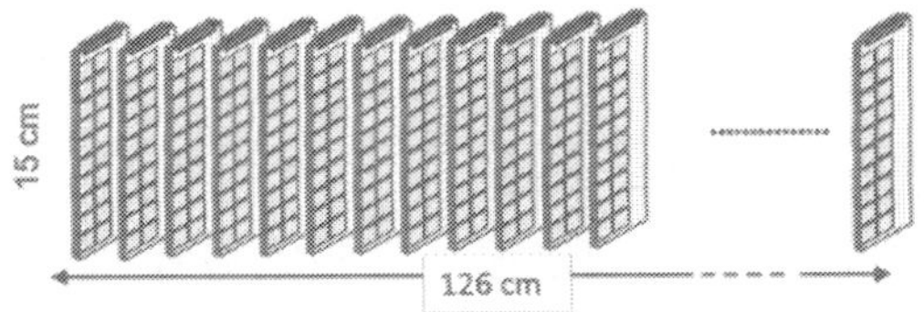

Fig. 5. Schematic layout of the PID-wall telescopes.

The R&D activity on the PID wall resulted in the choice of an array of two-stage telescopes of Silicon Carbide (SiC) [17] and Thallium-doped Cesium Iodide CsI(Tl) detectors. The active area of each element is 1.5 × 1.5 cm^2, with 0.2 mm dead space between adjacent cells. The SiC detector is 100-μm thick, while the CsI(Tl) scintillator is 5-mm thick and is coupled to a Hamamatsu S3590 photodiode with a 1 × 1 cm^2 area. The capabilities of the system in terms of radiation hardness and mass resolution are discussed in [12]. A scheme of the mechanical arrangement of the telescopes is presented in Fig. 5, as a result of an optimization of the detection efficiency and of the mechanical integration [11].

3 Conclusions

The NUMEN project has promoted a major upgrade of the INFN–LNS research facility allowing for a significant increase of the beam intensity, which is needed for the systematic exploration of the DCE transitions of interest for 0νββ. Several aspects of the technology involved in heavy ion collision experiments demand for challenging R&D.

In addition to the on-going improvement of the CS accelerator extraction technique for heavy ion beams in the regime of kW power and at energies from 15 to 70 MeV/u, crucial upgrades involve the MAGNEX spectrometer.

Detailed descriptions of such activities and of the proposed solutions are given in the NUMEN Technical Design Report [18]. Here an overview of the main results has been given, with a special focus on the target technology, on the focal plane detector tracker and particle identification wall.

This project has received funding from the European Research Council (ERC) under the European Union's Horizon 2020 research and innovation programme (grant agreement No 714625).

References

1. F. Cappuzzello et al., Eur. Phys. J. A **52**, 167 (2016)
2. F. Cappuzzello et al., Eur. Phys. J. A **54**, 72 (2018)
3. H. Lenske et al., Prog. Part. Nucl. Phys. **109** 103716 (2019)
4. E. Santopinto, et al., Phys. Rev. C **98**:061601 (2018)
5. S. Calabrese et al., Nucl. Inst.Method Phys. Res. A **980**:164500 (2020)
6. C. Agodi et al., Universe **7**, 72 (2021)
7. F. Iazzi, et al., WIT Trans. Engin. Sci. **116**, 61 (2017)
8. V. Capirossi et al., Acta Phys. Pol. B **51**, 661. (2020)
9. F. Pinna et al., Acta Phys. Pol. B **51**, 665 (2020)

10. D. Sartirana et al., Adv. Serv. Indust. Robot. **84**, 533 (2020)
11. F. Cappuzzello et al., Front. In Astr. And Space Sc. **8**, 668587 (2021)
12. P. Finocchiaro et al., Universe **6**, 129 (2020)
13. F. Cappuzzello et al., Nucl. Instr. And Meth. A **638**, 74 (2011)
14. M. Cavallaro et al., Eur. Phys. J. A **48**, 59 (2012)
15. D. Torresi et al., Nucl. Instr. Method A **989**, 164918 (2021)
16. F. Cappuzzello et al., Nucl. Instr. And Meth. A **621**, 419 (2010)
17. S. Tudisco et al., Sensors **18**, 2289 (2018)
18. F. Cappuzzello et al., Journ. of Modern Phys. A, submitted.

Recent experimental activity on heavy-ion induced reactions within the NUMEN project

Francesco Cappuzzello[1,2,*], *Luis* Acosta[3], *Clementina* Agodi[1], *Carmen* Altana[1], *Paulina* Amador-Valenzuela[4], *Ismail* Boztosun[5], *Giuseppe* A. Brischetto[1,2], *Salvatore* Calabrese[1,2], *Daniela* Calvo[6], *Vittoria* Capirossi[6,7], *Diana* Carbone[1], *Manuela* Cavallaro[1], *Efrain* R. Chávez Lomeli[3], *Irene* Ciraldo[1,2], *Mauro* Cutuli[1,2], *Franck* Delaunay[1,2,8], *Haris* Djapo[9], *Canel* Eke[5], *Paolo* Finocchiaro[1], *Suna* Firat[5], *Maria* Fisichella[1], *Marcilei* A. Guazzelli da Silveira[10], *Aylin* Hacisalihoglu[11], *Felice* Iazzi[6,7], *Gaetano* Lanzalone[1,12], *Laura* La Fauci[1,2], *Roberto* Linares[13], *Nilberto* H. Medina[14], *Mauricio* Moralles[15], *José* R. B. Oliveira[14], *Athena* Pakou[16], *Luciano* Pandola[1], *Horia* Petrascu[17], *Federico* Pinna[6,7], *Giuseppe* Russo[1], *Roberto* B. B. Santos[10], *Onoufrios* Sgouros[1], *Selçuk* O. Solakci[5], *George* Souliotis[18], *Vasileios* Soukeras[1,2], *Alessandro* Spatafora[1,2], *Domenico* Torresi[1], *Salvatore* Tudisco[1], *Aydin* Yildirin[5], *Vinicius* A. B. Zagatto[13]

for the NUMEN Collaboration

[1]Istituto Nazionale di Fisica Nucleare, Laboratori Nazionali del Sud, Catania, Italy
[2]Dipartimento di Fisica e Astronomia "Ettore Majorana", Università di Catania, Italy
[3]Instituto de Fısica, Universidad Nacional Autonoma de Mexico - Mexico City, Mexico
[4]Instituto Nacional de Investigaciones Nucleares - Ocoyoacac, Mexico
[5]Department of Physics, Akdeniz University - Antalya, Turkey
[6]Istituto Nazionale di Fisica Nucleare, Sezione di Torino, Italy
[7]DISAT, Politecnico di Torino, Italy
[8]LPC Caen, Normandie Universite, ENSICAEN, UNICAEN, CNRS/IN2P3 Caen, France
[9]Ankara University, Institute of Accelerator Technologies, Turkey
[10]Centro Universitario FEI - Sao Bernardo do Campo, Brazil
[11]Institute of Natural Sciences, Karadeniz Teknik University - Trabzon, Turkey
[12]Università degli Studi di Enna "Kore" - Enna, Italy
[13]Instituto de Fisica, Universidade Federal Fluminense - Niteroi, Brazil
[14]Instituto de Fisica, Universidade de Sao Paulo - Sao Paulo, Brazil
[15]Instituto de Pesquisas Energeticas e Nucleares IPEN/CNEN - Sao Paulo, Brazil
[16]Department of Physics and HINP, University of Ioannina - Ioannina, Greece
[17]IFIN-HH - Magurele, Romania
[18]Department of Chemistry, National and Kapodistrian University of Athens, Athens, Greece

Abstract. The possibility to use a special class of heavy-ion induced direct reactions, such as double charge exchange reactions, is discussed in view of their application to extract information that may be helpful to determinate the nuclear matrix elements entering in the expression of neutrinoless double beta decay half-life. The strategies adopted in the experimental campaigns performed at INFN - Laboratori Nazionali del Sud are briefly described, emphasizing the advantages of the multi-channel approach to nuclear reaction data analysis.

1 Introduction

Charge exchange (CE) reactions are nuclear processes in which the individual balance of neutrons and protons inside the projectile and the target is modified, while keeping the overall number of nucleons unchanged. In the isospin formalism the CE promote isovector excitations in both the projectile and target. CE are thus selective tools to emphasize the role of isospin in both nuclear structure and reaction studies. Relevant aspects of single charge exchange (SCE) induced by light projectile have been discussed in several articles, among which excellent reviews are in ref. [1] [2] [3]. For heavy-ion induced CE we refer to the recent review [4]. An especially appealing aspect is the possibility to use CE to investigate the spin-isospin-dependent response of nuclei.

Recently, particular interest has been given to double charge exchange (DCE) processes [5] [6] [7] as they can give access to the nuclear response to second order isospin operators, which also promote double beta decays.

SCE and DCE studies have a genuine nuclear physics interest, since they reflect a key aspect of nuclear dynamics. However, beyond that, there is an even broader impact of such investigations because of their relevance for nuclear beta decay, thus connecting strong and weak interactions.

In this context, the NUMEN and NURE projects at INFN-LNS laboratory in Catania [8] [9] aim at exploring heavy-ion induced single and double charge exchange (DCE)

* Corresponding author: cappuzzello@lns.infn.it

reactions in a systematic approach. The main goal is to extract information on nuclear matrix elements for SCE and DCE processes and to identify the possible connections with double beta decay.
Detailed information about these projects are found in recent articles, which we refer to [10] [11] [12] for both scientific and technological aspects.
In the present paper, the global strategy and methodologies of the above mentioned programmes, the challenges of the experimental activities together with the status of the already performed experimental runs are briefly outlined.

2 The motivation

Neutrinoless double beta decay ($0\nu\beta\beta$) is an hypothetical process in which a "parent" nucleus is spontaneously transformed into a "daughter" isobar by the conversion of two neutron/protons into two protons/neutrons. The process is accompanied by the emission of two electrons (or positrons), which guarantee also electric charge conservation. If detected, $0\nu\beta\beta$ would provide evidence for the Majorana nature of the neutrino (i.e. the equivalence of neutrino and antineutrino) and give access to the neutrino mass [13] [14] [15] [16]. Under the hypothesis that the neutrino, electrons and nuclear degrees of freedom can be separated in $0\nu\beta\beta$, the decay rate explicitly depends on the nuclear transition probability, accounted for in the so-called Nuclear Matrix Element (NME). The $0\nu\beta\beta$ NMEs are very sensitive to nuclear physics involved in decay, since they probe the nuclear wave functions of the initial and final states. In particular, the nuclear response to high-momentum second-order isospin operators is probed in $0\nu\beta\beta$, similarly to what happen in DCE reactions [17]. However, there are no experimental methods to directly measure $0\nu\beta\beta$ NMEs and accurate theoretical calculations are challenging, leading to discrepancy factors larger than two, as presently reported in literature [13]. Thus experimental inputs from SCE and DCE studies are useful to help evaluate the $0\nu\beta\beta$ NMEs and to constrain the calculations. In this context, an experimental campaign has started at the INFN-Laboratori Nazionali del Sud in Catania using the MAGNEX large acceptance magnetic spectrometer [18]. It focuses on the study of heavy-ion induced DCE reactions and competing reaction channels involving the nuclei of interest for $0\nu\beta\beta$ decay.
The possible connections of DCE measured cross-sections with double beta decay NMEs and, consequently, with the still unsolved neutrino puzzles are among the goals of the present studies.

2 Heavy-ion induced double charge exchange reactions

Challenging experimental issues must be addressed to measure heavy-ion induced DCE cross sections. Indeed one needs to detect heavy ions with good isotopic separation and energy resolution in a wide angular range, including zero-degree, in order to distinguish transitions to individual states and explore a wide momentum transfer range. In addition, the rather tiny DCE cross sections (down to few nb) demand for high experimental sensitivity, which strongly depends on the rejection capability against unwanted spurious events. These latter may be generated by competing reaction processes, some of which are more likely to occur than DCE. For this reason highly selective particle identification is a necessary prerequisite for these experiments. In addition, spurious data associated to a wrong determination of detection parameters could casually be misinterpreted as valid DCE data. Characterizing the relevant background sources can be subtle and require stringent conditions for the experimental set-up as well as a careful study of the detection response [19].
Recently, accurate cross sections for heavy-ion DCE reactions have been measured at the INFN-LNS laboratory, thanks to the use of the MAGNEX spectrometer.
The accurate description of heavy-ion induced DCE is a demanding task also for nuclear theory. From the reaction theory side initial and final state interactions of projectile and target systems need to be under control and the reaction amplitude must be described properly. In this context direct reaction theory can provide the suitable toolbox. Nuclear structure theory is asked to provide accurate input for modelling both projectile and target [20].
Recently, new theoretical developments for the reaction mechanism involved in heavy ion induced reactions have been achieved alongside. Presently this activity is in progress within the NUMEN project [21] [22] [23] [24].

3 The experiments

The NUMEN experiments are performed at INFN-LNS using the K800 Superconducting Cyclotron to accelerate beams and the MAGNEX large acceptance magnetic spectrometer for the detection of the ejectiles.
MAGNEX is a large acceptance magnetic device consisting of a large aperture vertically focusing quadrupole and a horizontally bending dipole magnet. A hybrid Focal Plane detector (FPD) featuring a gaseous tracker and a wall of silicon pad sensors is mounted at the focal plane to detect the reaction ejectiles [25] [26].
MAGNEX was designed to investigate processes characterized by very low yields, guaranteeing the identification of heavy ions with quite high mass ($\Delta A/A \sim 1/160$), angle ($\Delta\theta \sim 0.2°$) and energy resolutions ($\Delta E/E \sim 1/1000$), within a large solid angle ($\Omega \sim 50$ msr) and momentum range ($-14\% < \Delta p/p < +10\%$). High-resolution measurements for quasi-elastic processes, characterized by differential cross-sections as low as tens of nb/sr, were already performed with this setup [27] [28]. A crucial feature is the implementation of a technique of trajectory reconstruction, based on differential algebraic techniques and the accurate mapping of the spectrometer magnetic fields [29] [30] [31] [32], which solves the equation of motion of each detected particle to 10th order [33]. This is a unique characteristic of MAGNEX, which guarantees the above mentioned performances and its relevance in the worldwide scenario of heavy-ion physics.

The experimental activity with accelerated beams proposed and presently in progress consists of two main classes of experiments, aiming at the exploration of the two directions of isospin lowering $\tau^- \tau^-$ and rising $\tau^+ \tau^+$, characteristic of $\beta^-\beta^-$ and $\beta^+\beta^+$ decays, respectively.

In particular, the $\beta^+\beta^+$ direction in the target excitation modes is investigated using a $^{18}O^{8+}$ beam and measuring the (^{18}O,^{18}Ne) DCE transitions, together with other reaction channels involving same beam and target. Similarly, the $\beta^-\beta^-$ direction is explored via the (^{20}Ne,^{20}O) reaction, using a $^{20}Ne^{10+}$ beam and detecting the reaction products from the DCE channel and from the other open reaction channels.

In each case, the experiments are organized in order to explore together with the DCE reactions of interest, all the relevant quasi-elastic processes occurring in the projectile-target collision. In this way, we can get supplementary information from the detailed study of the reaction network, profiting of the strong selectivity of quasi-elastic processes. The nucleus-nucleus interaction can be extracted from elastic scattering data and applied to the evaluation of the distorted waves in each direct reaction channel. Inelastic scattering data give selective access to the role of the deformation in the coupled channel reaction network equations. Single nucleon transfer reactions shed light on the role of single particle configurations in the involved nuclear many body states. Two nucleon transfer reactions probe pairing correlations, while SCE give access to 1particle-1hole excitations. All of these information are provided in a consistent experimental framework and analysed in a multi-channel approach, based on quantum scattering theory of coupled channels. A unique opportunity is thus searched by NUMEN to build an over-constrained set of nuclear structure information, promoting the development of a new approach to $0\nu\beta\beta$ NMEs calculations.

Investigations of the two classes of experiments have been started in the recent years with the present facility. These studies highlighted the strengths and the limiting aspects of the adopted technique, providing first valuable information on relevant nuclear structure aspects for a number of double beta decay emitter candidates [8]. This activity has also given precious information to guide the major upgrade of the facility toward much higher beam intensity, which is presently going on.

3.1. Experiments with ^{18}O beam

For the experiments of this class, the reaction channels of our interest are listed below:

- (^{18}O,^{18}O) elastic and inelastic scattering
- (^{18}O,^{18}Ne) DCE reaction
- (^{18}O,^{18}F) SCE reaction
- (^{18}O,^{20}Ne) two-proton pickup reaction
- (^{18}O,^{19}F) one-proton pickup reaction
- (^{18}O,^{16}O) two-neutron stripping reaction
- (^{18}O,^{17}O) one-neutron stripping reaction

One of the main challenges of such experiments is the measurement at very forward angles, including zero-degree. This is performed by placing the spectrometer with its optical axis at +3° with respect to the beam axis. Thanks to its large angular acceptance, the range $-2° < \theta_{lab} < 9°$ is explored. The MAGNEX quadrupole and dipole magnetic fields are set in order that the incident beam, after passing through the magnets, reaches a region besides the FPD. For this class of experiments, in fact, the incident ion beam ($^{18}O^{8+}$) has higher magnetic rigidity (Bρ) than the ejectiles of interest (namely fully stripped ^{18}Ne, ^{18}F, ^{20}Ne, ^{19}F, ^{16}O, ^{17}O ions). The beam stops in a Faraday cup, placed in the high-Bρ region besides the FPD, which measures the collected charge for each run.

^{116}Sn, ^{76}Se and ^{48}Ti are the targets of interest for $0\nu\beta\beta$, which have already been explored by NUMEN, via (^{18}O,^{18}Ne) reaction at 15 and 22 AMeV. The purpose was to study the ^{116}Sn → ^{116}Cd, ^{76}Se → ^{76}Ge and ^{48}Ti → ^{48}Ca transitions and the competing channels as shown in Fig. 2. A dedicated run was also performed on ^{12}C target, in order to compare reaction models with an easier light system. The reduction and analysis of the collected data from these campaigns is presently in progress. First results have been already published in ref. [34] [35] [36] [37].

3.2. Experiments with ^{20}Ne beam ($\beta^-\beta^-$ direction)

In the class of experiments triggered by $^{20}Ne^{10+}$ beams, the reaction channels we are interested are the following:

- (^{20}Ne,^{20}Ne) elastic and inelastic scattering
- (^{20}Ne,^{20}O) DCE reaction
- (^{20}Ne,^{20}F) SCE reaction
- (^{20}Ne,^{18}O) two-proton stripping reaction
- (^{20}Ne,^{19}F) one-proton stripping reaction
- (^{20}Ne,^{22}Ne) two-neutron pickup reaction
- (^{20}Ne,^{21}Ne) one-neutron pickup reaction.

For these experiments, the incident beam ($^{20}Ne^{10+}$) has a lower magnetic rigidity compared to to the ejectiles of interest. Thus, for a fixed magnetic field setting, the beam is more bent than the ejectiles of interest. The spectrometer optical axis is typically placed at −3°, thus the covered angular range is $-8° < \theta_{lab} < +3°$. The quadrupole and dipole magnetic fields are set in order that the $^{20}Ne^{10+}$ beam reaches the low-Bρ region besides the FPD. The $^{20}Ne^{9+}$ and $^{20}Ne^{8+}$ components of the beam emerging for the target could be a source of background in these experiments. They are significantly reduced by the use of a supplementary stripper foil mounted after the target itself [38].

The systems already experimentally explored using the (^{20}Ne,^{20}O) reaction at 15 AMeV are the ^{116}Cd target (to study the ^{116}Cd → ^{116}Sn transition), the ^{130}Te (for the ^{130}Te → ^{130}Xe) and the ^{76}Ge (for the ^{76}Ge → ^{76}Se). The data reduction and analysis is in progress and the first results from some reaction channels have been recently published [39] [40] [41] [42].

A scheme of the transitions already studied in the experimental runs is shown in Fig. 2.

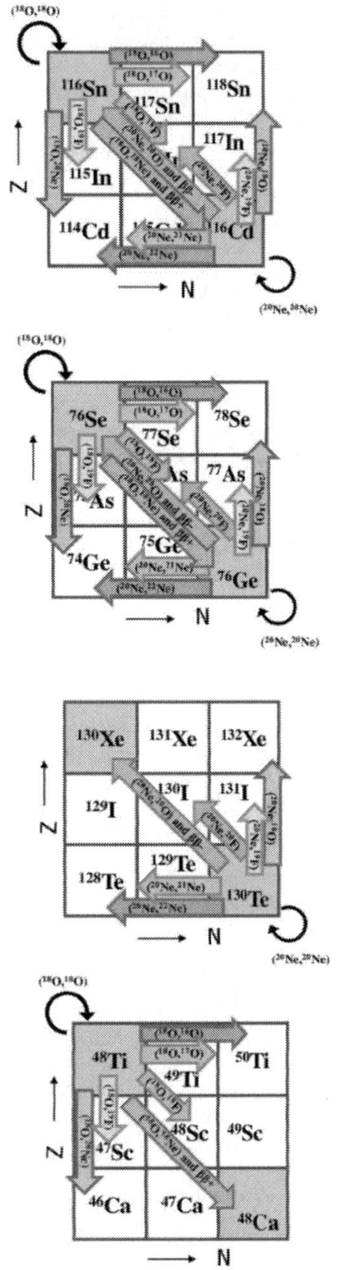

Fig. 2 Scheme of the reaction channels already studied within the NURE and NUMEN experiments.

Conclusions

A systematic study of heavy-ion induced DCE reactions and of the other reaction channels characterized by the same projectile and target is in progress at INFN-LNS. The main goal is to investigate the nuclear response to DCE reactions for all the isotopes candidate for $0\nu\beta\beta$ decay towards the determination of $0\nu\beta\beta$ NMEs.

Acknowledgments

This project has received funding from the European Research Council (ERC) under the European Union's Horizon 2020 research and innovation programme (grant agreement No 714625).

References

[1] Alford, W.P.; Spicer, B.M. Nucleon charge-exchange reactions at intermediate energy. Adv. Nucl. Phys. 1998, 1, 24.

[2] Osterfeld, F. Nuclear spin and isospin excitations. Rev. Mod. Phys. 1992, 64, 491..

[3] Taddeucci, T.N.; et al. The (p, n) reaction as a probe of beta decay strength. Nucl. Phys. A 1987, 469, 125..

[4] Lenske, H.; Cappuzzello, F.; Cavallaro, M.; Colonna, M. Heavy ion charge exchange reactions as probes for nuclear β-decay. Prog. Part. Nucl. Phys. 2019, 109, 103716..

[5] F. Cappuzzello et al., Eur. Phys. J. A 51, 145 (2015).

[6] Cappuzzello, F.; Cavallaro, M.; Agodi, C.; Bondi, M.; Carbone, D.; Cunsolo, A.;, Foti, A. Heavy-ion double charge exchange reactions: A tool toward 0νββ nuclear matrix elements. Eur. Phys. J. A 2015, 51, 145.

[7] Cappuzzello, F., and Agodi, C. (2021). The NUMEN Project: Shedding Light on Neutrinoless Double Beta Decay by Heavy-Ion Nuclear Reactions. Nucl, Phys. News, in press, DOI 10.1080/10619127.2021.1881368.

[8] Cappuzzello, F., Agodi, C., Cavallaro, M., Carbone, D., Tudisco, S., Lo Presti, D., et al.. (2018). The NUMEN project: NUclear Matrix Elements. Eur. Phys. J. A. 54, 72.

[9] Cavallaro, M., Aciksoz, E., Acosta, L., Agodi, C., Auerbach, N., Bellone, J. I., et al., (2017). "NURE: an ERC project to study nuclear reactions," in Proceedings, 55th International Winter Meeting on Nuclear Physics Bormio, (Bormio). doi: 10.22323/1.302..

[10] P. Finocchiaro et al., Universe 2020, 6, 129; doi:10.3390/universe6090129.

[11] C. Agodi et al., Universe 2021, 7, 72. https://doi.org/10.3390/universe7030072.

[12] F. Cappuzzello et al., The NUMEN Project: An Update of the Experimental Campaigns. Front. Astron. Space Sci. 8:668587. doi: 10.3389/fspas.2021.668587.

[13] H. Ejiri, J. Suhonen, K. Zuber, Physics Reports 797 (2019) 1.

[14] J. Engel, J. Menéndez, Rep. Progr. Phys. 60 (2017) 046301.

[15] S. Dell'Oro, S. Marcocci, and F. Vissani, Phys. Rev. D 90, 033005 (2014).

[16] J. Vergados, H. Ejiri, F. Šimkovic, Rep. Progr. Phys. 75 (2012) 106301.

[17] Cappuzzello, F., and Cavallaro, M. (2020). Nuclear Response to Second-Order Isospin Probes in Connection to Double Beta Decay. Universe. 6, 217.

[18] F. Cappuzzello et al. The MAGNEX spectrometer: Results and perspectives. Eur. Phys. J. A (2015) 51:145.

[19] S. Calabrese et al. Analysis of the background on cross section measurements with the MAGNEX spectrometer: The (20Ne,20O) Double Charge Exchange case. Nuclear Inst. and Methods in Physics Research, A 980 (2020) 164500.

[20] N. Shimizu, J. Menendez, and K. Yako, Phys. Rev. Lett. 120, 142502 (2018)..

[21] H. Lenske, J.I. Bellone, M. Colonna, J. A. Lay, Phys. Rev. C 98 (4), 044620 (2018)..

[22] E. Santopinto, H. García-Tecocoatzi, RIM Vsevolodovna, J. Ferretti, Phys. Rev. C 98 (6), 061601 (2018).

[23] J.I. Bellone, S. Burrello, M. Colonna, J. A. Lay, H. Lenske, Phys. Lett. B 807, 135528 (2020).

[24] Lenske, H.; Bellone, J.; Colonna, M.; Gambacurta, D. Nuclear Matrix Elements for Heavy Ion Sequential Double Charge Exchange Reactions. Universe 2021, 7, 98. https://doi.org/10.3390/universe7040098.

[25] Torresi, D., Sgouros, O., Soukeras, V., Cavallaro, M., Cappuzzello, F., Carbone, D., et al. (2021). An upgraded focal plane detector for the MAGNEX spectrometer. Nucl. Instr. and Meth. A. 989, 164918.

[26] M. Cavallaro et al. The low-pressure focal plane detector of the MAGNEX. Eur. Phys. J. A (2012) 48: 59. DOI 10.1140/epja/i2012-12059-8.

[27] J.R.B. Oliveira et al., Journal of Physics G: Nuclear and Particle Physics, 40 (2013) 105101.

[28] D. Pereira et al. Phys. Lett. B 710 (2012) 426.

[29] A. Lazzaro et al., Nucl. Instr. and Methods A 591, 394 (2008)..

[30] A. Lazzaro et al., Nucl. Instr. and Methods A 585, 136 (2008).

[31] A. Lazzaro et al., Nucl. Instr. and Methods A 570, 192 (2007).

[32] A. Lazzaro et al., Nucl. Instr. and Methods A 602, 494 (2009).

[33] F. Cappuzzello et al., Eur. Phys. J. A 52 (2016) 169.

[34] O. Sgouros et al., Phys. Rev. C, accepted.

[35] D. Carbone et al., Universe 7 (2021) 58..

[36] J. L. Ferreira et al., Phys. Rev. C 103 (2021) 054604.

[37] M. Cavallaro et al., Frontiers in Astron. and Space Science 8 (2021) 659815.

[38] M. Cavallaro et al., Results in Phys. 13 (2019) 102191.

[39] A. Spatafora et al., Phys. Rev. C 100 (2019) 034620.

[40] V. Soukeras et al., Results in Physics, (2021) submitted.

[41] S. Calabrese et al., Nucl. Instr. and Meth. A, 980 (2020) 164500.

[42] S. Calabrese et al., Acta Phys. Pol. B 49 (2018) 275.

Recent results for the one-proton transfer reaction in the ^{18}O+^{48}Ti collision at 275 MeV

O. Sgouros[1,*], *M.* Cavallaro[1], *F.* Cappuzzello[1,2], *D.* Carbone[1], *C.* Agodi[1], *C.* Altana[1], *G. A.* Brischetto[1,2], *S.* Burrello[3,4], *S.* Calabrese[1], *D.* Calvo[5], *V.* Capirossi[5,6], *E. R.* Chávez Lomelí[7], *I.* Ciraldo[1,2], *M.* Cutuli[1,2], *G.* De Gregorio[8,9], *F.* Delaunay[1,2,10], *H.* Djapo[11], *C.* Eke[12], *P.* Finocchiaro[1], *M.* Fisichella[1], *A.* Foti[13], *A.* Gargano[8], *A.* Hacisalihoglu[14], *F.* Iazzi[5,6], *L.* La Fauci[1,2], *R.* Linares[15], *J.* Lubian[15], *N. H.* Medina[16], *M.* Moralles[17], *J. R. B.* Oliveira[15], *A.* Pakou[18], *L.* Pandola[1], *F.* Pinna[5,6], *G.* Russo[2,13], *M. A.* Guazzelli[19], *V.* Soukeras[1], *G.* Souliotis[20], *A.* Spatafora[1,2], *D.* Torresi[1], *A.* Yildirim[21], and *V. A. B.* Zagatto[15] for the NUMEN collaboration

[1]INFN - Laboratori Nazionali del Sud, Catania, Italy
[2]Dipartimento di Fisica e Astronomia "Ettore Majorana", Università di Catania, Catania, Italy
[3]Université Paris-Saclay, CNRS/IN2P3, IJCLab, Orsay, France
[4]Technische Universität Darmstadt, Institut für Kernphysik, Darmstadt, Germany
[5]INFN - Sezione di Torino, Torino, Italy
[6]DISAT - Politecnico di Torino, Torino, Italy
[7]Instituto de Física, Universidad Nacional Autónoma de México, Mexico City, Mexico
[8]INFN - Sezione di Napoli, Napoli, Italy
[9]Dipartimento di Matematica e Fisica, Università della Campania "Luigi Vanvitelli", Caserta, Italy
[10]LPC Caen, Normandie Université, ENSICAEN, UNICAEN, CNRS/IN2P3, Caen, France
[11]Ankara University, Institute of Accelerator Technologies, Turkey
[12]Department of Mathematics and Science Education, Faculty of Education, Akdeniz University, Antalya, Turkey
[13]INFN - Sezione di Catania, Catania, Italy
[14]Institute of Natural Sciences, Karadeniz Teknik Universitesi, Trabzon, Turkey
[15]Instituto de Física, Universidade Federal Fluminense, Niterói, Brazil
[16]Instituto de Física, Universidade de São Paulo, São Paulo, Brazil
[17]Instituto de Pesquisas Energeticas e Nucleares IPEN/CNEN, São Paulo, Brazil
[18]Department of Physics, University of Ioannina and Hellenic Institute of Nuclear Physics, Ioannina, Greece
[19]Centro Universitario FEI, São Bernardo do Campo, Brazil
[20]Department of Chemistry, University of Athens and Hellenic Institute of Nuclear Physics, Athens, Greece
[21]Department of Physics, Akdeniz Universitesi, Antalya, Turkey

Abstract. The ^{18}O+^{48}Ti reaction was studied at the energy of 275 MeV for the first time under the NUMEN and NURE experimental campaigns with the aim to investigate the complete net of reaction channels potentially involved in the ^{48}Ca→^{48}Ti double charge exchange transition. Such a transition is of great interest because of its relevance to the extraction of ^{48}Ca→^{48}Ti double beta decay nuclear matrix element. The relevant experiment was carried out at the MAGNEX facility of INFN-LNS in Catania. Angular distribution measurements for the various reaction products were performed by using the MAGNEX large acceptance magnetic spectrometer. The present contribution is focused on the analysis of the one-proton transfer channel with emphasis on the particle identification technique and the estimation of background contaminations.

1 Introduction

The interest of the Nuclear Physics community for the Double Charge Exchange (DCE) reactions is still vivid due to their possible relation with the double beta ($\beta\beta$) decay [1]. Into this context, a novel idea has been recently conceived at Istituto Nazionale di Fisica Nucleare – Laboratori Nazionali del Sud (INFN-LNS), under the NUMEN (NUclear Matrix Elements for Neutrinoless double beta decay) project [2]. This is to use for the first time DCE reactions induced by heavy-ion beams as a mean for the determination of the $\beta\beta$ decay Nuclear Matrix Elements (NMEs). This unique opportunity is based on the fact that the two processes present some interesting similarities. Among them, both processes probe the same initial and final nuclear wave functions and the operators connecting them have a similar spin – isospin mathematical structure [2–4]. Thus, even if the two processes are mediated by different interactions, the involved NMEs could be connected and the determination of the DCE reaction cross-sections may provide an important piece of information on the $\beta\beta$ matrix elements. In the present study, which is a part of the

*e-mail: onoufrios.sgouros@lns.infn.it

NURE project [5], the $^{18}O+^{48}Ti$ reaction was investigated by measuring not only the DCE channel, but also Single Charge Exchange (SCE) and multi-step transfer reactions which may lead to the same final states as the DCE one. It is very important to quantify possible contributions from other reaction channels to the DCE one for a precise determination of the absolute DCE cross sections, which may be the key for accessing the information of the NMEs of the $\beta\beta$ decay. In the present contribution, the data analysis of the one-proton transfer reaction $^{48}Ti(^{18}O,^{19}F)^{47}Sc$ is presented. The experimental details are presented in Section 2, while the data reduction procedure is reported in Section 3. Our results are summarized in Section 4.

2 Experimental Setup

The experiment was performed at the MAGNEX facility [6] of INFN-LNS, where an $^{18}O^{8+}$ beam accelerated by the K800 Superconducting Cyclotron at the energy of 275 MeV impinged onto a TiO_2 target 510 $\mu g/cm^2$ thick. An aluminum foil with a thickness of 216 $\mu g/cm^2$ was used as a backing for the target material. In order to estimate the background originating from the different target components, two additional runs, one with a pure ^{27}Al target 226 $\mu g/cm^2$ thick and a WO_3 one (284 $\mu g/cm^2$) evaporated on a thin aluminium foil, were also performed. The various reaction products were momentum analyzed by the MAGNEX large acceptance magnetic spectrometer whose optical axis was set at θ_{opt}= 9° with respect to the beam direction, spanning an angular range between 4° and 15° in the laboratory reference frame. The vertical slits of the MAGNEX spectrometer, located 260 mm downstream the target position were set such as to cover a vertical angular range of ±2°. The various reaction products emerging from the beam-target interaction were detected by the MAGNEX Focal Plane Detector (FPD) [7, 8].

The MAGNEX FPD is consisted of a gas section followed by a wall of 60 single silicon detectors. The use of the gas detector is twofold. It serves as a proportional drift chamber providing the energy loss signal (ΔE) of the ions inside the gas, but also as a mean to map the ions track. The tracker is divided in six sections each one having at the top a proportional wire (DC) in which the (ΔE) signal is measured. The sum of the 6 signals, ΔE_{tot}, is used in the particle identification process. Above the DC wires, a set of 6 segmented anode strips is located where each strip is composed of 224 induction pads allowing the measurement of the horizontal position (X_{foc}) and thus, the determination of the horizontal angle (θ_{foc}). Moreover, the electron drift time measurements inside the gas allow the determination of the vertical position (Y_{foc}) and angle (ϕ_{foc}). With the present gas tracker, the maximum rate that the detector can sustain is limited to a few kHz. However, this will be insufficient for the future NUMEN experiments at the MAGNEX facility with high intensity beams. To this extent, the development of a new configuration [9, 10] for the MAGNEX FPD is in progress based on the THick GEM (THGEM) technology for the new tracker.

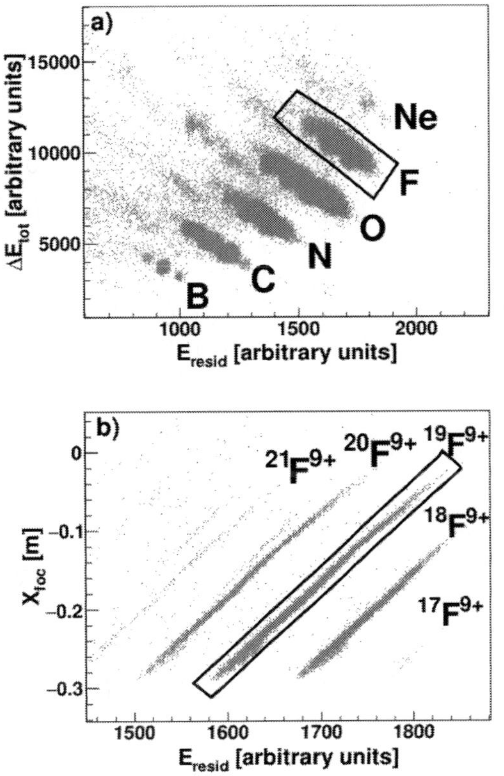

Figure 1. Typical particle identification spectra for the $^{48}Ti(^{18}O,^{19}F)^{47}Sc$ reaction at the energy of 275 MeV. a) ΔE_{tot} as a function the of residual energy, E_{resid}, measured by one silicon detector of the FPD. A graphical selection of the fluorine ion events is illustrated by the solid black line. b) The horizontal position at the MAGNEX focal plane, X_{foc}, as a function of the residual energy gated by the selected fluorine ions of panel (a). The different loci correspond to ions with different $\sqrt{m}/q$. A graphical selection on the $^{19}F^{9+}$ events is shown by the solid black line.

3 Data Reduction

The data analysis relies on the accurate identification of the reaction channel of our interest. The particle identification (PID) is the first step of the data analysis which was performed following the prescription of Ref. [11]. The various ion species were well-discriminated using the conventional ΔE–E technique. A typical ΔE_{tot}–E_{resid} spectrum is presented in the top panel of Figure 1 for a single silicon detector. It is evident that the fluorine isotopes, highlighted with the black contour, are well separated from any other ion family. However, we are interested in identifying the one-proton pick up reaction products i.e. $^{19}F^{9+}$ so, the Z separation is not enough. To this direction, after the identification of the fluorine ions, the mass separation is feasible via the X_{foc}–E_{resid} (horizontal position versus

https://doi.org/10.1051/epjconf/202125204002

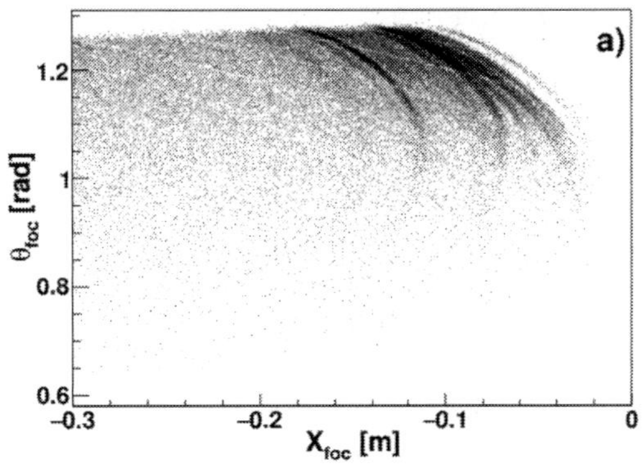

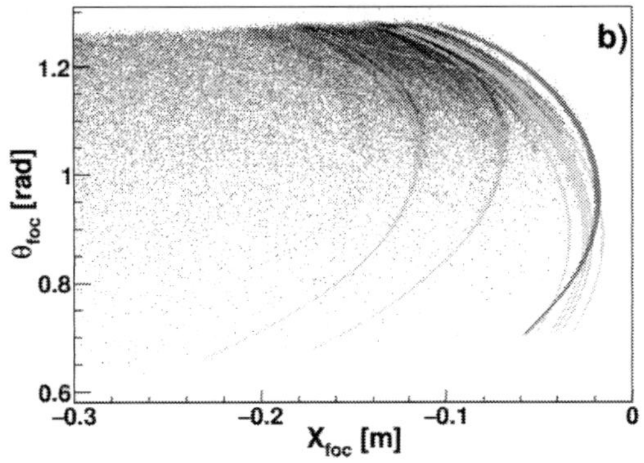

Figure 2. a) Correlation plot between the horizontal angle, θ_{foc}, and horizontal position, X_{foc}, measured by the focal plane detector of the MAGNEX spectrometer. b) A comparison between the experimental data of panel (a) with the simulated ones for three different reaction channels. The red, cyan and green points correspond to the simulated events originating from the one-proton transfer reaction on ^{27}Al, ^{48}Ti and ^{16}O target, respectively. For reasons of clarity, the simulations were performed only for a few selective transitions for the one-proton transfer reaction on each target.

residual energy measured for each silicon detector) spectra. In a spectrometer, the position along the dispersive axis is analogous to the momentum of the ion. So, the position and the energy are related according to the following expression:

$$X_{foc} \propto \frac{\sqrt{m}}{q} \sqrt{E_{resid}}, \quad (1)$$

where m, q and E_{resid} are the ions' mass, charge state and residual energy, respectively. Ions possessing a different mass over charge state ratio lie on a different region at the X_{foc}–E_{resid} representation as it is illustrated in the lower panel of Figure 1. Looking at Figure 1 it is obvious that the mass resolution of MAGNEX is excellent. Thus, the ΔE–E technique in conjunction with mass separation method presented above, provide the root for an accurate identification of the one proton pick-up reaction channel.

Once the $^{19}F^{9+}$ ions were identified, we have proceeded with the analysis of the the final phase space parameters. A typical θ_{foc}-X_{foc} spectrum obtained with the TiO_2+^{27}Al target is presented in Figure 2, where various loci are well-pronounced in the spectrum. Since a compound target was used in the current measurement, background events due to the reaction of the ^{18}O beam with the oxygen and aluminium components of the target were also present in our spectra. In order to identify the fingerprint of each reaction, dedicated Monte Carlo simulations by taking into account the reaction kinematics but also the complete geometry of the spectrometer and the spatial distribution of the dipole and quadrupole fields were performed. The results of the simulations are compared to the experimental data in the lower panel of Figure 2. It is seen that the simulations describe in an excellent way the experimental data giving further support to the validity of the dipole and quadrupole fields which were adopted in the simulations. The different slope met both in experimental and simulated events is attributed partially to the different kinematics for each reaction and partially to the chromatic aberrations which are present in large-acceptance optical devices like MAGNEX. However, the latter are effectively compensated when a high-order software trajectory reconstruction [12] is applied to the data.

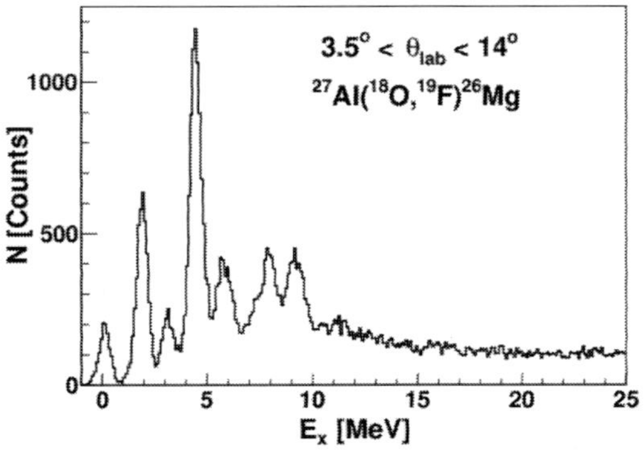

Figure 3. Reconstructed excitation energy spectrum for the ^{27}Al(^{18}O,^{19}F)^{26}Mg reaction at 275 MeV.

Having identified the background events, the analysis of the data obtained with the pure aluminum target and the WO_3+^{27}Al one was performed in order to estimate and subtract the contaminant events in the data set obtained with the TiO_2+^{27}Al target. A software ray reconstruction was applied to the data and the initial phase space parameters (e.g. θ_{lab}, kinetic energy) were reconstructed from the measured ones (e.g. θ_{foc}, ϕ_{foc}, X_{foc}). The excitation energy E_x was determined from the missing mass method [6] as:

$$E_x = Q_0 - Q, \quad (2)$$

where Q_0 is the ground state (g.s.) to g.s. transition. The reconstructed excitation energy spectrum for the data obtained with aluminum target is presented in Figure 3.

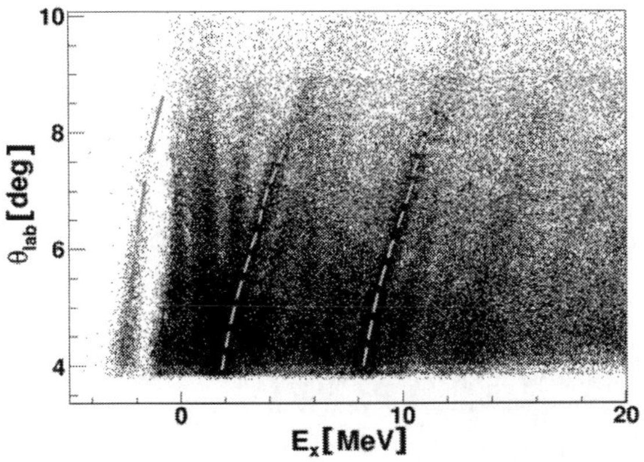

Figure 4. Bi-dimensional plot for the reconstructed scattering angle, θ_{lab}, as a function of the excitation energy, E_x, for the data obtained with the TiO_2+^{27}Al target. The inclined loci, highlighted with the dashed-green and long-dashed red curves, correspond to the one-proton transfer reaction of ^{18}O with the oxygen and aluminum components of the target, respectively. The vertical loci correspond to the reconstructed events from the reaction on the ^{48}Ti target.

Various peaks associated to transition to the states of ^{19}F and ^{26}Mg nuclei are well-formed in the spectrum up to the excitation energy of 10 MeV, where a rather continuum shape is observed due to the decay of ^{26}Mg nucleus (^{26}Mg$\rightarrow^{22}$Ne+α). A similar procedure was also followed for the data obtained with the WO_3+^{27}Al target and the excitation energy spectrum corresponding to the ^{16}O(^{18}O,^{19}F)^{15}N reaction was deduced. After completing the analysis of the background runs, the same trajectory reconstruction technique was applied to the data obtained with the TiO_2+^{27}Al target. The reconstructed θ_{lab}-E_x plot is shown in Figure 4. Like in case of Figure 2, we can immediately identify in the spectrum the presence of the reaction contaminants. For reasons of clarity, we have pointed-out a few selected transitions. The inclined locus, highlighted with the red long-dashed curve, correspond to the g.s. of the ^{27}Al(^{18}O,^{19}F)^{26}Mg reaction, while the ones illustrated with the green dashed curves correspond to the g.s. and the 6.324 MeV ($3/2^-$) state of ^{15}N from the ^{16}O(^{18}O,^{19}F)^{15}N reaction. The vertical oriented loci correspond to the ^{19}F and ^{47}Sc states populated in the ^{48}Ti(^{18}O,^{19}F)^{47}Sc reaction. In order to isolate the energy spectrum for the one-proton transfer reaction on ^{48}Ti, after projecting the data of Figure 4 onto the E_x axis, we have superimposed to the mixed spectrum the previously analyzed data with the aluminum and oxygen targets, appropriately normalized. Subsequently, the background spectra were subtracted from the mixed one and the energy spectrum corresponding to the ^{48}Ti(^{18}O,^{19}F)^{47}Sc reaction was deduced. The background subtraction procedure is presented in Figure 5. As it can be seen, the energy spectrum corresponding to the ^{48}Ti(^{18}O,^{19}F)^{47}Sc reaction is rather structure-less as a result of the high density of states of the ^{47}Sc nucleus. The data analysis for the reaction under study was recently completed and the results have been submitted for publication.

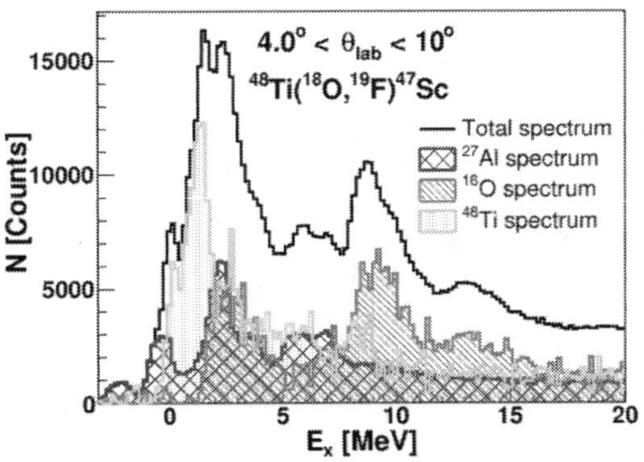

Figure 5. Decomposition of the excitation energy spectrum obtained with the TiO_2+^{27}Al target. The total spectrum is presented with the solid black line. The blue-hatched area corresponds to the normalized background originating from the ^{27}Al backing material, while the brown solid line corresponds to the background arising from oxygen. The green solid line is the obtained excitation energy spectrum for the ^{48}Ti(^{18}O,^{19}F)^{47}Sc reaction, after subtracting from the total spectrum the background contributions.

4 Summary

The study of the ^{18}O+^{48}Ti reaction at 275 MeV was performed, for the first time, at INFN-LNS under the NUMEN and NURE experimental campaigns. Angular distribution measurements for the ^{48}Ti(^{18}O,^{19}F)^{47}Sc reaction were performed by using the MAGNEX large acceptance magnetic spectrometer. Complementary measurements with aluminum and WO_3+^{27}Al targets of the appropriate thickness were performed in order to estimate the background arising from the use of a TiO_2 target evaporated onto a thin aluminum foil. After subtracting the background contamination, the excitation energy spectrum corresponding to the ^{48}Ti(^{18}O,^{19}F)^{47}Sc reaction was deduced. Given the high density of states of the populated ^{47}Sc nucleus and the finite experimental energy resolution (about 500 keV FWHM) the measured excitation energy spectrum was rather structure-less. The results of this analysis will be the subject of a forthcoming publication.

Acknowledgments

The research leading to these results was partially funded by the European Research Council (ERC) under the European Union's Horizon 2020 Research and Innovation Programme (Grant Agreement No 714625).

References

[1] H. Ejiri, J. Suhonen and K. Zuber, Phys. Rep. **797**, 1 (2019)
[2] F. Cappuzzello, C. Agodi, M. Cavallaro et al., Eur. Phys. J. A **54**, 72 (2018)
[3] E. Santopinto, H. García-Tecocoatzi, R. I. Magaña Vsevolodovna et al., Phys. Rev. C **98**, 061601(R) (2018)
[4] H. Lenske, F. Cappuzzello, M. Cavallaro et al., Prog. Part. Nucl. Phys. **109**, 103716 (2019)
[5] M. Cavallaro, E. Aciksoz, L. Acosta et al., POS **302**, (BORMIO2017) 015 (2017)
[6] F. Cappuzzello, C. Agodi, D. Carbone et al., Eur. Phys. J. A **52**, 167 (2016)
[7] M. Cavallaro, F. Cappuzzello, D. Carbone et al., Eur. Phys. J. A **48**, 59 (2012)
[8] D. Torresi, O. Sgouros, V. Soukeras et al., Nucl. Instrum. Methods Phys. Res. A **989**, 164918 (2021)
[9] C. Agodi, A. D. Russo, L. Calabretta et al., Universe **7**, 72 (2021)
[10] F. Cappuzzello, L. Acosta, C. Agodi et al., Front. Astron. Space. Sci. **8**, 668587 (2021)
[11] F. Cappuzzello, M. Cavallaro, A. Cunsolo et al., Nucl. Instrum. Meth. A **621**, 419 (2010)
[12] F. Cappuzzello, D. Carbone, M. Cavallaro, Nucl. Instrum. Methods Phys. Res. A **638**, 74 (2011)
[13] M. Cavallaro, F.Cappuzzello, D.Carbone et al., Nucl. Instrum. Meth. A **637**, 77 (2011)

Low energy reactions of halo nuclei

Ismael Martel[1*]

[1]Departamento de Ciencias Integradas, Universidad de Huelva, 21071 Huelva, Spain

Abstract. Halo nuclei are extreme nuclear states consisting of one or more weakly-bound valence nucleons spatially decoupled from a tightly bound nuclear core. The weakly bound nature of the halo dominates the reaction probability, but the specific reaction mechanisms depend also on the core and target nuclei. Despite of the inherent complexity of the reaction process, simple two-body models and direct reaction theories can be used to extract useful information of the structure of the halo nucleus and its dynamics. These ideas are discussed using selected experiments of Coulomb barrier reactions with one- and two- neutron halo systems.

1 Introduction

Nuclear systems, such as ^{6}He, ^{11}Li, ^{11}Be, and ^{15}C are known to have an extended neutron distribution, the so-called neutron halo [1]. The appearance of haloes is strongly correlated with the existence of weakly bound valence nucleons, which increases the tunnelling probability through the effective nuclear potential. This effect tends to de-localise the valence nucleons, extending the nuclear density up to large distances and producing the so-called "neutron halo". Among the peculiar properties of this exotic systems, there are the appearance of soft dipole modes, with a pronounced maximum in the dipole strength B(E1) close to the neutron breakup threshold, very narrow transverse momentum distributions and large interaction cross-sections [2]. Known halo systems accommodate at most one bound state, and often form three-body Borromean systems, like ^{6}He and ^{11}Li, where none of the subsystems obtained by removing one of the valence neutrons happens to be bound.

Coulomb barrier energies (CBE) are interesting to study halo dynamics as the relative velocity is comparable to the nucleon motion in the impinging nuclei, which enhances the correlation between the relative motion and the internal degrees of freedom. Strong couplings can be induced between the elastic channel and the inelastic, transfer, breakup and fusion reactions and, for halo nuclei, the states lying in the continuum play a very important role [3]. Therefore, CBE provide a good energy range to probe the influence of the nuclear halo on reaction dynamics, testing few-body nuclear models and investigating the role of nucleon-nucleon correlations.

The elastic scattering is a peripheral process, and therefore the angular distribution of the elastic cross sections can be used to probe the extended tail of the halo wave function. The contributions of the couplings to the relevant reaction channels, often due to the dipole component of the interaction (dipole polarizability) produce a strong reduction of the elastic yield around the grazing angle, leading to the disappearance of the Coulomb rainbow, a typical feature of the scattering of stable nuclei. At CBE the elastic channel exhibits large yields, so the measurement of the angular distribution of the elastic cross section can be an important tool to get a first information of halo systems. Values down to few tenths of mb can be measured with radioactive beams with intensities down to 10^{4} pps on target.

2 Scattering of ^{6}He

The exotic nucleus ^{6}He has halo "Borromean" structure with a core of ^{4}He and two valence nucleons and has no bound states. It is by far the most investigated halo nucleus at CBE, and more than fifty data sets are currently available at the EXFOR data base.

Our collaboration has investigated the dynamics of the system ^{6}He + ^{208}Pb in a wide range of energies around the barrier, from 14 - 22 MeV at the Cyclotron Research facility at Louvain-la-Neuve, Belgium [4, 5]. The large Z = 82 of the lead target generates an intense Coulomb field, strongly repelling the ^{4}He core. Thus, considering the low CBE projectile velocities, the two neutrons are dragged through the trajectory by the core with essentially only the strong (attractive) nuclear force. When approaching the target this effect will distort the shape of the ^{6}He density distribution, provoking the excitation of the low- lying states of the continuum. As a result, strong couplings are induced between the elastic channel and breakup and transfer channels, mainly due to the dipole component involved in the deformation process. On the experimental side, this effect can be observed in the shape of the angular distribution of the elastic cross sections and its correlation with the yields of ^{4}He arising from breakup and neutron transfer mechanisms.

Figure 1a shows with black dots the measured angular distributions of the elastic scattering over Rutherford at the lab energy of 22 MeV, just above the Coulomb barrier (~ 20 MeV) [4, 5]. It is observed a strong absorption pattern up to very small scattering angles

* Corresponding author: imartel@uhu.es

https://doi.org/10.1051/epjconf/202125204003

down to $\theta \sim 40°$, much below the grazing angle expected at $\theta \sim 110°$. In semiclassical approach this corresponds to a very large projectile-target distance-of-closest approach as far as r_{max} = 22 fm, which brings out the appearance of long-range reaction mechanisms (this is more clearly shown in the data of Figure 1b, which includes also intermediate scattering angles). Note the disappearance of the nuclear rainbow, a typical signature of the heavy ion scattering found in stable nuclei such e.g., ^{4}He + ^{208}Pb or ^{6,7}Li + ^{208}Pb at CBE.

In Figure 1a we also show three Optical Model (OM) calculations: a prediction using a volume complex potential obtained from systematic studies of stable nuclei [6] (dashed line, BOM); a fit of the data (solid line, FOM) obtained by varying this potential; a fit adding a semiclassical dipole polarization potential (dot-dashed line, DPP) to include the Coulomb part of the dipole coupling [7]; and a calculation obtained by removing from the last one the Coulomb dipole coupling (NCD) (dotted line). We can see that prediction based on BOM is far from the data, resulting in a pronounced nuclear rainbow and reduced absorption, and therefore halo-nuclei cannot be described by directly extrapolating parameters obtained from systematics of stable nuclei. The data are well reproduced with the FOM calculations using a shallow imaginary potential with extremely large imaginary diffuseness ~ 1.8 fm, about three times that of stable nuclei. This is a typical signature of CBE halo scattering, which arises from long-range reaction mechanisms. By comparison with the DPP calculation, it is seen that the effect of the Coulomb dipole part of the interaction produces only part of the absorption, mainly at large scattering angles.

The effect of the couplings on the reaction dynamics is more detailed investigated using Coupled Discretised Coupled Channel Calculations (CDCC) by including the effects of coupling to the continuum. The results are shown in Figure 1b [8, 9]. In this case ^{6}He has been modelled as a ^{4}He core plus a di-neutron system (2n), which highlights the strong correlations among halo neutrons. Three calculations are shown: a full calculation (FC) that reproduces the data (red-solid line); a calculation (NC) obtained by removing the coupling to the continuum (purple-dashed line); and a calculation (ND) obtained by removing *only* the dipole component of the *nuclear* part of the coupling to the continuum (green-dashed line). It should be noticed that the FC can nicely reproduce the angular distribution using this simple model for ^{6}He using a standard d-^{208}Pb potential. In fact, this 2n halo model can simultaneously reproduce the angular and energy distributions of the ^{4}He fragments by assuming neutron transfer to the continuum; this idea will be discussed later in some more detail. What it is interesting about the CDCC results is that although the system is dominated, as expected, by the coupling to continuum states (compare FC with NC), the *nuclear* part of the dipole couplings plays the dominant role, as can be seen by comparing NC with ND. In fact, it is the coherent interference between nuclear and Coulomb couplings that allows to reproduce the data.

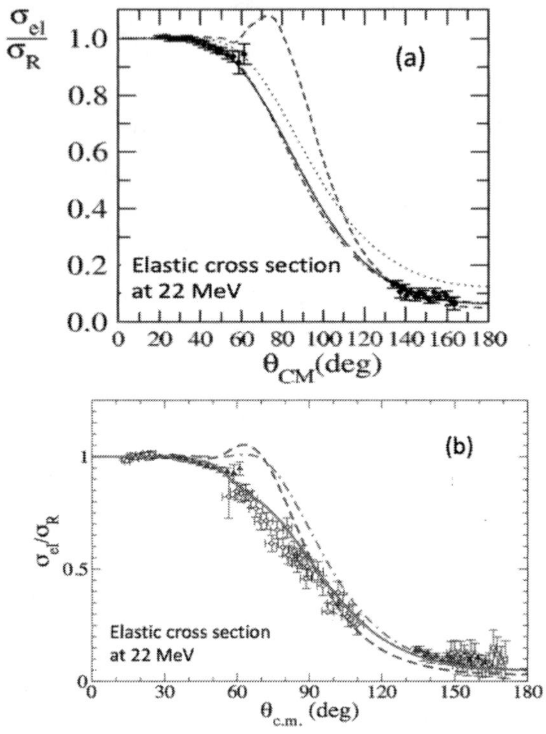

Fig. 1. Elastic scattering of the system ^{6}He + ^{208}Pb at 22 MeV from ref [3] (up) and ref. [4] (down). The various curves are scattering calculations; see text for details.

The angular distributions of the ^{4}He events, normalised to the elastic scattering, is shown in Figure 2(a). We observe large ratios even at energies well below the Coulomb barrier, still about 10% at 14 MeV collision energy, while the contribution of the Coulomb dipole component of the coupling, relative to the nuclear part, is increased [10]. Using the data at 18 MeV a value of the polarizability of ^{6}He is found to be 1.2 fm^3 [11], in good agreement with that given in Ref. [12] using high energy reactions.

To further investigate the reaction mechanism, we carried out additional Distorted Wave Born Approximation (DWBA) calculations for one and two neutron stripping using the 2n model with binding energy ε_b = 0.9 MeV. The result is compared with the previous CDCC calculations and the data in figure 2(b) and 2(c). The solid black and blue lines (TC) depict the calculated 2n transfer angular and energy distributions, the dash-dot line (NT) the single neutron stripping, and the dashed line the direct breakup distributions (DBU) calculated with CDCC. The black arrow in panel 2(c) marks the energy of ~ (4/6) E_{Lab} where the events arising from direct breakup should be most probably expected. The TC calculation reproduces well both the angular and energy distribution of the ^{4}He fragments at backward angles, assuming the direct transfer of a di-neutron cluster to high excited states in ^{210}Pb just above the two-neutron separation energy. This peculiar transfer process was named "transfer to the continuum". Single neutron stripping becomes relevant around 80° Lab, whereas the direct breakup yield at much forward angles ~30° Lab. In a semiclassical picture, the CBE dynamics of ^{6}He+^{208}Pb can be interpreted as a combined action of the repulsive Coulomb interaction core - target at large distances, generating breakup events, and the attractive nuclear part of the interaction producing transfer of neutrons at the vicinity of the target, close to the semiclassical turning point, with almost zero relative velocity and Q-value. A post-acceleration effect on ^{4}He fragments after the 2n transfer, arising from the Coulomb repulsion by the ^{208}Pb target, can be observed.

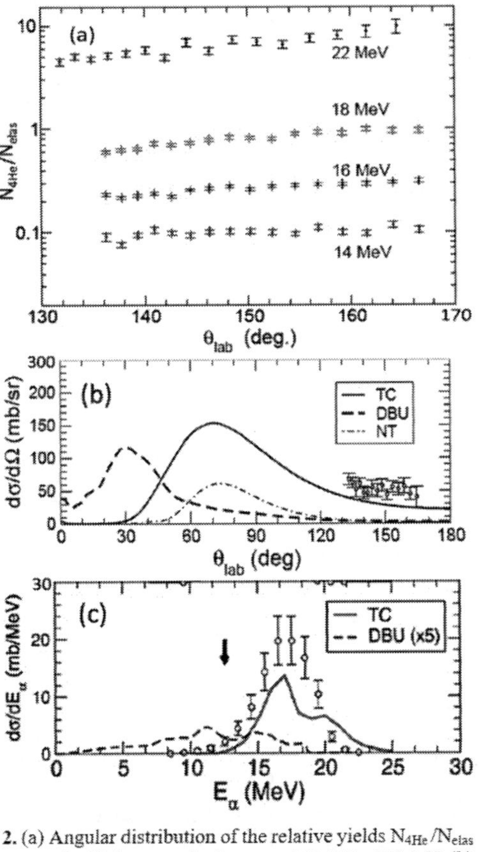

Fig. 2. (a) Angular distribution of the relative yields N_{4He}/N_{elas} in a range of collision energies between 14 MeV - 22 MeV. (b) Angular distribution of the cross section for ^{4}He fragment production. (c) Energy distribution of ^{4}He fragments for the angular range 132°-164°. The black arrow in panel 2(c) marks the energy of ~ (4/6) E_{Lab} where the events arising from direct breakup should be expected. See text for discussion.

The CBE of ^{8}He system constitutes an ideal testbench for improving dynamic polarization potentials to describe coupling to breakup and continuum excitations [13, 14], and for developing di-neutron and four body models [15-17].

The vast amount of data available makes it possible to develop systematic models of CBE scattering by introducing suitable scaling parameters. Concerning the total reaction cross sections σ_{Reac}, the authors of ref. [18] find a systematic behaviour with the collision energy, by using a scaling factor for the collision energy proportional to the Coulomb-barrier height, $ZpxZt/(Ap^{1/3}+At^{1/3})$ and a cross-section scaling factor $(Ap^{1/3}+At^{1/3})^2$ proportional to the size of the scattering system. The reduced cross sections σ_R for ^{6}He and ^{4}He scattering is represented with respect to the reduced collision energy E_{Red} in Figure 3, for a wide range of targets from ^{12}C to ^{209}Bi. The data can be fitted using the simple analytical model of Wong [19], where the reaction cross section is calculated by the quantum tunnelling of an effective "reaction barrier" of parabolic shape. The results of the fit are shown in the figure with black solid lines.

The consequence of this study is of very much importance. It demonstrates that (1) ^{6}He reactions at CBE can be described by a "universal" reaction function, (2) the appearance of two separate halo effects, one due to the size and other due to its peculiar dynamics and (3), a clear evidence for core-halo decoupling in reactions induced by ^{6}He at CBE. A more detailed comparison of light-ion CBE scattering highlighting bound, weakly-bound and halo systems is given in ref. [20].

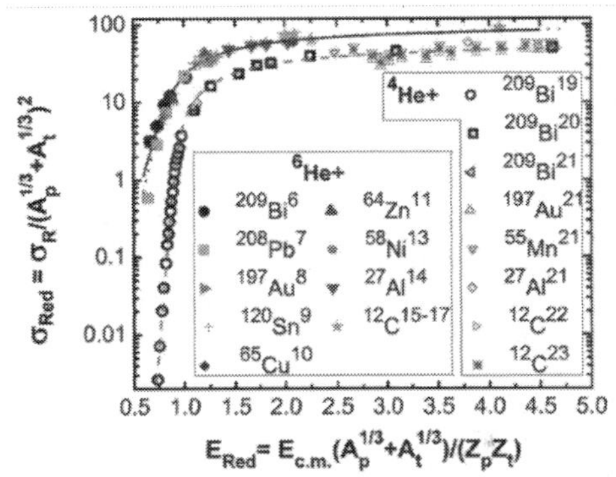

Fig. 3. Reaction cross sections for ^{6}He and ^{4}He at several collision energies and targets [19].

The regularities in the scattering observables can be further investigated by using the angular distributions of the elastic cross sections. In a semiclassical approach, the ratio between the elastic cross sections and the Rutherford cross section P_{el} can be interpreted as a probability for the occurrence of the elastic process. When plotted versus the distance of closest approach r_{max} in a Coulomb trajectory, the shape of the distribution $P_{el}(r_{max})$ exhibits the characteristic distance at which the absorption takes place. This is shown in Figure 4 (Top) for the scattering data of ^{6}He+^{208}Pb in a range of energies from 14-27 MeV, just around the Coulomb barrier [21, 22]. The overall trend of the data, which covers several energies, can be well described by using a semiclassical model with only a diffuse imaginary proximity potential (red solid line) (i.e., no real potential) with a set of parameters similar to the complex potential obtained from the OM calculation [1]. This reveals that the collision is dominated by a very strong absorption starting at large distances (~ 22 fm in the figure), a signature of the presence of long-range reaction mechanisms. However, what is somehow surprising is that the data of such a wide range of energies can be described by the *same potential*, meaning that the absorption depends strongly on the distance of closest approach.

On the other hand, as ^{6}He is Borromean, weakly bound and has no excited states, thus any reaction should mainly lead to the production of ^{4}He fragments. In the semiclassical approach, this reaction probability can be also estimated as the ratio between the angular distribution of the cross section for ^{4}He production and the Rutherford cross section, and the result is shown in Figure 4, as a function of the distance of closest approach. The reaction probability can be estimated as $Pr(r_{max}) = 1 - Pel(r_{max})$, and this is shown with a dashed black line. It is surprising the good agreement obtained, despite the fact that the trajectory of the alphas should be modified due to the loss of the two-neutrons [23]. This suggests that the dominant dynamical process is actually a "soft" neutron transfer with reduced Q-value and momentum transfer in the vicinity of the distance of closest approach, producing a small disruption on the Coulomb trajectory. On the other hand, it shows that almost all the reaction cross section is exhausted by the alpha production mechanism at large distances, leaving very little room for complete fusion. This was demonstrated in a series of experiments with ^{208}Pb and ^{238}U targets [24].

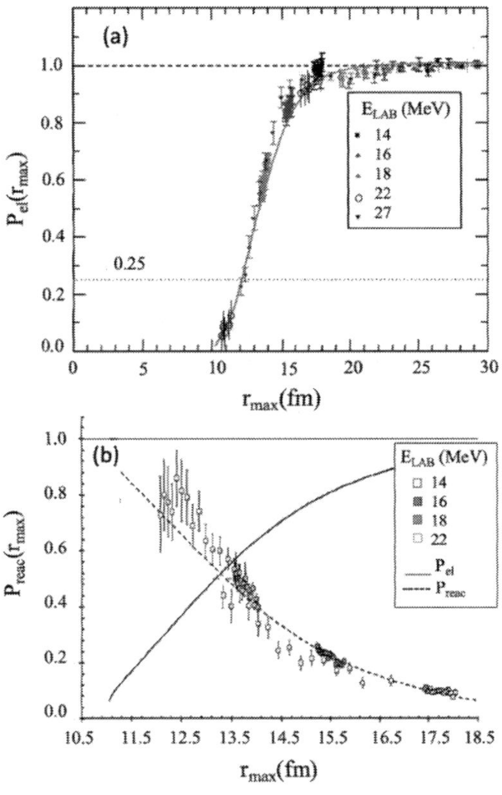

Fig. 4. (a) Ratio of the elastic cross sections over Rutherford $Pel(r_{max})$, as a function of the distance of closest approach in a Coulomb trajectory for the reaction ^{6}He+^{208}Pb. (b) Same as before for the ratio of the cross section for alpha production over Rutherford, $P_{reac}(r_{max})$.

3 Scattering of ^{8}He

^{8}He is the most neutron-rich particle-stable nucleus with N/Z = 3 and it has a Borromean neutron structure ^{4}He+4n, as the system ^{7}He is unbound. The main interest is about the study of the dynamics of the neutron skin. The neutron skin is characterised by a concentration of the neutron density around the nuclear core, whereas for a neutron halo the density is diffuse and extends to large distances from the core. ^{8}He is also unique as it can produce neutron transfer and breakup to ^{6}He and ^{4}He. Therefore, the CBE of ^{8}He makes possible the study the competition of multi-neutron transfer and breakup. On the other hand, the comparison of ^{8}He and ^{6}He scattering can reveal a different dynamics for skin and halo. The neutron separation energy is much higher in ^{8}He than in ^{6}He, so the breakup channel should be reduced.

In Figure 5 (a) it is shown the experimental angular distribution of the elastic scattering cross section over Rutherford for ^{8}He (black filled circles) and ^{6}He (white circles) with a ^{208}Pb target at E_{Lab}= 22 MeV [25]. The solid and dashed lines are OM calculations for ^{8}He and ^{6}He, respectively. Note that for ^{8}He and ^{6}He, the collision energies in centre of mass (Ecm) are 21.2 MeV and 21.4 MeV, respectively, and the Coulomb barriers (Vb) are 23.8 MeV and 24.4 MeV. Thus the observed differences in the angular distributions are mostly due to the dominant reaction mechanisms, and not to the kinematics. The data of ^{8}He presents a sharp fall-off around 60° pointing to a localised absorption, whereas for ^{6}He the absorption starts at very small angles ~ 45° and decreases smoothly with the scattering angle. The different behaviour is more clearly seen in Figure 5 (b), where the cross sections are plotted as a function of the distance of closest approach r_{max} instead of the scattering angle. In the case of ^{6}He scattering, a 2% reduction in the cross sections is reached at distances as long as 19.5 fm, exhibiting a smooth decay as the distance is reduced, whereas in the case of ^{8}He goes down to 16.2 fm with a sharp absorption. The OM gives similar reaction cross sections for both isotopes ~ 1500 mb, but in the case of ^{8}He, the imaginary optical potential has much larger radius and smaller imaginary diffusivity than in the case of ^{6}He.

In Figure 5a it is shown the experimental angular distribution of the elastic scattering cross section over Rutherford for ^{8}He (black filled circles) and ^{6}He (white circles) with a ^{208}Pb target at ELab= 22 MeV [25]. The solid and dashed lines are OM calculations for ^{8}He and ^{6}He, respectively. Note that for the lead target, the collision energies in centre of mass and the corresponding Coulomb barriers are very similar. Thus the observed differences in the angular distributions are due to the reaction mechanisms, and not to the kinematics.

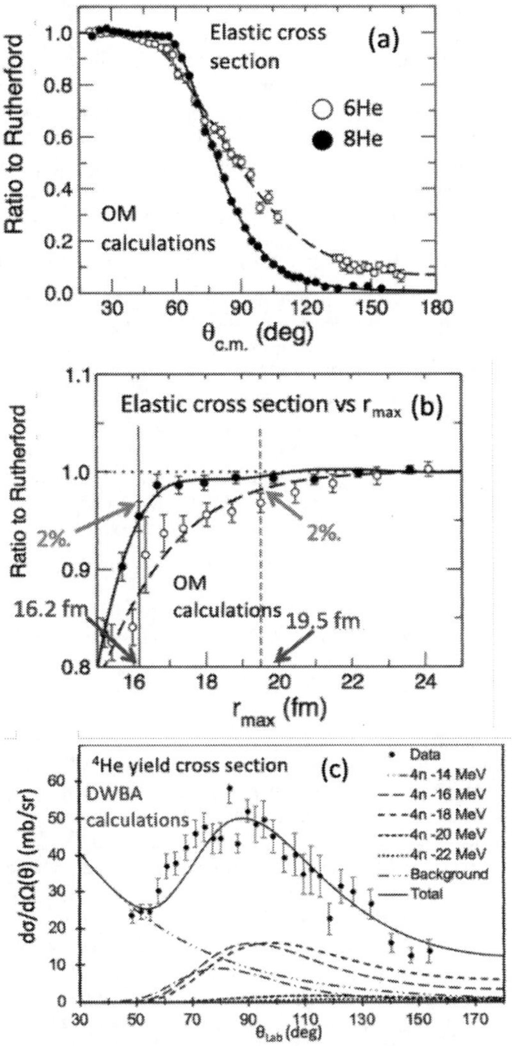

Fig. 5. Experimental angular distribution of the elastic scattering cross section over Rutherford for ^{8}He (black filled circles) and ^{6}He (white circles) with a ^{208}Pb target at E_{Lab}= 22 MeV [25]. The solid and dashed lines are OM calculations for ^{8}He and ^{6}He, respectively. (b) As before, but as a function of the distance of closest approach r_{max}. (c) Experimental angular distribution of the cross section for ^{4}He production in the scattering of ^{8}He + ^{208}Pb at E_{Lab}= 22 MeV. The data can be reproduced by DWBA calculations assuming 4n cluster transfer to high excited states in ^{212}Pb by fitting the relevant spectroscopic factors [26]. The solid line is thus a fit of the data, and the other lines the partial contributions from different excited states in ^{212}Pb.

The data of ^{8}He presents a sharp fall-off around 60° whereas for ^{6}He the absorption starts at very small angles ~ 45° and decreases smoothly with the scattering angle. The different behaviour is more clearly seen in Figure 5b, where the cross sections are plotted as a function of the distance of closest approach r_{max} instead of the scattering angle. In the case of ^{6}He scattering, a 2% reduction in the cross sections is reached at distances as long as 19.5 fm, exhibiting a smooth decay as the distance is reduced, whereas in the case of ^{8}He goes down to 16.2 fm. The OM gives similar reaction cross sections for both isotopes ~ 1500 mb, but in the case of ^{8}He, the imaginary optical potential has much larger radius and smaller imaginary diffusivity than in the case of ^{6}He. The different behaviour can be attributed to the differences in skin and halo. In the case of the halo, the combination of the extended neutron distribution, the weak binding and the dipole polarizability make the system break from very large distances to the target, and although there is a neutron transfer mechanism occurring in the vicinity of the target, it populates highly excited levels of ^{210}Pb lying in the continuum. In the case of ^{8}He, the neutron skin is more tightly bound and localised, and therefore the breakup and neutron stripping mechanism are induced closer to the target, and the breakup is weaker due to the higher binding energy.

In a recent work [26] the cross sections and energy distribution for the ^{8,6}He fragments produced in the ^{8}He+^{208}Pb reaction [25], were analysed in detail. By using DWBA calculations it is found that the neutron transfer process is dominated by (1) single neutron stripping to low excited states of ^{209}Pb (^{6}He in the exit channel) and (2) the transfer of ^{4}n clusters to excited states in ^{212}Pb (^{4}He in the exit channel). The total cross sections obtained are quite large ~ 900 mb and 400 mb, respectively. The angular distribution of the cross sections for ^{4}He production is shown in Figure 5 (c). The experimental data are shown with black circles. The data can be well reproduced by DWBA calculations assuming the ^{4}n transfer to various excited sates in ^{212}Pb (black solid line), by fitting the spectroscopic factors. The partial contributions of the transfer to various excited states are also shown. In order to confirm this results, it would be interesting to measure the exclusive ^{6,4}He angular distributions in coincidence with gammas (and neutrons) produced in the decay of lead isotopes populated in the stripping process.

4 Scattering of ^{11}Be

The isotope ^{11}Be is known to exhibit a single neutron halo structure with a ^{10}Be core. It also has one bound excited state at 320 keV with largest known B(E1) transition strength to the ground state [27]. To explain this phenomenon it was actually noticed for first time the need to integrate the matrix elements up to large distances, presumably due to the extension of the gs wave function. The theoretical treatment of ^{11}Be might appear to be simpler as compared to Borromean systems like ^{6}He and ^{8}He, but in this case the deformation of the ^{10}Be plays an important role in the reaction dynamics, which is absent in helium isotopes.

One of the first investigations of CBE dynamics using ^{11}Be beams was the system ^{11}Be + ^{209}Bi, which was measured between 35 - 50 MeV at the RIKEN RIPS facility [28]. The idea was to study the effect of the neutron excess on the complete fusion. Due to the halo structure, ^{11}Be should exhibit a large hindrance factor as compared to the stable isotope ^{9}Be. It should be also an effect in ^{10}Be due to the deformation and coupling to the first excited state.

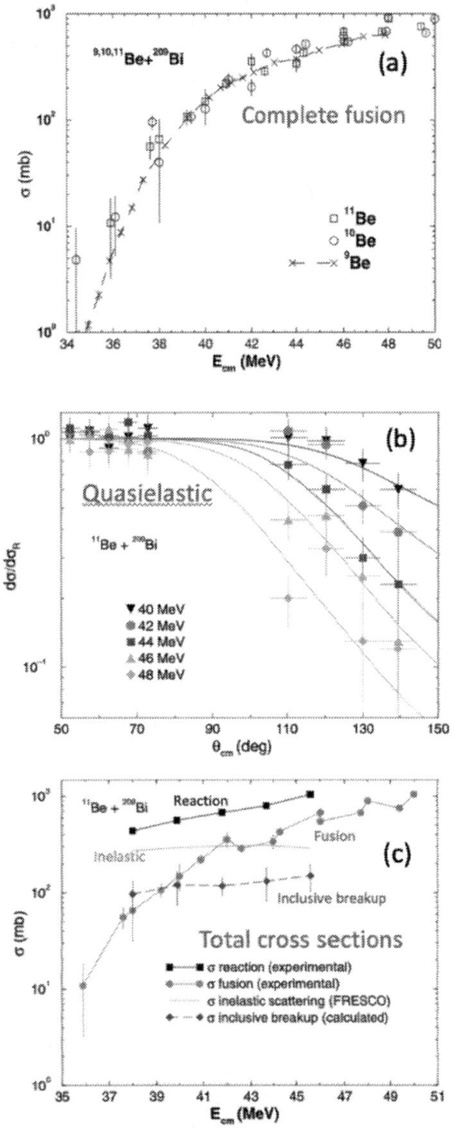

Fig. 6. (a) Fusion cross sections of the isotopes ^{11,10,9}Be with a ^{209}Bi target [28]. (b) Angular distribution of the quasielastic cross sections for ^{11}Be + ^{209}Bi for E = 40 - 48 MeV [29]. (c) Energy dependence of the reaction cross sections, fusion, inelastic and breakup. See text for discussion.

The results of the experiment, shown in Figure 6 (a), were quite puzzling, as they found almost identical energy distribution of the complete fusion for the three isotopes, except at very deep sub-barrier energies, where the data showed a slightly larger fusion cross sections for ^{11}Be than ^{9}Be that could be attributed to a halo effect. The data could be reproduced using CCFULL and CDCC calculations, which suggested a strong role of the breakup channel in the suppression of complete fusion for the halo system ^{11}Be. The halo would break at large distances before reaching the target, reducing the fusion probability. This was also observed in the case of ^{6}He [24]. The system ^{11}Be + ^{209}Bi was also investigated in [29], where the angular distribution of the quasielastic cross sections was measured between 40 - 48 MeV. The result is shown in Figure 6b. The data were well described by OM, DWBA and Coupled-Channel calculations, assuming a rotational model with a deformation parameter obtained from the experimental B(E1) value.

Using the measured reaction cross sections, the calculated inelastic cross sections and the fusion cross sections of the previous experiment [28], the energy distribution of the total cross sections for breakup could be deduced. The results are shown in Figure 6c. The reaction cross sections were quite large ~ 800 mb, about ten times the fusion cross sections; however the calculated breakup cross sections were relatively small ~ 100 - 150 mb, even below the fusion yield, which was somehow surprising for such a weakly bound halo system.

The CBE of ^{11}Be was later investigated by using a ^{120}Sn at E_{Lab}= 32 MeV at the RIB facility REX-ISOLDE at CERN, where the angular distribution of both quasielastic and ^{10}Be fragments were measured for the first time [30]. The angular distribution of the quasielastic cross section is shown in Figure 7a (black dots) and is compared with the scattering of ^{9}Be+^{120}Sn at similar collision energy [31] (yellow circles). The Coulomb-nuclear interference is strongly damped and the deviation from Coulomb occurs around ~30° cm, a clear signature of the neutron halo, and the presence of long-range reaction mechanisms.

The dynamics was also investigated with CC calculations using a simple vibrational model for ^{11}Be (deformation length derived from experimental B(E1), including the coupling to the bound excited state at 320 keV and the first two resonances at 1.78 MeV and 3.41 MeV, and two "fictitious" dipole states at ε= 0.55 MeV and I=1/2-,3/5-, just above the neutron separation energy, to simulate the effect of the dipole couplings to the continuum. The calculation reproduces well the quasielastic angular distribution. The effect of removing the coupling to the "fictitious" dipole states is shown by the red-dotted and blue-dashed lines. The former contains only the coupling to the 1st excited state, and the latter adding the resonances with negligible effect. This angular distributions are very similar to the experimental angular distribution of the stable nucleus ^{9}Be (yellow circles). This result brings out the importance of the coupling to the low-energy continuum dipole states in the CBE dynamics of ^{11}Be. On the other hand, although the cross section for excitation is relatively large, the effect of the coupling has a small effect on the shape of the angular distribution of the quasielastic cross section.

 https://doi.org/10.1051/epjconf/202125204003

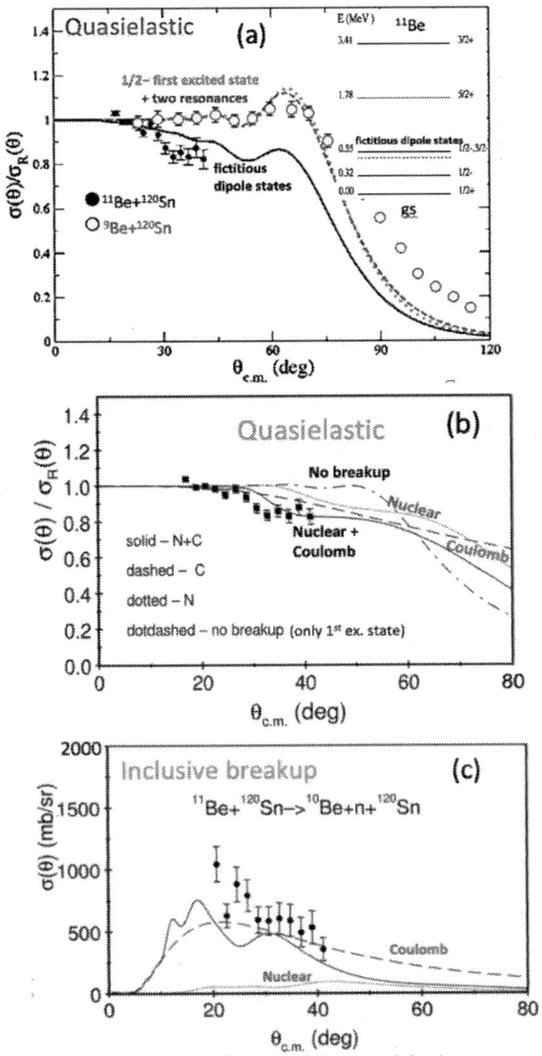

Fig. 7. (a) Angular distribution of the quasielastic cross sections for ^{11}Be + ^{120}Sn at E = 32 MeV [30] (black circles), and for the system ^{9}Be + ^{120}Sn at similar collision energy. (b) CDCC calculations are compared with the experimental quasielastic cross section (black squares) [31]. (c) Angular distribution of the breakup cross sections. Various calculations are shown with solid, dashed and dotted lines; see text for explanation.

The coupling to the continuum is investigated with more detail in [32] using CDCC calculations. In Figure 7b the calculations (solid black line, N+C) are compared with the data of the quasielastic cross sections (black circles). The dot-dashed line is a calculation obtained by removing the coupling to the continuum leading to breakup but keeping the coupling to the 1st excited state. This demonstrates, as just discussed, that the dynamics is dominated by dipole couplings to the continuum. The dashed line (C) results from removing the Coulomb dipole couplings and the dotted line (N) from removing the nuclear dipole couplings. A strong Coulomb-nuclear interference effect is observed where Coulomb and nuclear contributions interfere destructively.

In Figure 7c the predictions for breakup cross sections (solid black line) are compared with the experimental data (black circles). The calculation reproduces well the shape of the data, with a total breakup yield about 50% of the total cross sections. The dipole Coulomb coupling produces most of the breakup yield (dashed line) and the nuclear component is small (dotted line).

Although the Coulomb dipole coupling to the continuum dominates the reaction dynamics, the nuclear part plays a non-trivial role due to coherent interference effects, which are important to reproduce the measured angular distribution of both quasielastic and breakup. Another interesting feature arises from the energy distribution of ^{10}Be fragments, where the effect of Coulomb post-acceleration can be observed with an energy shift of ~ 1.8 MeV [32] suggesting breakup taking place at distances ~ 15 fm from the target. This is consistent with the long-rage absorption mechanisms.

A more recent experiment performed at REX-ISOLDE/CERN was the CBE of the system ^{11}Be + ^{64}Zn at E= 28.7 MeV [34, 35] where the angular distribution of the cross sections for quasielastic and ^{10}Be fragments (breakup) were measured. The use of the smaller Z = 30 target ^{64}Zn allows a deeper access to the nuclear part of the nuclear interaction producing coupling to the continuum and breakup. The angular distribution of the quasielastic scattering cross sections is shown in Figure 8a using black squares and is compared with the scattering of non-halo isotopes ^{10}Be and ^{9}Be at the same centre of mass collision energy. As in the previous case, as compared to the stable isotope ^{9}Be, the Coulomb-nuclear interference is strongly damped and there is a clear deviation from Coulomb scattering from about θ ~ 25° due to long range reaction mechanisms.

The inelastic cross section was calculated by semiclassical Coulomb excitation theory and is shown in the inset of the figure (dashed lines). In order to describe the data, an imaginary surface term with large diffuseness (~ 3.5 fm) had to be introduced. This potential simulates the dynamic polarization potential accounting for the dipole couplings to the continuum, as explained before. The large breakup cross sections are about ~ 40% of the reaction cross sections, almost twice the values found for ^{9,10}Be. The contributions of elastic and non-elastic BU calculations (EBU and NEB, respectively) are investigated in [35]. In Figure 8b, it is shown the result of EBU using XCDCC calculations (red solid line), which includes the effect of the core excitation in the reaction process, and standard CDCC calculation (black dashed line) which does not. Both prescriptions produce identical results and describe well the experimental data (yellow circles), meaning that the quasielastic cross sections are not sensitive to core deformation. Although both calculations underpredict the angular distribution of ^{10}Be fragments by ~20% see figure 8c (red solid and dashed lines), the distribution appears to be dominated by the elastic breakup mechanism. The NBU gives an important contribution (blue dot-dashed line) to the breakup yield and must be added incoherently to reproduce the data (TBU, black solid line). The NBU was calculated using the so-called participant-spectator IAV model [36]. Both EBU and NEB are consistent also with the energy distributions of ^{10}Be fragments. As in the previous case, Coulomb post-acceleration is observed, with a kinetic energy excess of ~ 1 MeV giving a breakup distance of ~ 15 fm angle-independent, consistent with the result obtained for ^{120}Sn.

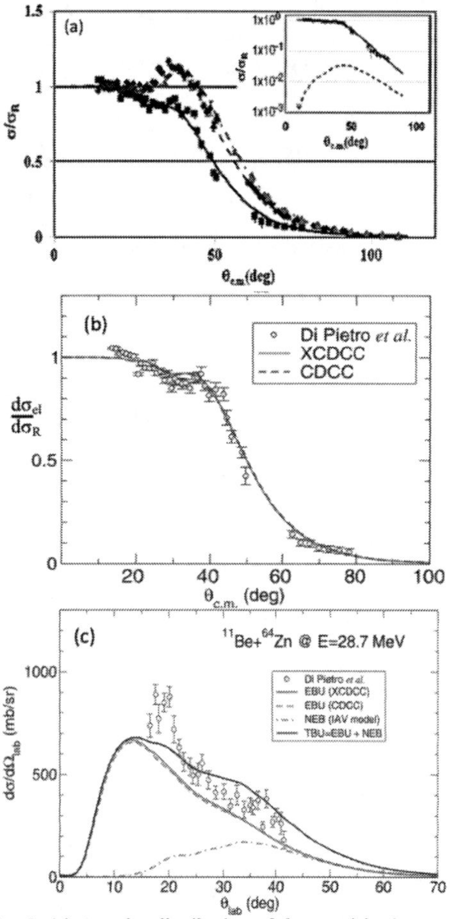

Fig. 8. (a) Angular distribution of the quasielastic cross sections for ^{11}Be +^{64}Zn at E = 28.7 MeV [34] (black circles), and for the system ^{10,9}Be + ^{64}Zn at similar collision energy (blue and red squares). (b) CDCC and XCDCC calculations are compared with the experimental quasielastic cross section (yellow circles) [35]. (c) Angular distribution of the breakup cross sections (yellow circles). Various calculations are shown with solid, dashed and dotted lines; see text for explanation.

5 Summary and conclusions

This contribution intends to highlight some peculiar aspects of CBE scattering of the neutron halo systems ^{6}He (2n-halo), ^{8}He(2n/4n-skin) and ^{11}Be (1n-halo). Common features are the large reaction and breakup cross sections, the disappearance of the Coulomb nuclear interference and the coupling to dipole states lying in the low energy continuum. Halo neutrons do not produce more fusion, as the Coulomb interaction breakups the system at distances far from the target. In the case of ^{6}He it is found a transfer mechanism of two neutrons to the continuum, whereas for ^{8}He the transfer seems to proceed through single neutron stripping to excited states in ^{209}Pb and 4-neutron cluster transfer to ^{212}Pb. ^{11}Be is dominated by the breakup channel. Elastic and inelastic breakup and core deformation are essential ingredients.

Systematic studies of reactions cross sections induced by halo nuclei at several energies around the Coulomb barrier are useful to understand the dynamical behaviour, extract scaling laws and universal rules. Simpler three body models can describe the gross properties of the scattering, but few-body models are needed for accurate descriptions and to probe the nature of nucleon-nucleon correlations, often leading to new interesting discoveries.

References

1. I. Tanihata et al., Phys. Rev. Lett., **55**, 2676 (1985); P.G. Hansen and B. Johnson, Europhys. Lett **4**, 409 (1987)
2. T. Aumann, et al., Eur. Phys. J. A **26**, 441 (2005)
3. N. Keeley et al. Prog. Part. Nucl. Phys. 63, 396 (2009)
4. A.M. Sánchez-Benítez, et al. Nucl. Phys. A **803**, 30 (2008)
5. L. Acosta et al., Phys. Rev. C **84**, 044604 (2011)
6. J. Cook. Nucl. Phys. A **388,** 153-172 (1982)
7. M. V. Andrés and J. Gómez-Camacho. Phys. Rev. Lett. **82**, 1387 (1999)
8. N. Keeley et al., Phys. Rev. C **68,** 054601 (2003).
9. K. Rusek et al., Phys. Rev. C **72**, 037603 (2005)
10. D. Escrig et al, Nucl. Phys. A **79**, 2 (2007)
11. V. Parkar et al., Acta Phys. Pol. B **42**, 761 (2011)
12. J.A. Lay et al., Phys. Rev. C **82**, 024605 (2010)
13. M. Rodríguez-Gallardo et al., Phys. Rev. C **77**, 064609 (2008)
14. R.S. Machintosh and N. Keley, Phys. Rev. C **79** (2009) 014611
15. N. Keeley, K.W. Kemper, K. Rusek. Phys. Rev. C **88**, 017602 (2013)
16. A.M. Moro, et al, Phys. Rev. C **75**, 064607 (2007)
17. M. Rodríguez-Gallardo et al., Phys. Rev. C **80**, 051601(R)(2009)
18. V. Morcelle et al, Phys. Lett. B **732**, 228 (2014)
19. E.F. Aguilera et al., Phys. Rev. C **83**, 021601(R) (2011)
20. C. Y. Wong, Phys. Rev. Lett. **31**, 766 (1973)

21. J.J. Kolata and E.F. Aguilera et al., Phys. Rev. C **83**, 027603 (2009)
22. A.M. Sánchez-Benítez et al., Acta Phys. Pol. B **37** (2006) 1.
23. I. Martel, et al. EPJ Web of Conferences **17**, 13002 (2011)
24. R. Wolski et al., Eur. Phys. Jour. A **47**, 111 (2011)
25. G. Marquínez-Durán et al., Phys. Rev. C **94,** 064618 (2016), Phys. Rev. C **95**, 024602 (2018)
26. I. Martel et al., Phys. Rev. C **102**, 34609 (2020)
27. D. J. Millener et al, Phys. Rev. C **28**, 497 (1983)
28. C. Signorini et al., Nucl. Phys. A **735**, 329 (2004)
29. M. Mazzocco et al., Eur. Phys. J. A **28**, 295 (2006)
30. L. Acosta et al., Eur. Phys. J. A **42**, 461 (2009)
31. A. Arazi et al., Phys. Rev. C **97**, 044609 (2018)
32. K. Rusek et al. Act. Phys. Pol. B. **43**, 129 (2012)
33. L. Acosta. PhD Thesis, University of Huelva, unpublished (2010)
34. A. Di Pietro et al. Phys. Rev. Lett. **105**, 022701 (2010)
35. A. Di Pietro et al., Phys. Lett. B **798**, 134954 (2019)
36. M. Ichimura, N. Austern, C.M. Vincent, Phys. Rev. **32**, 431 (1985)

Reaction dynamics of exotic and stable weakly-bound nuclei using a four-body continuum-discretized coupled-channels formalism

Manuela Rodríguez-Gallardo[1,2,*] and *Jesús* Casal[1]

[1]Departmento de Física Atómica, Molecular y Nuclear, Universidad de Sevilla, Spain
[2]Instituto Carlos I de Física Teórica y Computacional, Universidad de Sevilla, Spain

Abstract. Reactions induced by four weakly-bound nuclei on a lead target, at energies around the Coulomb barrier, are described using the same four-body continuum-discretized coupled-channels formalism. The nuclei considered are the halo nuclei ^{11}Li and ^{6}He, the stable nucleus ^{9}Be and the proton-rich nucleus ^{10}C. The differences and similarities on the reaction dynamics are addressed.

1 Introduction

A unified description of the structure and reactions, for the all nuclei at the Segrè chart, is still a problem to be solved in nuclear physics [1]. Weakly-bound nuclei can be the nexus between stable and exotic nuclei. Weakly-bound (stable or exotic) nuclei have a low breakup threshold, making likely the breakup of the system when colliding with another nucleus.

Among weakly-bound nuclei, an interesting case is the so-called Borromean nuclei, i.e. three-body composite systems with no binary bound states. With this structure, halo nuclei such as ^{11}Li and ^{6}He are weakly-bound, spatially extended systems, comprising a core and two valence nucleons. On the other hand, the nucleus of ^{9}Be is a stable weakly-bound nucleus with a Borromean structure consisting of two alpha particles and one weakly-bound neutron. Going a step further, the Brunnian nucleus ^{10}C has a four-body structure with no bound binary subsystems. This nucleus can be studied approximately in a three-body model as ^{8}Be$+p+p$.

The description of the reactions induced by these four nuclei, in order of increasing separation energy and using the same formalism, is the objective of this work. For that, the reactions on a same target, ^{208}Pb, at energies around the Coulomb barrier for each system are considered. The formalism used to study these reactions is the four-body continuum-discretized coupled-channels formalism [2, 3].

In Sec. 2 we introduce the continuum-discretized coupled-channels formalism. In Sec. 3 we compare the theoretical results with the available experimental data for the reaction of ^{11}Li, ^{6}He, ^{9}Be and ^{10}C on ^{208}Pb at energies around the Coulomb barrier. Finally in Sec. 4 we present the main conclusions.

2 Continuum-Discretized Coupled-Channels (CDCC) formalism

The coupling to breakup channels in reactions involving weakly-bound nuclei can be formally described within the CDCC formalism [4], in which the total scattering wave function is expanded in internal states of the projectile in a given few-body model. This is typically referred to as three-body CDCC, for two-body projectiles [4], or four-body CDCC, for three-body projectiles [2,, 5]. The corresponding coupling form factors are generated following a multipole expansion of the projectile-target interaction, using as input suitable fragment-target optical potentials. Details of this procedure can be found in Eqs. (12-14) of Ref. [2]. Each term of the expansion is referred to as a multipole of order Q.

In the present work, for ^{11}Li, ^{6}He, ^{9}Be and ^{10}C nuclei induced reactions we perform four-body CDCC calculations using the three-body structure models (^{9}Li $+ n + n$, $\alpha + n + n$, $\alpha + \alpha + n$ and ^{8}Be$+p+p$) and fragment-target optical potentials presented in Refs. [3, 6–8]. These reaction calculations have no free-parameters, once the projectile structure is adjusted to reproduce the experimentally known energy spectrum.

The states of ^{11}Li and ^{6}He are computed using a binning procedure (Ref. [3]) from the true three-body continuum states, whereas ^{9}Be and ^{10}C states are generated using a pseudostate approach, the analytical Transformed Harmonic Oscillator method presented in Ref. [9]. Both procedures use hysperspherical coordinates for describing the projectile nucleus. The binning procedure provides a more precise description of the low-lying continuum and this is essential for halo nuclei (^{6}He, ^{1}1Li). Nevertheless, the implementation of the binning procedure for three-body systems and, in particular, for three-body nuclei with more than one charged particle is complicated. For that reason, a pseudostate approach is assumed for ^{9}Be and ^{10}C, providing converged results for the reactions considered.

*e-mail: mrodri@us.es

Once the projectile structure is obtained, the CDCC equations are solved up to convergence in the number of partial waves and excitation energy above the respective breakup thresholds, including continuum couplings.

Within the CDCC formalism, the influence of the projectile binding energy on the reaction dynamics is implicitly contained in the corresponding coupling potentials. If the ground state is closer to the threshold, its wave function explores larger distances, increasing the coupling to the low-lying continuum states. This, in first order, produces larger breakup cross sections as less bound is the system.

3 Results

Elastic differential cross sections for the reactions $^{11}Li+^{208}Pb$ at 29.8 MeV, $^{6}He+^{208}Pb$ at 22 MeV, $^{9}Be+^{208}Pb$ at 44 MeV and $^{10}C+^{208}Pb$ at 66 MeV are studied comparing theoretical calculations and experimental data. For the first two reactions, the differential cross section of fragment ($^{9}Li/^{4}He$) production, for which experimental data exist, are also addressed. The order for presenting the results is starting with the less bound nucleus, ^{11}Li, and finishing with the nucleus that is more bound, ^{10}C.

3.1 ^{11}Li

The Borromean nucleus ^{11}Li is bound only by 0.369 MeV with respect to the $^{9}Li+n+n$ threshold. To describe the ^{11}Li nucleus, a three-body structure is assumed with an inert and spinless ^{9}Li core surrounded by two valence neutrons. In that way, the $J^{\pi} = 3/2-$ ground state becomes a $J^{\pi} = 0^{+}$ state. For details on the structure calculations, see Refs. [6, 10].

In order to study the reaction of ^{11}Li on ^{208}Pb target at 29.8 MeV, states up to a maximum energy of 5 MeV above the breakup threshold and with angular momenta $J^{\pi} = 0^{+}, 1^{-}, 2^{+}, 3^{-}$ are included. The elastic differential cross section for this reaction is very sensitive to the $B(E1)$ strength close to the threshold and, in particular, to the presence of a dipole resonance. Since the precise location of this resonance has not yet been firmly established, we adjusted the position of this resonance to the energy that best reproduces the elastic and breakup cross sections. This gives an energy of 0.32 MeV above the breakup threshold. This value is in agreement with the latest experimental results published Refs. [11, 12].

Fig. 1 shows the elastic differential cross section, relative to Rutherford cross section, as a function of the center-of-mass (c.m.) scattering angle for the reaction of ^{11}Li on ^{208}Pb at 29.8 MeV. Experimental data [10] are plotted by yellow circles. The four-body CDCC calculation including only the ground state is represented by the pink dashed line and the full four-body CDCC calculation is the red solid line. It can be seen that the inclusion of the coupling to the continuum produces a very large reduction of the elastic cross section that describes very well the data. The reduction of the cross section with respect to the Rutherford scattering is attributed to the strong dipole coupling

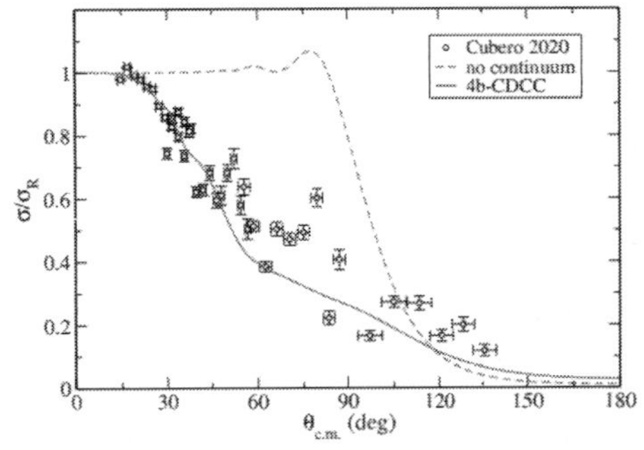

Figure 1. Elastic differential cross section relative to Rutherford as a function of the c.m. scattering angle for the reaction $^{11}Li+^{208}Pb$ at 29.8 MeV.

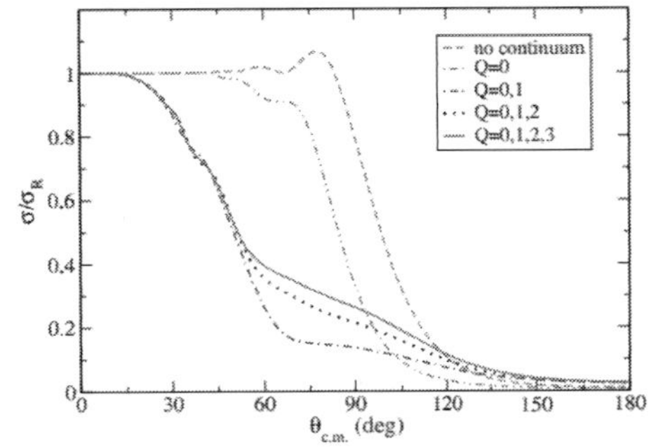

Figure 2. Effect of the different Q multipoles on the elastic differential cross section relative to Rutherford as a function of the c.m. scattering angle for the reaction $^{11}Li+^{208}Pb$ at 29.8 MeV.

between the ground and the continuum states in ^{11} Li, as it can be seen in Fig. 2. In this figure the contribution of the different multipoles to the elastic differential cross section, with respect to Rutherford cross section, is shown.

Fig. 3 shows differential cross section of ^{9}Li fragments, in the laboratory frame. Experimental data [6, 13] are plotted by yellow circles. The four-body CDCC calculation is the red solid line. This calculation reproduces the data for almost the whole angular range. This means that the *elastic* breakup is the process dominating in this reaction. It is also included in the figure, with a blue dashed line, the contribution from the transfer of the two halo neutrons to the continuum of the target (for details see Ref. [13]). This process gives a relevant contribution for the very backward angles only.

3.2 ^{6}He

The Borromean nucleus ^{6}He is bound by 0.973 MeV with respect to the $^{4}He+n+n$ threshold. Apart from the ground state, with angular momentum $J^{\pi} = 0^{+}$, a 2^{+} resonance is experimentally known at 0.825 MeV over the breakup threshold. To describe the ^{6}He nucleus, a three-body structure is assumed with an inert ^{4}He core surrounded by two

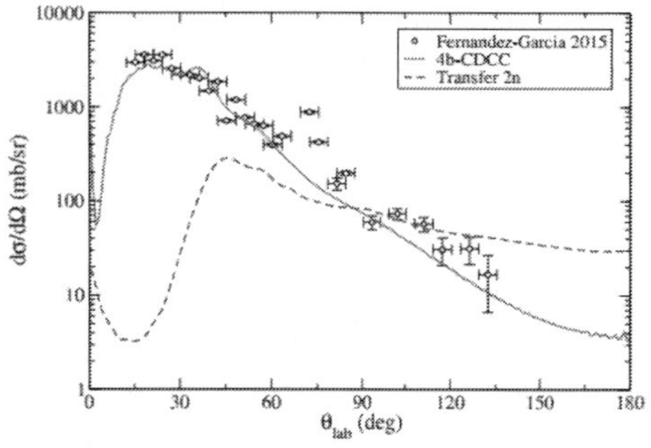

Figure 3. Differential cross section of the ^{9}Li fragments, in the laboratory frame, for the reaction ^{11}Li+^{208}Pb at 29.8 MeV.

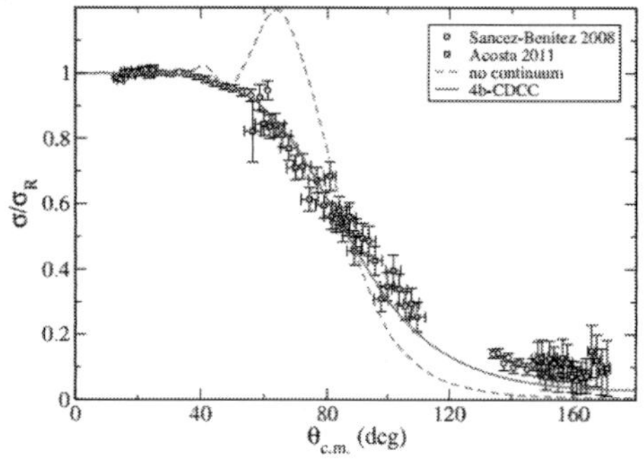

Figure 4. Elastic differential cross section, relative to Rutherford, as a function of the c.m. scattering angle for the reaction ^{6}He+^{208}Pb at 22 MeV.

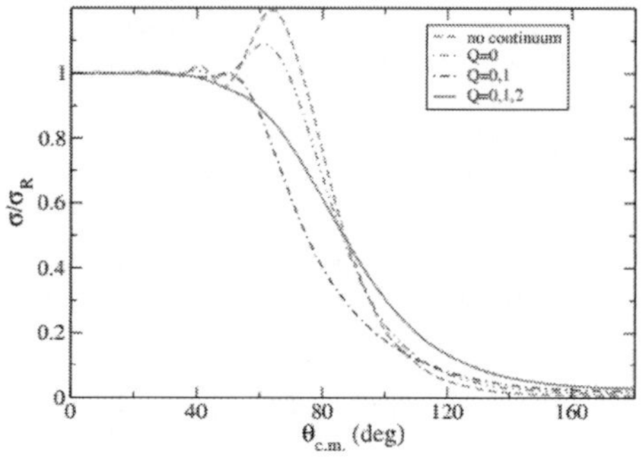

Figure 5. Effect of the different Q multipoles on the elastic differential cross section, relative to Rutherford, as a function of the c.m. scattering angle for the reaction of ^{6}He on ^{208}Pb at 22 MeV.

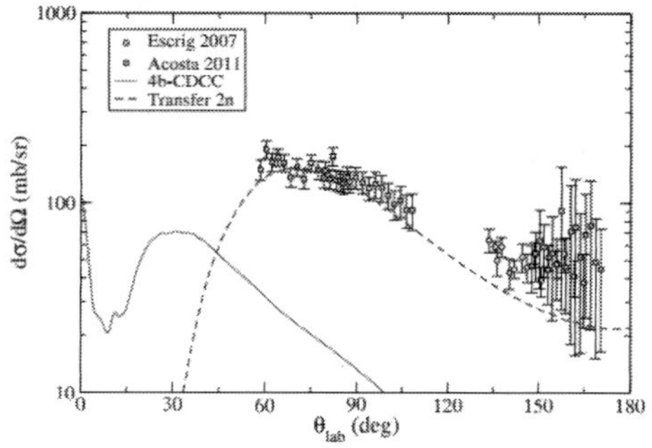

Figure 6. Differential cross section of α-particle production, in the laboratory frame, for the reaction ^{6}He+^{208}Pb at 22 MeV.

valence neutrons. For details on the structure calculations see Ref. [3].

In order to study the reaction of ^{6}He on ^{208}Pb target at 22 MeV, states up to 8 MeV above the breakup threshold and with angular momenta $J^\pi = 0^+, 1^-, 2^+$ are included. The elastic cross section angular distribution with respect to Rutherford is shown in Fig. 4. The experimental data are plotted by yellow circles [14] and green squares [15]. The pink dashed line is the four-body CDCC calculation without including the continuum and the red solid line is full four-body CDCC calculation. As in the ^{11}Li case, the inclusion of the continuum part of the spectrum produces an important, although smaller, reduction of the elastic cross section that describes very well the data. In particular, the rainbow at around 60 degrees disappears. This reduction is due mostly to the coupling to the dipolar states, although in the ^{6}He there is no a fully-fledged dipolar resonance. The effect of the different multipoles to the elastic distribution is shown in Fig. 5.

Fig. 6 shows the differential cross section of α-particle production, in the laboratory frame, for the reaction of ^{6}He on ^{208}Pb at 22 MeV. Experimental data from two experiments are plotted by yellow circles [16] and green squares [15]. The red solid line is the full four-body CDCC calculation and the blue dashed line is a calculation considering a scheme in which the two halo neutrons are transferred to the continuum of the target (see [16] for details). Contrary to the ^{11}Li case, the *elastic* breakup (obtained from CDCC) only dominates the angular region below 30 degrees. In the region where experimental data are available (above 30 degrees) the transfer of the two halo neutrons to the target is the process that dominates this reaction. This process is one of the contributions to the *non-elastic* breakup [17].

3.3 ^{9}Be

The Borromean nucleus ^{9}Be is bound by 1.5736 MeV with respect to the ^{4}He+^{4}He+n threshold. The low-energy spectrum of ^{9}Be shows, apart from the $J^\pi = 3/2^-$ ground state, several resonances with angular momenta $J^\pi = 1/2^+, 5/2^-, 1/2^-, 5/2^+, 3/2^+$, and possibly others, whose positions and widths have been investigated by many authors in different experiments [18–20]. The spin-parity assignment for some of these states is still under discussion, although there are no implications for the lowest $1/2^+$ and $5/2^-$ resonances, that are fairly-well reproduced in the theoretical model. Details on the structure calculations can be found in Ref. [7].

In order to study the reaction of ^{9}Be on ^{208}Pb target at 44 MeV, states up to 8 MeV above the breakup threshold

 https://doi.org/10.1051/epjconf/202125204004

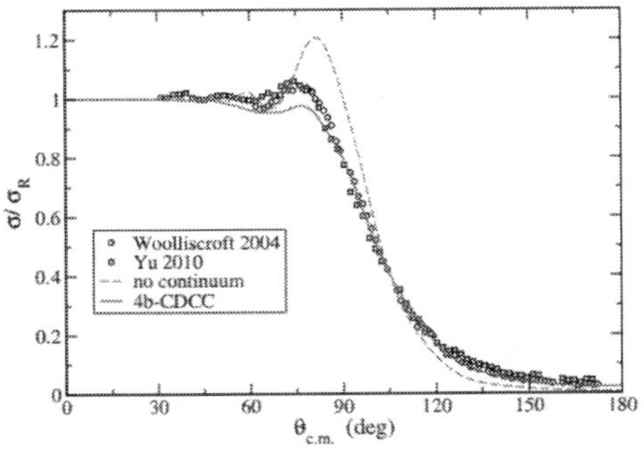

Figure 7. Elastic differential cross section, relative to Rutherford, as a function of the c.m. scattering angle for the reaction $^9Be+^{208}Pb$ at 44 MeV.

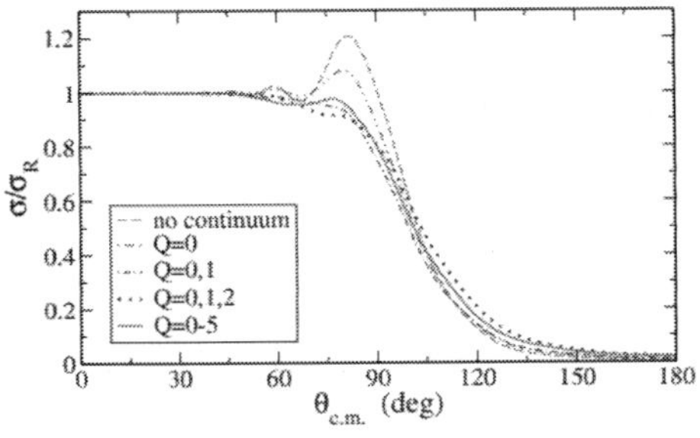

Figure 8. Effect of the different Q multipoles on the elastic differential cross section, relative to Rutherford, as a function of the c.m. scattering angle for for the reaction $^9Be+^{208}Pb$ at 44 MeV.

and with angular momenta $J^\pi = 1/2^\pm, 3/2^\pm, 5/2^\pm$ are included. The elastic differential cross section, with respect to Rutherford, as a function of the c.m. scattering angle is shown in Fig. 7. The experimental data are plotted by yellow circles [21] and green squares [22]. The pink dashed line is the four-body CDCC calculation without including the continuum and the red solid red is the full four-body CDCC calculation. It can be seen that the coupling to the continuum is still important to reproduce the experimental data, specially the effect of decreasing the rainbow at 80-90 degrees.

Fig. 8 shows the contribution of the different multipoles to the elastic differential cross section. In this case, the effect of the monopolar and dipolar contributions are of the same order of magnitude. This difference with respect to the halo nuclei, ^6He and ^{11}Li can be clearly seen comparing Figs. 2, 5 and 8 at the rainbow area for each reaction.

3.4 ^{10}C induced reactions

The Brunnian nucleus ^{10}C is bound by 3.820 MeV with respect the $^4He+^4He+p+p$ threshold. The ground state is a $J^\pi = 0^+$ state and a 2^+ excited bound state appears close to the breakup threshold. Here, ^{10}C is described within a $^8Be+p+p$ configuration. Details of the structure calculation can be found in Ref. [8].

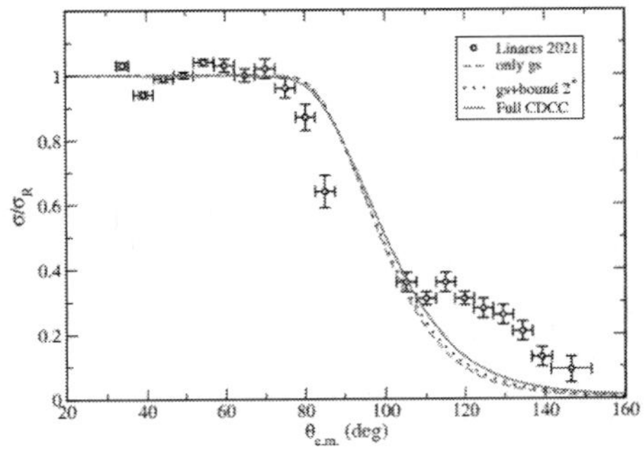

Figure 9. Elastic differential cross section, relative to Rutherford, as a function of the c.m. scattering angle for the reaction of $^{10}C+^{208}Pb$ at 66 MeV.

In order to study the reaction of ^{10}C on ^{208}Pb target at 66 MeV, states up to 8 MeV above the breakup threshold and with angular momenta $J^\pi = 0^+, 1^-, 2^+$ are included. The elastic cross section with respect to Rutherford is shown in Fig. 9. The experimental data [8] is plotted by yellow circles. The dashed pink line is the four-body CDCC calculation including only the ground state, the dotted blue line is the four-body CDCC calculation including the ground state and the bound excited state and the solid red line is the full four-body CDCC calculation. In this case it is observed that the calculation with only the ground state has no rainbow peak and the coupling to the continuum is very small. Even the coupling to the bound excited state hardly affects the elastic cross section.

4 Conclusions

The study of the reactions of four weakly-bound nuclei on a lead target at energies around the Coulomb barrier for each system has been presented. The nuclei considered run from exotic nuclei, the halos ^{11}Li and ^6He, to the stable nucleus ^9Be and proton-rich nucleus ^{10}C. All the reactions are studied using the four-body CDCC formalism, that reproduces the elastic experimental data for all the cases.

Comparing the four reactions, it can be see how the coupling to the continuum is crucial to reproduce the elastic data for ^{11}Li, ^6He and even for ^9Be, decreasing this effect with the binding energy. For the case of ^{10}C, the most bound nucleus, the effect is practically negligible. It is also seen that the coupling that produces the reduction of the rainbow is mainly due to dipolar ($Q = 1$) couplings for the halo nuclei ^{11}Li and ^6He, meanwhile for ^9Be monopolar and dipolar coupling effects are of the same order.

Finally, the cross sections of fragment production for the reactions induced by ^{11}Li and ^6He, have been addressed showing different mechanisms of reaction. For ^{11}Li on ^{208}Pb at 29.8 MeV, the *elastic* breakup obtained by the four-body CDCC calculations reproduce the experimental distribution for almost the whole angular range. For ^6He on ^{208}Pb at 22 MeV, the *elastic* breakup is only dominating the forward angles, while at angles larger than

50 degrees the process is dominated by the transfer of the two halo neutrons to the continuum of the target.

Acknowledgments

This work was supported by the Spanish Ministry of Economy and Competitiveness, the European Regional Development Fund (FEDER), under Project No. FIS2017-88410-P and by the European Union's Horizon 2020 research and innovation program, under Grant Agreement No. 654002.

References

[1] M.A.G. Alvarez, M. Rodríguez-Gallardo, J.P. Fernández-García, J. Casal, J.A. Lay, Phys. Rev. C 103, 054614 (2021)

[2] M. Rodríguez-Gallardo, J.M. Arias, J. Gómez-Camacho, R.C. Johnson, I.J. Thompson, J.A. Tostevin, Phys. Rev. C 77, 064609 (2008).

[3] M. Rodríguez-Gallardo, J.M. Arias, J. Gómez-Camacho, I.J. Thompson, J.A. Tostevin, Phys. Rev. C 80, 051601(R) (2009).

[4] M. Yahiro, Y. Iseri, H. Kameyama, M. Kamimura, M. Kawai, Prog. Theor. Phys. Suppl. 89, 32 (1986).

[5] T. Matsumoto, T. Egami, K. Ogata, Y. Iseri, M. Kamimura, M. Yahiro, Phys. Rev. C 73, 051602(R) (2006).

[6] J. P. Fernández-García, M. Cubero, M. Rodríguez-Gallardo et al., Phys. Rev. Lett. 110, 142701 (2013).

[7] J. Casal, M. Rodríguez-Gallardo, J.M. Arias, Phys. Rev. C 92, 054611 (2015).

[8] R. Linares, Mandira Sinha, E.N. Cardozo et al., Phys. Rev. C 103, 044613 (2021).

[9] J. Casal, M. Rodríguez-Gallardo, J. M. Arias, Phys. Rev. C 88, 014327 (2013).

[10] M. Cubero, J. P. Fernández-García, M. Rodríguez-Gallardo et al., Phys. Rev. Lett. 109, 262701 (2012).

[11] R. Kanungo, A. Sanetullaev, M. Tanaka et al., Phys. Rev. Lett. 114, 192502 (2015).

[12] M. Tanaka, R. Kanungo, M. Alcorta et al., Phys. Lett. B 774, 268 (2017).

[13] J. P. Fernández-García, M. Cubero, L. Acosta, Phys. Rev. C 92, 044608 (2015).

[14] A.M. Sánchez-Benítez, D. Escrig, M.A.G. Alvarez et al., Nucl. Phys. A 803, 30 (2008).

[15] L. Acosta, A.M. Sánchez-Benítez, M.E. Gomez et al., Phys. Rev. C 84, 044604 (2011).

[16] D. Escrig, A.M. Sánchez-Benítez, A.M. Moro et al., Nucl. Phys. A 792, 2 (2007).

[17] J. Lei, A.M. Moro, Phys. Rev. C 92, 044616 (2015).

[18] K. Sumiyoshi, H. Utsunomiya, S. Goko, and T. Kajino, Nucl. Phys. A 709, 467 (2002).

[19] C.W. Arnold, T.B. Clegg, C. Iliadis et al., Phys. Rev. C 85, 044605 (2012).

[20] O. Burda, P. von Neumann-Cosel, A. Richter, C. Forssén, B.A. Brown, Phys. Rev. C 82, 015808 (2010).

[21] R. J. Woolliscroft, B. R. Fulton, R. L. Cowin et al., Phys. Rev. C 69, 044612 (2004).

[22] N. Yu, H.Q. Zhang, H.M. Jia et al., J. Phys. G: Nucl. Part. Phys. 37, 075108 (2010).

EPJ Web of Conferences **252**, 04005 (2021) https://doi.org/10.1051/epjconf/202125204005
HINPw6

Linking structure and dynamics with two-neutron halos

Jesús Casal[1,*], *Mario* Gómez-Ramos[1], and *Antonio M.* Moro[1]

[1]Dpto. de Física Atómica, Molecular y Nuclear, Facultad de Física, Universidad de Sevilla, Apartado 1065, E-41080 Sevilla, Spain

Abstract. We present a theoretical description of (p, pn) reactions induced by two-neutron halo nuclei at intermediate energies. For this purpose, we use a reaction framework that incorporates a full three-body model for the projectile, and we focus on the cases of ^{11}Li and ^{14}Be. Our results provide a direct link between the structure properties of these nuclei and different reaction observables, such as relative-energy spectra or momentum distributions. We discuss also on the link between dineutron correlations and opening-angle distributions.

1 Introduction

Two-neutron halo nuclei are very exotic systems characterized by a diffuse matter distribution arising from the very weakly bound nature of two valence neutrons around a more compact core [1–3]. These (core + n + n) systems are called Borromean, i.e., they contain no bound binary subsystems and thus present a signature three-body structure [4, 5]. Typical examples are ^{6}He, ^{11}Li or ^{14}Be. Many different techniques have been used to explore the properties of two-neutron halo nuclei, such as Coulomb Dissociation [6, 7], neutron removal [8, 9], or low-energy scattering [10, 11], which provide invaluable insight into the limits of nuclear stabilily. Recently, two-neutron halo nuclei have been studied via quasifree knockout reactions with proton targets. The analysis of the relative-energy spectrum and momentum distributions after one-neutron removal provide information about the halo properties and the spectrum of the corresponding unbound subsystems [12–14]. Even more recently, the opening angle as a function of the intrinsic neutron momentum has been also studied to characterize dineutron correlations in the halo [15].

Three-body models are a well-established approach to describe two-neutron halo nuclei and have shown that, while a good knowledge of the unbound core + n subsystems is key to understand their properties, the correlation between the valence neutrons plays also a fundamental role [16, 17]. In this contribution, we provide a summary of some recent developments in the description of (p, pn) reactions induced by two-neutron halos using a full three-body model for the projectile. We first study the relative-energy spectrum and transverse momentum distributions at intermediate energies using the Transfer to the Continuum reaction framework [18]. Then, we adopt a sudden model approximation to analyze momentum and angular distributions, and we link the reaction observables with structure properties such as the partial-wave content and dineutron correlations.

*e-mail: jcasal@us.es

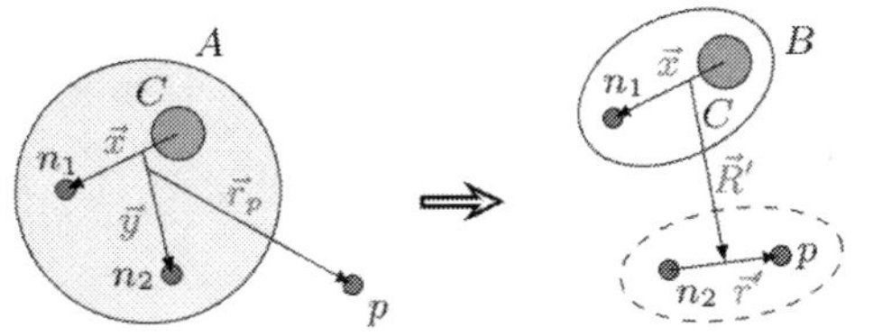

Figure 1. Schematic representation of the (p, pn) process induced by a two-neutron halo projectile.

2 Theoretical description

2.1 Three-body model and structure overlaps

Two-neutron halo nuclei can be described within the hyperspherical formalism [4]. In this work, as in Refs. [14, 19], we need the ground-state wave function of the Borromean projectile, $\Phi_{g.s.}(\boldsymbol{x}, \boldsymbol{y})$. Here, $\{\boldsymbol{x}, \boldsymbol{y}\}$ are the usual Jacobi coordinates (see Fig. 1). In order to build this state, we diagonalize the three-body Hamiltonian in a discrete basis, a procedure which is referred to in the literature as a pseudostate method. This method provides not only the ground state but also a discrete representation of the continuum (i.e., the pseudostates), but for the present calculations only the former is required. Different bases can be used, but we employ an analytical transformed harmonic oscillator basis [20]. Details about the three-body formalism can be found, for instance, in Ref. [21] and references therein.

The description of the knockout process requires the evaluation of a transition matrix involving both the ground-state of the three-body projectile and the scattering states, $\phi(\boldsymbol{k}_x, \boldsymbol{x})$, of the unbound core + n subsystem after the removal of a halo neutron (k_x is the intrinsic momentum). If the transition operator does not change the state of this residue, then the key structure ingredient for reaction calculations is the corresponding overlap

$$\varphi(k_x, \boldsymbol{y}) = \langle \phi(\boldsymbol{k}_x, \boldsymbol{x}) | \Phi_{g.s.}(\boldsymbol{x}, \boldsymbol{y}) \rangle_x, \quad (1)$$

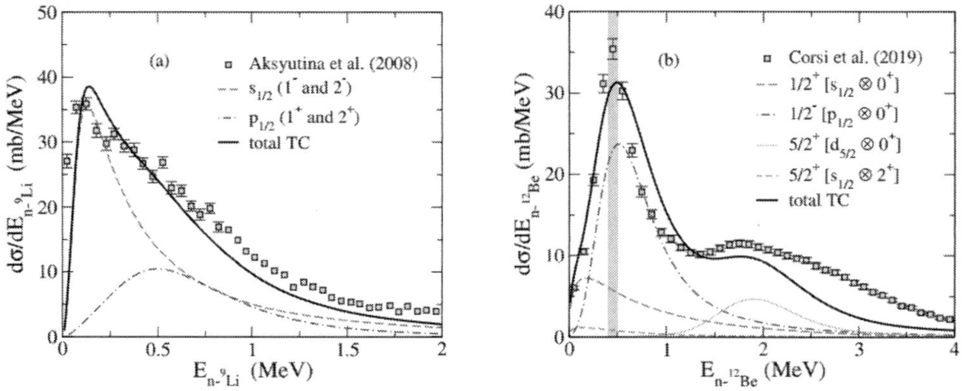

Figure 2. Relative-energy spectra of a) ^{10}Li and b) ^{13}Be after one-neutron knockout from ^{11}Li and ^{13}Be, respectively, and compared with the data of Refs. [12, 14] after convolution with the experimental resolution. The shaded area in the right panel is the region integrated to generate the momentum distribution in Fig. 3.

which can be split into different single-particle configurations of the knocked-out neutron using the appropriate coupling scheme for $\Phi_{g.s.}$ and ϕ [19, 22].

2.2 Reaction framework

We described the (p, pn) process within the Transfer to the Continuum (TC) framework. Using the structure overlaps defined above, the transition amplitude in prior form can be written as

$$\mathcal{T} = \sqrt{2}\langle \Upsilon_f^{(-)}(\boldsymbol{R}', \boldsymbol{r}')|V_{pn} + U_{pB} - U_{pA}|\varphi(k_x, \boldsymbol{y})\chi_{pA}^{(+)}\rangle, \quad (2)$$

where χ_{pA} is the distorted wave generated by the auxiliary optical potential U_{pA}, and $\Upsilon_f^{(-)}$ is the $B + p + n$ wave function in the final state, with B the unbound core $+\, n$ residue. This expression assumes a participant-spectator approximation, in which the transition operator does not change the state of B, and the factor $\sqrt{2}$ arises from the two identical neutrons in the halo. To solve this problem, in the TC approach the final wave functions $\Upsilon_f^{(-)}$ are expanded in p-n continuum states following the binning procedure employed in the continuum-discretized coupled-channels method [23]. The dependence of the T-matrix on the momentum k_x of the binary subsystem after knockout enables the description of the corresponding relative-energy spectrum. Note that the TC method provides absolute cross sections, so no arbitrary scaling is needed to compare with (p, pn) experimental data.

3 Results

Recently, we carried out calculations to describe the ^{11}Li$(p, pn)^{10}$Li and ^{14}Be$(p, pn)^{13}$Be reactions [14, 19]. By adopting different structure models for the core + n unbound nucleus, we explored the sensitivity of the relative-energy spectrum after knockout to the position of resonances and virtual states in these subsystems, as well as to the resulting partial-wave content in the two-neutron halo wave function. In the case of ^{14}Be, our structure model included the coupling to the first 2^+ excited state of the ^{12}Be core, and we also analyzed the transverse momentum distributions. Here we summarize and compare some of these results. In Fig. 2, we show the ^{9}Li-n (left panel) and ^{12}Be-n (right panel) relative-energy spectra obtained within our reaction model at 280 MeV/nucleon and 265 MeV/nucleon, respectively, and compared with the data of Refs. [12, 14].

For ^{10}Li, our model includes the finite $(3/2^-)$ spin of the ^{9}Li core, so that the single-particle configurations $s_{1/2}$ and $p_{1/2}$ appear as doublets. For simplicity, here we present the summed contributions. As can be seen in the figure, the experimental data are consistent with parity inversion in ^{10}Li, with the $s_{1/2}$ contribution (coming mostly from a virtual state) being lower in energy than the $p_{1/2}$ resonances. The agreement with the data is remarkable, especially since no fitting is performed. Our three-body model for ^{11}Li is governed by s-wave components, with the p-waves contributing only about $\sim 31\%$ of the norm. Note that the same structure model has been also successful in describing ^{11}Li(p, d) transfer data [22].

The results for ^{13}Be are shown in the right panel of Fig. 2. In this case, the presence of core excitations in ^{14}Be leads to components in which the knocked-out neutron has a different angular momentum than the neutron than remains unperturbed in ^{13}Be. This is the case of the $5/2^+$ $[s_{1/2} \otimes 2^+]$ contribution, which corresponds to an $s_{1/2}$ single-particle component coupled to the 2^+ state of the core, but leading to a $d_{5/2}$ knockout. More details are given in Ref. [14]. As can be seen in the figure, our calculations are compatible with the spectrum being governed by a $p_{1/2}$ resonance around 0.5 MeV. This state carries about 60% of the ground-state norm. The agreement at low ^{13}Be energies is reasonable, while the calculations tend to underestimate the data at higher energies. This may be due to some missing components (or ^{13}Be resonances) in the wave function. Note that the ^{14}Be model used here does not include the excitation to all bound states of the ^{12}Be

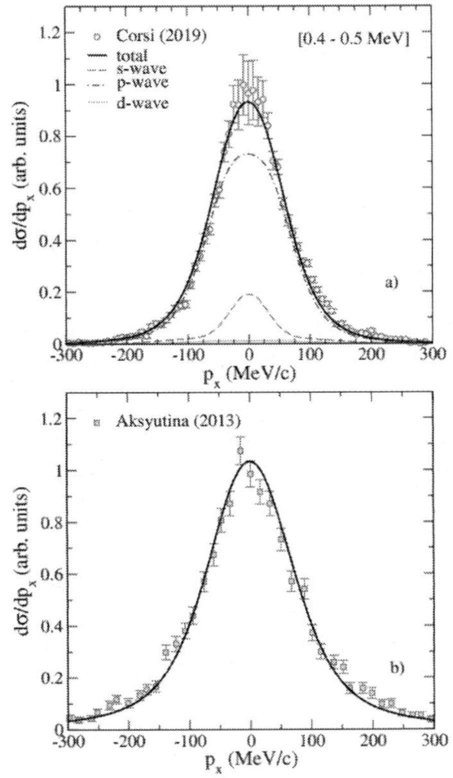

Figure 3. a) Transverse momentum distribution of the knocked-out neutron in the ^{14}Be(p, pn)^{13}Be reaction at 265 MeV/nucleon [14]. The ^{13}Be energy range (0.4-0.5 MeV) corresponds to the shaded area in Fig. 2, right panel. Calculations have been convoluted with the experimental resolution. b) The same but for the reaction at 304 MeV/nucleon [13, 25].

core, which will require further developments in the three-body description of the projectile.

The nature of the low-lying peak in the ^{13}Be spectrum has been long debated. In Ref. [24], the analysis of (p, pn) data at 69 MeV/nucleon was also interpreted with the population of a p-wave resonance. In the measurement of Refs. [13, 25] at 304 MeV/nucleon, however, the relative-energy spectrum could be described with a dominant s-wave. As we discussed in Ref. [14], our p-wave resonance is also consistent with the observed transverse momentum distributions. This is shown in Fig. 3 (top panel), integrated in the energy region corresponding to the shaded area in Fig. 2. The relative weights of s, p and d waves are given by the relative-energy spectrum, and the total distribution is rescaled to match the data through a χ^2 fit. It is clear from Fig. 3 that the width of the momentum distribution of Ref. [14] is totally compatible with a p-wave resonance. Interestingly, the same structure model and reaction calculations for ^{14}Be reproduce also the transverse momentum distribution for the (p, pn) data reported in Ref. [25], as shown in Fig. 3 (bottom panel). This highlights the consistency of the present results, which points to a dominant p-wave contribution in the low-lying spectrum of ^{13}Be considering various (p, pn) observables.

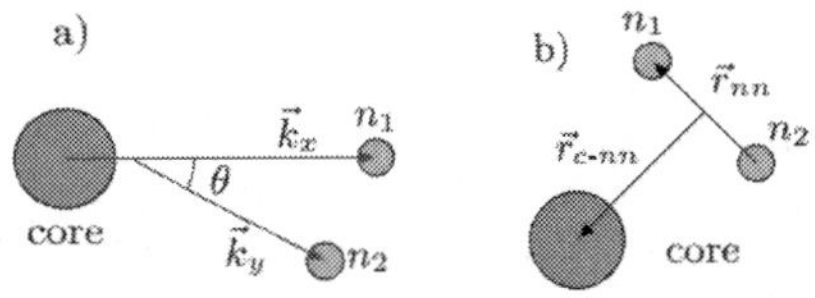

Figure 4. a) Momenta and opening angle in the Jacobi-Y representation; b) nn and core-nn distances in the Jacobi-T representation.

3.1 Opening angle and dineutron correlations

In addition to exploring the properties of the unbound core + n subsystems, (p, pn) reactions with Borromean two-neutron halo nuclei can also be used to study dineutron correlations. This was proposed in Ref. [26] and realized experimentally in Ref. [15], where the measurement of the opening angle as a function of the intrinsic neutron momentum in ^{11}Li was linked to the surface localization of the dineutron. The mixing between different-parity components in the ground-state wave function favors dineutron correlations in coordinate space [27], thus leading to large opening angles (> 90 degrees) between the neutrons in momentum space.

The theoretical model in Ref. [26] assumes a quasi-free sudden approximation, where the reaction mechanism involves a zero-range V_{pn} interaction and absorption effects are given by the eikonal S-matrix between the proton target and the core of the Borromean nucleus. Within such model, the transition amplitude can be written as

$$\mathcal{T} \propto \langle \phi(\boldsymbol{k}_x, \boldsymbol{x}) \otimes e^{\boldsymbol{k}_y \cdot \boldsymbol{y}} | S(y) \Phi_{g.s.}(\boldsymbol{x}, \boldsymbol{y}) \rangle, \quad (3)$$

where the S-matrix dependence on its impact parameter is approximated by the y coordinate (see Fig. 1). The previous equation involves the same structure overlaps introduced in Sec. 2.2 and can be seen as a distorted Fourier transform. By evaluating this transition amplitude, we can easily obtain the intrinsic momentum distributions in k_x, k_y and the opening angle between $\boldsymbol{k}_x$ and $\boldsymbol{k}_y$, in the so-called Jacobi-Y representation as depicted in Fig. 4 (left panel). Details on how to obtain the relevant expressions, as well as the absorption effects, will be presented elsewhere [28].

Here we restrict ourselves to the correspondence between dineutron correlation in coordinate space and opening angle in momentum space. In Fig. 5 (top panel), we present preliminary results for the the opening-angle distribution (integrated for all momenta) of ^{11}Li using the same three-body model introduced above. The result exhibits a clear asymmetry towards large angles ($\cos(\theta) < 0$). This is consistent with the recent experimental results in Ref. [15]. The large opening angle is linked to the spatial localization of the dineutron, which can be seen by plotting the two-dimensional ground-state density as a function of the nn and core-nn distances in the so-called Jacobi-T representation (see Fig. 4). This is also shown in Fig. 5

 https://doi.org/10.1051/epjconf/202125204005

(bottom panel), where the probability density has a clear maximum at small r_{nn} distances, i.e., small angles between the neutrons in coordinate space. Calculations of the intrinsic momentum distributions and the average opening angle as a function of the intrinsic neutron momentum are in progress.

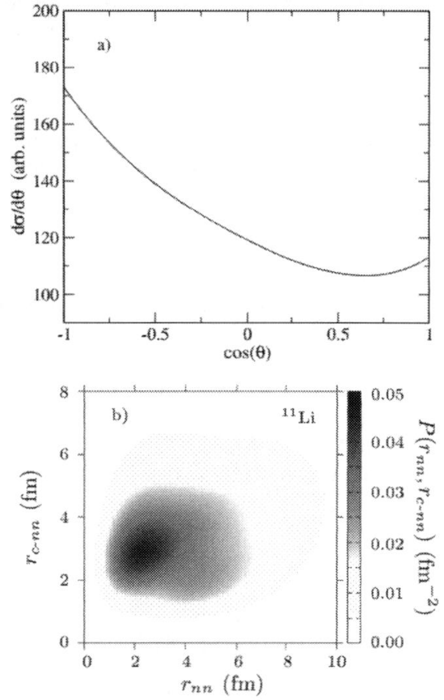

Figure 5. a) Opening angle distribution for the ^{11}Li(p, pn) reaction. b) Ground-state probability density of ^{11}Li as a function of r_{nn} and $r_{c\text{-}nn}$. Results are preliminary.

Acknowledgements

This work has been partially funded by the Spanish Ministerio de Ciencia, Innovación y Universidades (project No. FIS2017-88410-P), by the European Union research and innovation programme under Marie Skłodowska Curie Actions (grant agreement 101023609), and by the European Social Fund and Junta de Andalucía (PAIDI 2020, grant number DOC-01006).

References

[1] I. Tanihata *et al.*, Phys. Rev. Lett **55**, 2676 (1985)
[2] P. G. Hansen and B. Jonson, Europhys. Lett. **4**, 409 (1987)
[3] K. Riisager, Rev. Mod. Phys. **66**, 1105 (1994)
[4] M. V. Zhukov *et al.*, Phys. Rep. **231**, 151 (1993)
[5] E. Nielsen, D. Fedorov, A. Jensen, and E. Garrido, Phys. Rep. **347**, 373 (2001)
[6] T. Nakamura *et al.*, Phys. Rev. Lett. **96**, 252502 (2006)
[7] K. J. Cook *et al.*, Phys. Rev. Lett. **124**, 212503 (2020)
[8] H. Simon *et al.*, Phys. Rev. Lett. **83**, 496 (1999)
[9] H. Simon *et al.*, Nucl. Phys. A **791**, 267 (2007)
[10] L. Acosta *et al.*, Phys. Rev. C **84**, 044604 (2011)
[11] M. Cubero *et al.*, Phys. Rev. Lett. **109**, 262701 (2011)
[12] Y. Aksyutina *et al.*, Phys. Lett. B **666**, 430 (2008)
[13] Y. Aksyutina *et al.*, Phys. Lett. B **718**, 1309 (2013)
[14] A. Corsi *et al.*, Phys. Lett. B **797**, 134843 (2019)
[15] Y. Kubota *et al.*, Phys. Rev. Lett. **125**, 252501 (2020)
[16] K. Hagino and H. Sagawa, Phys. Rev. C **72**, 044321 (2005)
[17] J. Casal and E. Garrido, Phys. Rev. C. **102**, 051304(R) (2020)
[18] A. M. Moro, Phys. Rev. C. **92**, 044605 (2015)
[19] M. Gómez-Ramos, J. Casal, and A. M. Moro, Phys. Lett. B **772**, 115 (2027)
[20] J. Casal, M. Rodríguez-Gallardo and J. M. Arias, Phys. Rev. C **88**, 014327 (2013)
[21] J. Casal *et al.*, Phys. Rev. C **102**, 064627 (2020)
[22] J. Casal, M. Gómez-Ramos, and A. M. Moro, Phys. Lett. B **767**, 307 (2017)
[23] N. Austern *et al.*, Phys. Rep. **154**, 125 (1987)
[24] Y. Kondo *et al.*, Phys. Lett. B **690**, 245 (2010)
[25] Y. Aksyutina *et al.*, Phys. Rev. C **87**, 064316 (2013)
[26] Y. Kikuchi, K. Ogata, Y. Kubota, M. Sasano, and T. Uesaka, Prog. Theor. Exp. Phys. **2016**, 103D03 (2016)
[27] F. Catara, A. Insolia, E. Maglione, and A. Vitturi, Phys. Rev. C **29**, 1091 (1984)
[28] J. Casal and M. Gómez-Ramos, in preparation (2021).

 https://doi.org/10.1051/epjconf/202125204006

Searching for treasures at sub-barrier energies: the case of ^{8}B and ^{7}Be

A. Pakou[1,2,*], *P. D.* O'Malley[3], *L.* Acosta[4], *A. M.* Sánchez-Benítez[5], *J. J.* Kolata[3], *K.* Palli[6], *O.* Sgouros[7], *V.* Soukeras[7], and *G* Souliotis[6]

[1]Department of Physics, The University of Ioannina, Ioannina GR45110, Greece
[2]HINP, The University of Ioannina, Ioannina GR45110, Greece
[3]Department of Physics, University of Notre Dame, Notre Dame, Indiana 46556, USA
[4]Instituto de Física, Universidad Nacional Autónoma de México, A.P. 20-364, Mexico City 01000, Mexico.
[5]Centro de Estudios Avanzados en Física, Matemáticas y Computación (CEAFMC), Department of Integrated Sciences, University of Huelva, 21071 Huelva, Spain
[6]Department of Chemistry, National and Kapodistrian University of Athens, Athens, Greece
[7]INFN Laboratori Nazionali del Sud, via S. Sofia 62, 95125, Catania, Italy

Abstract. Unexpected and challenging experimental results, at below barrier energies for weakly bound nuclei, are briefly reviewed in this article. The emphasis will be on our recent breakup results for ^{8}B+^{208}Pb at deep sub-barrier energies, indicating a dominance of direct mechanisms at this low energy regime. We will also present, a preliminary analysis of the ^{4}He and ^{3}He - particle production events for the ^{7}Be + ^{208}Pb reaction. These data were collected at the same experiment and at similar deep sub-barrier energies, exhibiting large yields compatible with cluster transfer processes. To confirm these results a new experiment to be performed at *TriSol* is planned, as soon as the upgrading of this facility will be completed.

1 Introduction

The investigation of reaction dynamics at near barrier energies for weakly bound nuclei was pursued systematically the last 20 years. It proved to be a fruitful play ground in relation with channel coupling effects affecting, either elastic scattering and probing a new type of the standard potential threshold anomaly -see e.g Refs. [1–3], or/and the suppression and enhancement of fusion cross sections above and below barrier energies [4–8]. On the other hand research at deep sub-barrier energies, in relation with reaction mechanisms, is mainly devoted to stable systems investigating fusion hindrance -see e.g ref. [9–11]. Recently in Ref. [12] a phenomenological description of the ratios direct to total reaction cross sections is reported, predicting an increasing trend at near barrier energies as well as a quenching in favor of the direct part for below barrier energy ratios. Moreover, a simultaneous description of elastic scattering and backscattering results [13], predicts not only an unusual behavior of the optical potential for the weakly bound nucleus ^{6}Li, with an increasing behavior at near barrier energies, but also a drop of the imaginary potential, not at near, but at deep sub-barrier energies. The last issue indicates the persistence of reaction channels at deep sub-barrier energies (see Fig. 1). These facts intrigued our interest at this low energy regime and initiated our new experiment [14] at the Notre Dame University, *TwinSol* facility [15].

*e-mail: apakou@uoi.gr

In the following, we will make a brief review of our breakup results at 30 MeV for ^{8}B +^{208}Pb [14], while we will also give new results of our preliminary analysis for ^{7}Be +^{208}Pb, in relation with the ^{4}He and ^{3}He - particle production. We should note here that another important observation, for neutron rich stable and radioactive weakly bound nuclei, is the excessive α - production [16]. Research in this direction is mainly related to ^{6,7}Li and ^{6}He projectiles. An overview of existing data is given in Ref. [17]. In Ref. [18] exclusive α - production data are comprehensively analyzed and the various mechanisms are reported with emphasis on breakup following transfer. Recently inclusive measurements are also reported for the proton rich radioactive nucleus ^{7}Be on a ^{28}Si [17] and a ^{58}Ni target [19]. These measurements have been performed at near barrier energies with the conclusion that standard Continuum Discretized Coupled-Channel calculations (CDCC) and Distorted Wave Born Approximation (DWBA) calculations for breakup and transfer can not accommodate the very large measured α - production cross sections, in excess of the ^{3}He ones. A rather large ^{3}He stripping process may be inferred from these measurements.

This paper includes in section 2 details of our experiment and the results, for both secondary beams that is ^{8}B and ^{7}Be. A concluding discussion with summary will be given in section 3.

 https://doi.org/10.1051/epjconf/202125204006

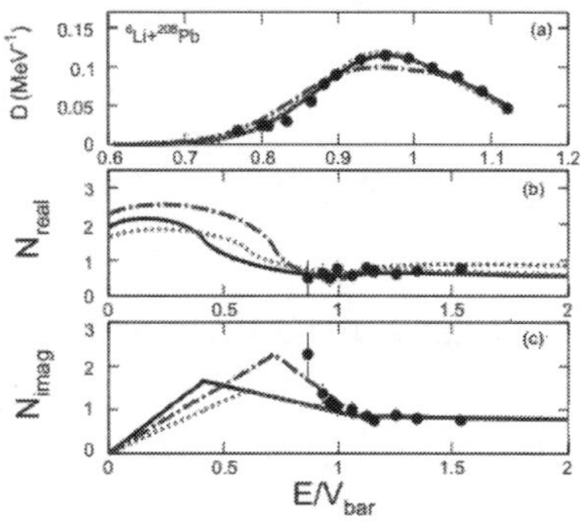

Figure 1. Determining the optical potential for ^{6}Li + ^{208}Pb via elastic scattering and backscattering measurements. (a) Barrier distributions obtained in the backscattering measurements- see Ref. [13]. (b) and (c) the real and imaginary part of the optical potential, determined in the analysis of elastic scattering data, reported in Ref. [1]. The lines represent trial potentials in a dispersive relation scheme, taking into account simultaneously the elastic scattering and backscattering measurements. Figure taken from Ref. [13].

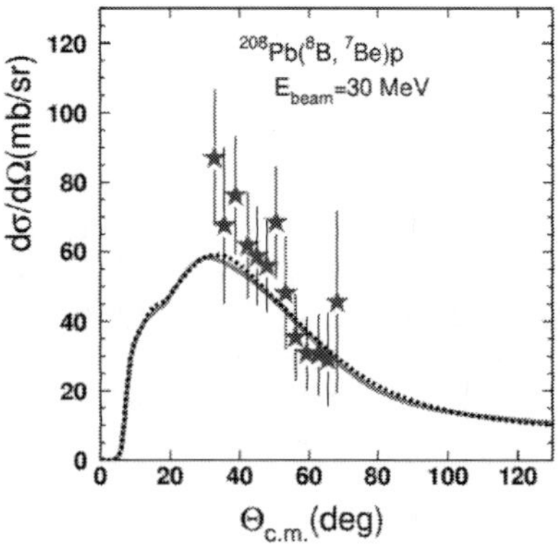

Figure 2. Angular distribution of the observed $^{8}B \rightarrow {}^{7}Be + p$ breakup yield on a lead target at a ^{8}B incident energy of $\approx$ 30 MeV (middle of the target). The solid and dashed curves denote the summed breakup angular distributions from the full (Coulomb plus nuclear potentials) and Coulomb potentials only CDCC calculations, respectively. Figure taken from Ref. [14].

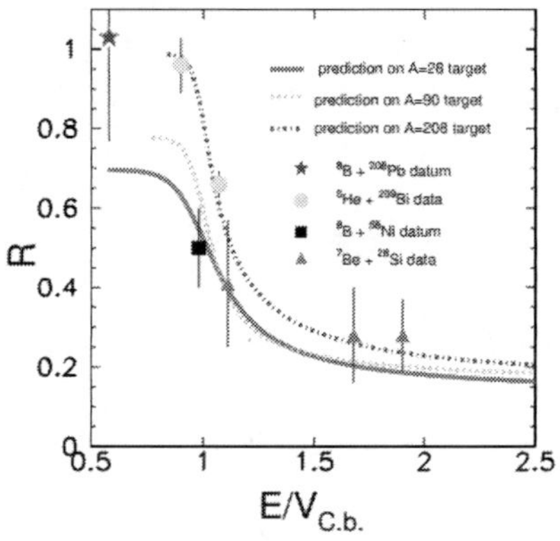

Figure 3. Ratio's, R, direct to total reaction cross section. Lines correspond to predictions obtained earlier [12], from experimental data of weakly bound neutron rich nuclei on various targets, appropriately reduced (red-solid: A=28, green-dotted A=90, blue-dotted-dashed A=208). The experimental datum for ^{8}B + ^{208}Pb [14], is designated with the blue star. A previous datum ^{8}B + ^{58}Ni with a black box [20, 21], while previous data for ^{7}Be + ^{28}Si [17] are designated with the red filled arrows, and ^{6}He + ^{209}Bi data with cyan filled circles [22].

2 Experimental details- Results

Comprehensive experimental details are given in Ref. [14]. Few pertinent points will be given below. The experiment was performed at the *TwinSol* facility [15] of the University of Notre Dame (UND). A cocktail secondary beam, composed of ^{8}B at the energy of 30.5 MeV, of ^{7}Li at 13.1 MeV, and of ^{7}Be at 22.5 MeV, was produced by using the ^{6}Li + ^{3}He reaction with beam fluxes in the range of 3000 to 5000 pps. The ^{8}B and ^{7}Be beams were very well separated by imposing a TOF (Time Of Flight) requirement between the occurrence of an energy signal in our detection system and the RF timing pulse from the beam buncher. The reaction products were detected in two telescopes of the SIMAS array (Sistema Móvil de Alta Segmentación), of LEMA (Laboratorio Nacional de Espectrometría de Masas con Aceleradores), the National Laboratory of the Physics Institute at the Autonomous National University of Mexico. Each telescope included a Double Sided Silicon Strip Detector (DSSSD) with nominal thickness of $\approx$ 20 μm (the effective thickness of each strip was estimated during the run and calibrations to be 24 $\pm$4 μm), backed by a silicon pad 150 μm thick. One telescope was set beam left covering an angular range between $\approx 25.8^0$ to 69.2^0, being 59.9 mm far away from the lead target, while the second telescope was set beam right covering an angular range between $\approx 28.5^0$ to 66.5^0 being 69.95 mm away from the target. The observed reaction products were ^{7}Be for the ^{8}B+^{208}Pb studied reaction, and ^{4}He and ^{3}He - particles for the ^{7}Be+^{208}Pb one, and were detected in inclusive measurements. It is assumed that for the boron induced reaction, all ^{7}Be fragments originate from a breakup reaction mechanism, since our DWBA calculations predict a negligible proton stripping channel. Our experimental angular distribution is demonstrated in Fig. 2, together with Continuum Discretized Coupling Channel (CDCC) calculations, exhibiting a very good agreement with them. The measured cross section, obtained by integrating the angular distribution results, is $\sigma_{break}{}^{exp}$ = 326$\pm$84 mb, to be compared with the CDCC result of $\sigma_{break}{}^{theory}$ =300 mb. The very good agreement between experiment and theory, justifies the use of a theoretical value for total reaction

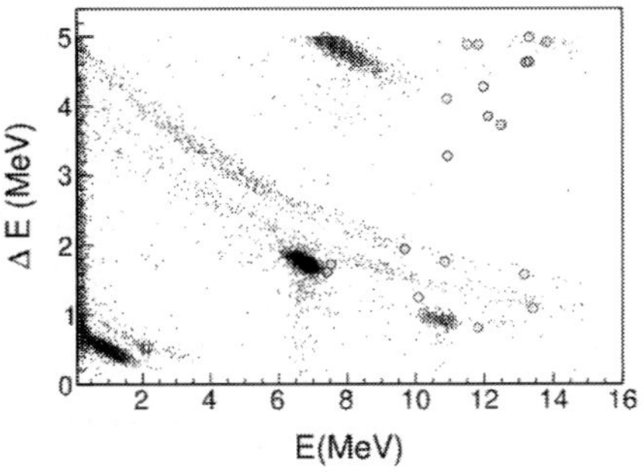

Figure 4. ΔE - E spectrum for ^{7}Be + ^{nat}Pb at 22.5 MeV. The spectrum, events denoted with black dots, is zoomed for ΔE ≤ 5 and is taken without any time window. The open circles in red and blue, denote reaction events with a time window to the ^{7}Be beam with different time bounces. The data refer to the three lower pixels of strip 27 of the DSSSD detector installed beam right, corresponding to $\theta_{lab} \approx 41°$.

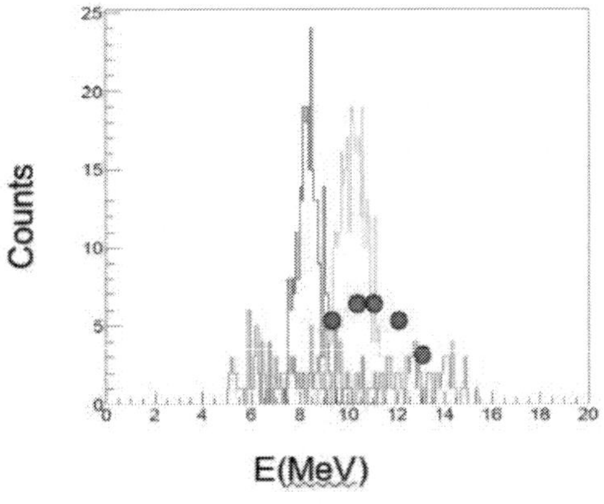

Figure 5. Simulated energy spectra for ^{3}He reaction products of the reaction ^{7}Be + ^{208}Pb at 22.5 MeV. The peaks with the black and green lines correspond to events due to elastic breakup and ^{4}He transfer, respectively. The red line correspond to non elastic breakup. The red spots correspond to our experimental energy spectra.

cross section via CDCC of $\sigma_{tot}{}^{theory}$= 316 mb, since an experimental total reaction cross section at these energies is unobtainable. In this respect, a ratio of direct to total is determined as R=1.03±0.27. This value together with other ratios for various projectiles and targets are presented in Fig. 3. The experimental values are compared well with our phenomenological predictions [12]. The present result confirms the ratio quenching to 100% for the heavy targets, demonstrating the dominance of direct mechanisms below barrier for weakly bound projectiles and heavy targets.

We will proceed now with our preliminary analysis for the ^{7}Be + ^{208}Pb reaction at 22.5 MeV. We should note the following

α - production can originate under the following four processes.

- Breakup : ^{7}Be + ^{208}Pb → ^{4}He + ^{3}He + ^{208}Pb, S_α =-1.586 MeV.
- ^{3}He - stripping : ^{7}Be + ^{208}Pb → ^{4}He + ^{211}Po, Q=4.03 MeV and Q_{opt} = -10.60 MeV.
- n - stripping : ^{7}Be + ^{208}Pb → ^{6}Be + ^{207}Pb, Q= -6.74 MeV, $Q_{opt} \approx 0$ MeV.
- n - pickup : ^{7}Be + ^{208}Pb → ^{8}Be + ^{207}Pb, Q= +11.53 MeV, $Q_{opt} \approx 0$ MeV.

The ^{3}He - production can originate under the following two processes:

- Breakup : ^{7}Be + ^{208}Pb → ^{4}He + ^{3}He + ^{208}Pb, S_α =-1.586 MeV.
- ^{4}He - stripping : ^{7}Be + ^{208}Pb → ^{3}He + ^{212}Po, Q=-10.544 MeV and Q_{opt} = -10.60 MeV.

In Fig. 4, we present a two dimension ΔE versus E plot, zoomed in the helium area. These data are due to one strip of the DSSSD detector, corresponding to $\theta_{lab} \approx 41°$. We can observe in black, two well distinguished curves

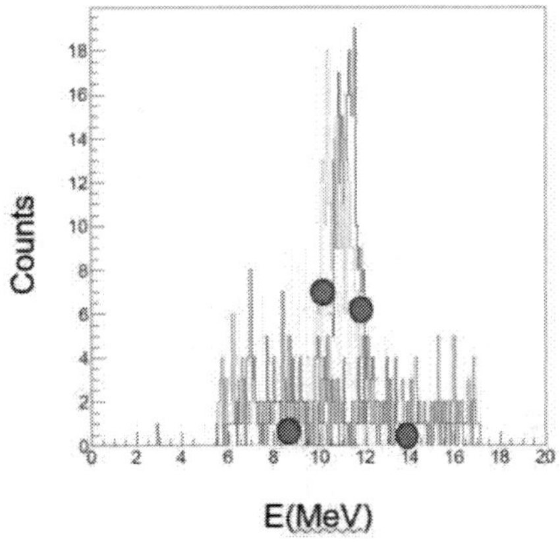

Figure 6. Simulated energy spectra for ^{4}He reaction products of the reaction ^{7}Be+^{208}Pb at 22.5 MeV. The peaks with the black and green lines correspond to events due to elastic breakup and ^{3}He transfer, respectively. Other lines correspond to all other processes underlined in the text. The red spots correspond to our experimental energy spectra.

corresponding to α- and ^{3}He - particle production. After applying the time windows corresponding solely to the ^{7}Be beam, and applying also restrictions for interstrip rejection, the only remaining events are the three ones designated with blue and red open circles (two successive rf timing pulses) for ^{4}He -particles and only one, designated with the red open circle for ^{3}He. Some events are also remaining around the elastically scattered ^{7}Be beam particles, a small part of which appears in black on the top of the Figure. The two black spots below our curves, have been identified as beam helium particles which pos-

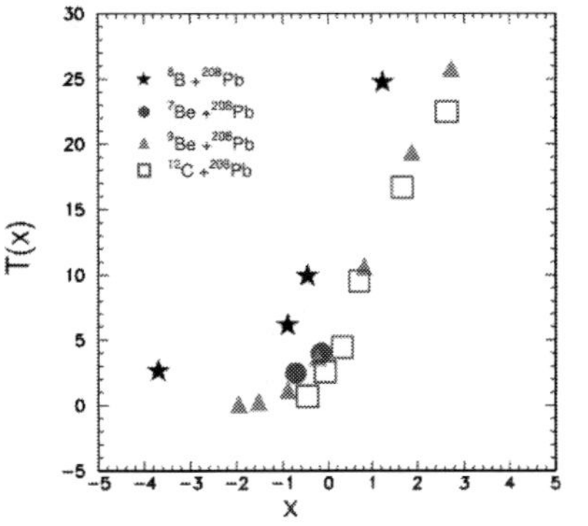

Figure 7. Reduced total reaction cross sections as a function of a reduced energy. The reduction is according to Wong relations as reported before in Ref. [23]. The cross sections are experimental or/and theoretical values from Refs. [4, 14] for ^{8}B, [24] for ^{7}Be, [25] for ^{9}Be and [26], for ^{12}C.

sibly are scattered under an angle and not perpendicular to the target and therefore accumulate in the ΔE detector less energy. Elastically scattered particles due to these beam particles, appear around the black spots.

To possibly clarify the origin of the events in the ^{3}He and ^{4}He curves by kinematics, we have performed detailed simulations. The results are included in Figures 5 and 6 for the ^{3}He and ^{4}He - particles respectively. Obviously from Figure 5 we can see that the experimental energy spectra, designated with the red spots, probe as a mechanism the cluster transfer, that is ^{4}He stripping. The situation is more complicated as expected for the ^{4}He case, where several mechanisms contribute and where the peaks due to breakup and transfer coincide. We should note here that these energy spectra is the result of events in all strips of both detectors. Also, the observed yields are quite large as we will discuss in the following section.

3 Discussion - Summary

We have outlined experimental results in relation with reaction mechanisms at deep sub-barrier energies. Two nuclei are under our investigation, by considering their induced reactions with a lead target. These are the proton halo nucleus ^{8}B, with a binding energy of only 136 keV, and the radioactive ^{7}Be nucleus with a dominant cluster structure of (^{4}He + ^{3}He) and binding energy of 1.59 MeV. The question is how these nuclei with very different structure will behave at the same deep sub-barrier energy conditions? The answer for ^{8}B was reported in Ref. [14] and it is also reviewed within this paper. In summary, the measurement of a strong and dominant breakup channel for ^{8}B on a lead target at 30 MeV, an energy corresponding to a distance of closest approach of D$\approx$ 22 fm to be compared with a sum of radii for the two colliding nuclei of D$\approx$ 10fm, is observed and quantified. This issue may be of paramount importance for both reaction mechanism problems but also for the structure of this challenging nucleus. At this distance of closest approach, the only interaction experienced by projectile and target nuclei, is the Coulomb one. In a simplistic approach, the proton of the halo around the compact ^{7}Be core, is tunnelling between a forbidden and allowed quantum mechanically region and at some instance of the collision, it faces the huge Coulomb potential of the lead, and scatters (breakups) to the continuum. In the conventional CDCC theory this procedure is very well accommodated and compatible results of both experiment and theory have been obtained, justifying a ratio of direct to total cross sections almost equal to one. This result is found to be in excellent agreement with our phenomenological predictions [12]. The same prediction however should apply to ^{7}Be, that is the direct cross section should almost exhaust the total reaction cross section. Our analysis is in progress in this direction. Preliminary experimental results, indicate large cross sections of the order of $\approx$ 100 mb. Moreover for the ^{3}He production, this mechanism is ^{4}He -stripping. Transfer mechanisms instead of a breakup process, may be well understood for ^{7}Be, from the point of view that at these large distances of closest approach, where we have no halo structure as is the case of ^{8}B, we need long range interactions, for inducing any type of reaction. Can a whole cluster of nuclei like a ^{3}He or ^{4}He nucleus tunnel through the strong Coulomb barrier of the lead target? In a conventional theoretical framework, preliminary calculations, based on elastic breakup (EBU) and non elastic breakup (NEB) mechanisms have been performed [27]. EBU is calculated into a CDCC approach, while for the NEB calculations a model proposed by Ichimura et al. [28], is employed within the distorted-wave Born approximation (DWBA). These calculations indicate total reaction cross sections of the order of $\approx$ 40 mb, with most of the probability to be connected with an inelastic excitation of ^{7}Be in the first excited state and only a very small probability below 1 mb, to be related with other direct mechanisms. Moreover looking to systematics of total reaction cross sections for ^{8}B and ^{7}Be, as well as of ^{9}Be and ^{12}C, presented in Figure 7, we comment the following. For stable projectiles, the total cross sections drops to zero around barrier. For the weakly bound nuclei ^{8}B and ^{9}Be around barrier starts a constant cross section quenching which for boron is close to 300 mb, while for ^{9}Be close to 10 mb. For ^{7}Be the information is limited to two points. Under an extrapolation, we expect total reaction cross sections between 30 to 170 mb, depending of the type of extrapolation and following the trend of ^{9}Be. Under the above motivation, and our preliminary results indicating large ^{3}He and ^{4}He yields, searching for a new "treasure" at deep sub-barrier energies, we plan a new experiment for ^{7}Be + ^{208}Pb in Notre Dame as long as the upgrading of *Twinsol* to *Trisol* will be complete. With the new conditions, the optimization of the secondary beam on ^{7}Be, is plausible and a clean beam of other particles, as helium beam nuclei, is expected to be produced with a high flux of the order of 10^{6} pps. Such a beam could make possible the determination of cross sections of the order of 1 mb and below. Further more another measurement will focus on the inelastic excitation of ^{7}Be by gamma spectroscopy [29].

Acknowledgements This work was partially supported by the U.S. National Science Foundation: PHY-1713857 and PHY14-01343 and by CONACyT 299073 and CB-01-254619 and DGAPA-PAPIIT IN107820 and AG101120 of Mexico. The authors want to warmly thank to A. Huerta-Hernández and M. Pérez-Vielma for their technical support in the construction of the SIMAS array.

References

[1] A. Pakou, N. Alamanos, G. Doukelis, A. Gillibert et al., Phys. Rev. C **69**, 054602(2004).
[2] N. Keeley et al., Nucl. Phys. **A571**, 326 (1994).
[3] I. Martel, J. Gómez-Camacho,, K. Rusek, G. Tungate; Nucl. Phys. **A605**, 417 (1996).
[4] N. Keeley, R. Raabe, N. Alamanos, and J. L. Sida; Prog. Part. Nucl. Phys. **59**, 579 (2007).
[5] N. Keeley, N. Alamanos, K. W. Kemper and K. Rusek, Phys. Rev. C **82**, 034606(2010).
[6] L. F. Canto, P. R. S. Gomes, R. Donangelo, and M. S. Hussein; Phys. Rep. **424**, 1 (2006).
[7] B. B. Back, H. Esbensen, C. L. Jiang, and K. E. Rehm; Reviews of Modern Physics **86**,317(2014)
[8] J. J. Kolata, V. Guimaraes and E. F. Aguilera, Eur. Phys. J. A 52, 123 (2016).
[9] C. L. Jiang et al.; Phys. Rev. Lett. **89**, 052701(2002).
[10] C. L.Jiang et al.; Phys. Lett. **B 640**, 18 (2006).
[11] G. Montagnoli et al.; Phys. Rev. C **87**, 014611(2013).
[12] A. Pakou, D. Pierroutsakou, M. Mazzocco, L. Acosta et al., ; Eur. Phys. J. **A 51**, 55 (2015).
[13] K. Zerva, A. Pakou, N. Patronis, P. Figuera, Eur. Phys. J. A **48**, 102 (2012).
[14] A. Pakou, L. Acosta, P. D. O'Malley, S. Aguilar et al.; Phys. Rev. C **102**, 031601(R)(2020).
[15] M. Y. Lee, F. D Becchetti, T. W. O' Donnell, D A. Roberts et al., Nucl. Instrum. Methods Phys. Res., Sect. **A 422**, 536 (1999).
[16] A. Pakou et al., Phys. Rev. Lett. **90**, 202701 (2003).
[17] O. Sgouros et al., Phys. Rev. C **94**, 044623 (2016).
[18] D. H. Luong et al., Phys. Rev. C **88**, 034609 (2013).
[19] M. Mazzocco et al., Phys. Rev. C **92**, 024615 (2015).
[20] V. Guimaraes et al., Phys. Rev Lett. **84**, 1962 (2000).
[21] J. J. Kolata et al.,Phys. Rev. C **63**, 024616(2001).
[22] E. F. Aguilera et al., Phys. Rev. Lett. **84**, 5058 (2000).
[23] L.F. Canto, P.R.S. Gomes, J. Lubian, L.C. Chamon, E. Crema, J. Phys. G 36, 015109 (2009).
[24] M. Mazzocco, N. Keeley, A. Boiano, C. Boiano et al. Phys. Rev. C **100**, 024602(2019).
[25] A. Gomez Camacho, E. F. Aguilera, A. M. Moro; Nuclear Physics A **762**, 216 (2005).
[26] S. Santra, P. Singh, S. Kailas, A. Chatterjee et al. **64**, 024602 (2001).
[27] Antonio Moro; private communication
[28] M. Ichimura, N. Austern, and C. M. Vincent, Phys. Rev. C **32**, 431 (1985).
[29] S. L. Henderson, T. Ahn, M. A. Caprio, P. J. Fasano et al. ; Phys. Rev. C **99**, 064320 (2019).

Study of the ^{4}He(^{4}He,^{4}He)^{4}He* inelastic scattering at the MAGNEX facility

V. Soukeras[1,2*], *F.* Cappuzzello[1,2], *M.* Cavallaro[1], *D.* Carbone[1], *A.* Hacisalihoglu[1,3], *M.* Fisichella[1], *C.* Agodi[1], *H.-W.* Becker[4], *G. A.* Brischetto[1,2], *S.* Calabrese[1,2], *C.* Ciampi[5], *M.* Cicerchia[6], *M.* Cinausero[6], *I.* Ciraldo[1,2], *M.* D′ Andrea[7], *D.* Dell′ Aquila[8], *S.* Firat[9], *C.* Frosin[5], *M.* Hilcker[4], *L.* La Fauci[1,2], *I.* Lombardo[7], *T.* Marchi[6], *O.* Sgouros[1], *A.* Spatafora[1,2], *D.* Torresi[1], *M.* Vigilante[10], and *A.* Yildirim[9]

[1]INFN – Laboratori Nazionali del Sud, Catania, Italy
[2]Dipartimento di Fisica e Astronomia "Ettore Majorana", Università di Catania, Catania, Italy
[3]Institute of Natural Science, Karadeniz Teknik Universitesi, Trabzon, Turkey
[4]Ruhr-Universitat Bochum, Bochum, Germany
[5]INFN – Sezione di Firenze, Florence, Italy
[6]INFN – Laboratori Nazionali di Legnaro, Legnaro, Italy
[7]INFN – Sezione di Catania, Catania, Italy
[8]Rudjer Bošković Institute, Zagreb, Croatia
[9]Akdeniz University, Antalya, Turkey
[10]INFN – Sezione di Napoli and Università degli Studi di Napoli "Federico II", Napoli, Italy

Abstract. The ^{4}He(^{4}He,^{4}He)^{4}He* inelastic scattering was revisited in a new coincidence measurement at the MAGNEX facility of Istituto Nazionale di Fisica Nucleare – Laboratori Nazionali del Sud. The ^{4}He + ^{4}He → ^{4}He + ^{4}He* → ^{4}He + ^{3}H + ^{1}H and ^{4}He + ^{4}He → ^{4}He + ^{4}He* → ^{4}He + ^{3}He + n reactions were measured simultaneously by detecting the ^{4}He particles at the MAGNEX magnetic spectrometer in coincidence with the ^{3}H and ^{3}He at the OSCAR silicon telescope. The main concept of the experiment is described and the data reduction strategy is reported.

1 Introduction

The ^{4}He is a well bound nucleus with a breakup threshold of 19.8 MeV for the ^{3}H – ^{1}H mode and 20.7 MeV for the ^{3}He – n one. No bound states are existing in its level scheme however it exhibits, very close to the breakup threshold, a pronounced resonance with the same spin and parity (0^+) as the ground state. Several investigations have been performed so far aiming at shedding light on the characteristics of this resonance either by using a ^{4}He(e,e′)^{4}He* [1-4] or a ^{4}He(^{4}He,^{4}He)^{4}He* [5-6] reaction.

Additionally, a recent ab-initio calculation of the monopole transition form factor of ^{4}He with realistic nuclear forces pointed to a strong dependence on the different realistic potential used and revealed a significant disagreement with respect to all existing electron scattering data when a method based on modern Hamiltonians from chiral perturbation theory was adopted [7]. The inconsistencies met between the recent ab-initio form factor and all existing data from ^{4}He(e,e′)^{4}He* studies call for further investigation.

In a recent theoretical work [8], the transition densities to the 0^+ excited state of ^{4}He were calculated and the form factors were constructed by folding procedure. The cross section angular distributions were also calculated for this state by using both macroscopic and microscopic models for the form factors.

From both the experimental and theoretical point of view the ^{4}He(e,e′)^{4}He* and ^{4}He(^{4}He,^{4}He)^{4}He* studies are challenging. The first ones give a direct access to the form factor, which is governed by a pure electromagnetic interaction. However, only the protons are involved in the interaction. Other disadvantages of (e,e′) include its very low cross section, the fact that it consists of a mixed isoscalar and isovector probes and finally, higher multipolarities are not possible to be excluded. On the other hand, the ^{4}He(^{4}He,^{4}He)^{4}He* studies benefit by a pure isoscalar probe, a significantly higher cross section and the higher multipolarities are easier to be unfolded. Nevertheless, the ^{4}He induced scattering is governed by the nuclear force, making the extraction of the electric monopole form factor less direct. In addition, an optical model analysis is needed to describe the distortions in the incoming and outcoming waves due to the initial and final state nucleus-nucleus interactions.

Taking into account the positive and negative features of the (^{4}He,^{4}He) probe, we performed inelastic scattering measurements for the ^{4}He + ^{4}He reaction at 53 MeV incident energy ($E_{c.m.}$ = 26.5 MeV) in order to extract the characteristics of the 0^+ resonance of ^{4}He in a new measurement and resolve previous

* Corresponding author: vasileios.soukeras@lns.infn.it

inconsistencies. Our study includes also a careful description of the significant phase space for both the ^{3}H – ^{1}H and ^{3}He – n modes. Finally, a global interpretation together with the elastic scattering channel, which was also measured, ensures a precise optical model analysis, allowing the extraction of the form factors.

2 The experiment

The ^{4}He beam was accelerated by the K800 Superconducting Cyclotron of Istituto Nazionale di Fisica Nucleare – Laboratori Nazionali del Sud (INFN – LNS) at 53 MeV and impinged on a ^{4}He target implanted on a thin aluminum foil. Measurements with a 2 MeV ^{4}He were performed at Ruhr – Universitat Bochum before the implantation in order to test the thickness and the uniformity of the aluminum foils while the ^{4}He scattering centres after the implantation were measured by a measurement with a 2.2 MeV proton beam at the same laboratory. The density of scattering centres of the ^{4}He as well as the other target materials, namely aluminum, oxygen and carbon are presented in Table 1. The measurement was repeated with a target composed of aluminum, oxygen and carbon in the appropriate proportions for the background estimation. However, the majority of the background events was excluded by performing the coincidence measurement described below.

Table 1. The density of scattering centres of helium, aluminum, oxygen, and carbon atoms present in the target.

Material	Thickness (at/cm^2)	Thickness (μg/cm^2)
He	1.92 E+17	1.29
Al	4.64 E+18	210
O	1.00 E+17	2.66
C	1.20 E+16	0.24

2.1 Elastic scattering measurement

The ^{4}He ejectiles were momentum analysed by the MAGNEX magnetic spectrometer [9] whose optical axis was set at θ_{opt} = 6.6° spanning an angular range between 2° and 13°. The different ions were detected by the Focal Plane Detector (FPD) [10-11] of MAGNEX, which consists of a gas tracker followed by a wall of 60 silicon pad detectors. The gas tracker includes six sections, each one having at the top a proportional wire. The energy loss (ΔE) of the ions inside the gas is measured in each of the six wires from the charge signal generated by the proportional multiplication of the primary electrons close to the wires. The residual energy (E_{resid}) is measured by the silicon pad detectors. The charge generated by the electron avalanches close to the multiplication wires is also induced into a segmented anode featuring a pattern of 223 independent pads. The anode is mounted above each DC wire allowing the measurement of the horizontal position (X_{foc}) and angle (θ_{foc}). The measurement of the electron drift time inside the gas allows the determination of the vertical position (Y_{foc}) and angle (φ_{foc}).

2.2 Inelastic scattering measurement

No bound states are reporting in ^{4}He structure [12] therefore the ^{4}He* excited states can be reconstructed by the detection of its fragments. In particular, the two different decay modes ^{4}He + ^{4}He → ^{4}He + ^{4}He* → ^{4}He + ^{3}H + ^{1}H and ^{4}He + ^{4}He → ^{4}He + ^{4}He* → ^{4}He + ^{3}He + n were reconstructed by two coincidence measurements. In both modes ^{4}He nuclei were detected by the MAGNEX FPD [11], while the ^{3}H and ^{3}He fragments were detected by OSCAR (hOdoscope of Silicons for Correlations and Analysis of Reactions) telescope [13]. OSCAR consists of two detection stages: a Single Sided Silicon Strip Detector (SSSSD) 20 μm thick as ΔE, followed by 16 silicon pads (4x4) 300 μm thick, providing the measurement of the residual energy E. OSCAR telescope was mounted inside the scattering chamber of MAGNEX, 15 cm from the target spanning the angular range between 19° and 38°. The 16 strips of the SSSSD allowed a measurement with an angular resolution of about 1°. The detection threshold in OSCAR was 2.5 MeV for ^{3}H, 5.0 MeV for ^{3}He and 5.5 MeV for ^{4}He, taking into account the SSSSD thickness and the electronics threshold for both ΔE and E.

3 Data Reduction

The particle identification (PID) of the data collected by the MAGNEX FPD was performed as described in Ref. [14]. The first step is the selection of the data belonging to the Z of interest ($Z = 2$) in a $\Delta E - E$ plot (Fig. 1). The mass separation is feasible by applying the appropriate graphical selection conditions to the $X_{foc} - E_{resid}$ spectra, where the different loci correspond to ions with different $\sqrt{m}/q$. Such a spectrum for a single silicon detector is illustrated in Fig. 2. Having identified the events of interest, a software 10th order ray – reconstruction [15] is applied to the data in order to obtain the kinetic energy, scattering angle, and excitation energy spectra.

The standard $\Delta E - E$ technique was applied for a precise PID at the OSCAR telescope and a typical $\Delta E - E$ plot is shown in Fig. 3.

The experimental inelastic scattering data collected by the two coincidence measurements were compared with kinematical simulations based on the multipurpose Monte Carlo simulation algorithm MULTIP [16]. This algorithm is a powerful tool for two –, three – and four – body reactions kinematics and its validity has been tested in several works and for many reactions [17-24]. In the present work, the inelastic scattering reaction ^{4}He + ^{4}He → ^{4}He + ^{4}He* → ^{4}He + ^{3}H + ^{1}H was simulated for both the resonant (0$^+$) and the non – resonant breakup, while the ^{4}He + ^{4}He → ^{4}He + ^{4}He* → ^{4}He + ^{3}He + n was simulated only for the non – resonant breakup since its threshold is higher than the 0$^+$ resonant region. The transfer processes ^{4}He + ^{4}He → ^{3}H + ^{5}Li → ^{4}He + ^{3}H +

^{1}H and ^{4}He + ^{4}He → ^{3}He + ^{5}He → ^{4}He + ^{3}He + n leading to the same final channels were also simulated. The experimental data compared with the kinematical simulations are illustrated in Figs. 4 – 5 for the two different modes showing an excellent agreement themselves. It should be noted that except the importance at the preparation of the experiment, the simulations were also useful in order to estimate the energy efficiency due to the coincidence measurements. This efficiency is mainly dominated by the detection threshold of OSCAR, mentioned in section 2.2. The energy efficiency was extracted by taking the ratio between the simulated events with all the experimental conditions in angle, kinetic energy and excitation energy and the simulated events without the kinetic energy conditions. Since the efficiency may change for different excitation energy regions and angles, it was extracted for a step of dE_x = 0.8 MeV and $d\theta_{lab}$ = 0.5°, ensuring a precise estimation. Finally, the solid angle of both MAGNEX and OSCAR was deduced by geometrical simulations taking into account for the case of OSCAR the possible "dead regions" of its E – stage.

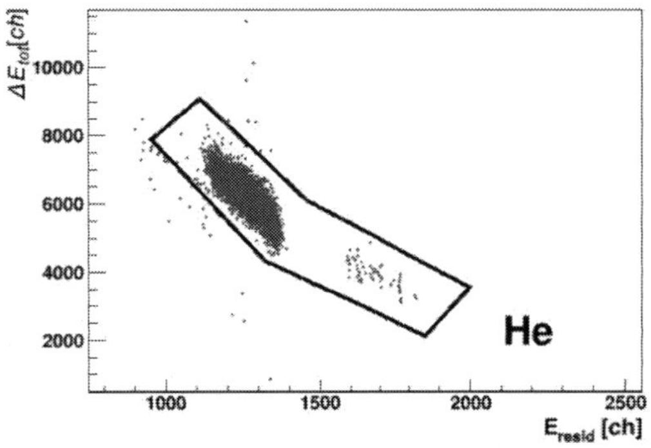

Fig. 1. A typical ΔE_{tot} versus residual energy (E_{resid}) spectrum for a single silicon detector of the FPD.

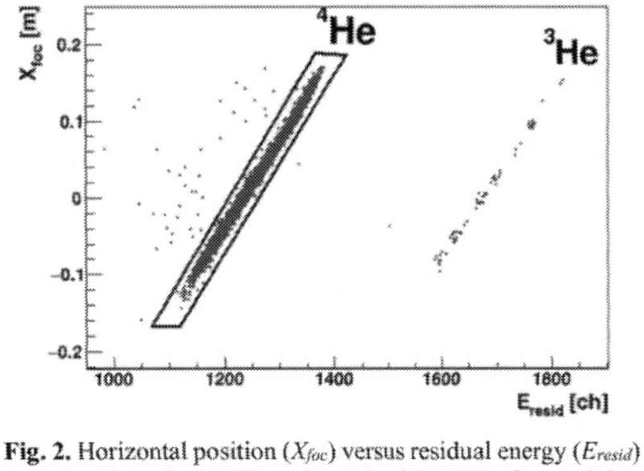

Fig. 2. Horizontal position (X_{foc}) versus residual energy (E_{resid}) spectrum for a single silicon detector of the FPD after applying the graphical selection on *Z* = *2* highlighted in Fig.1. The selection condition of ^{4}He events is illustrated by the solid black line.

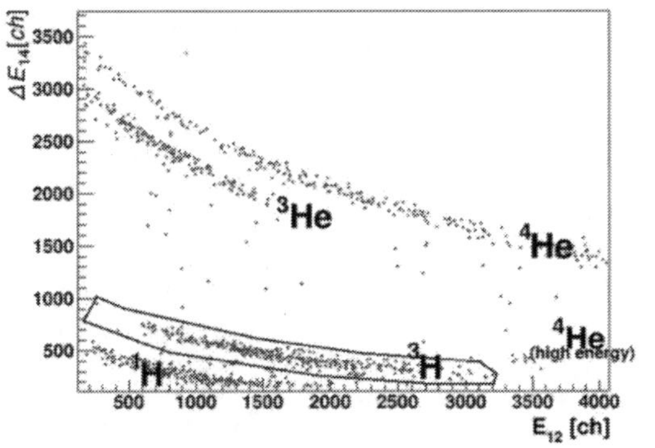

Fig. 3. A typical $\Delta E - E$ spectrum for a ΔE strip correlated with a single *E* pad of the OSCAR telescope. One can see the loci of protons, tritons, ^{3}He and ^{4}He while, the selection of tritons is depicted with the solid black line.

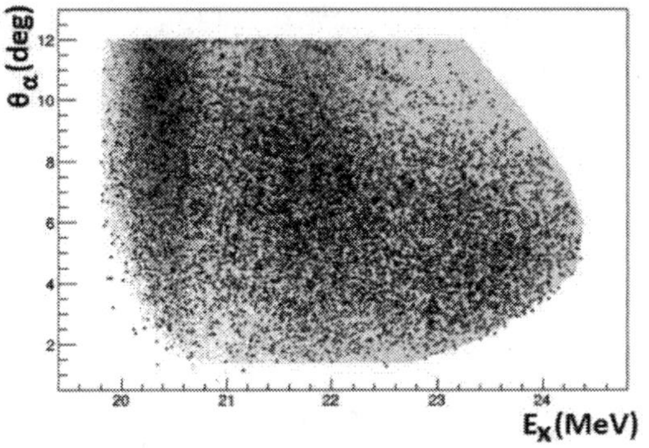

Fig. 4. A θ_α versus excitation energy (E_x) spectrum for the ^{4}He + ^{4}He → ^{4}He + ^{3}H + ^{1}H mode. The experimental data points, presented in black, were acquired in a ^{4}He - ^{3}H coincidence measurement. Preliminary simulations for the resonant (0^+) and the non-resonant continuum are presented with the red and green points, respectively, while the simulated events for the transfer are presented with the blue points. See text for more details.

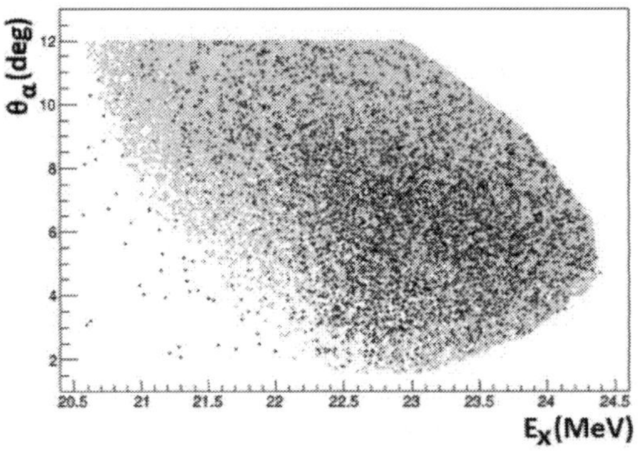

Fig. 5. A θ_α versus excitation energy (E_x) spectrum for the ^{4}He + ^{4}He → ^{4}He + ^{3}He + n mode. The experimental data points, presented in black, were acquired in a ^{4}He - ^{3}He coincidence measurement. Preliminary simulations for the non-resonant breakup and transfer are presented with the green and cyan blue points, respectively. See text for more details.

4 Summary

The $^4He + ^4He \rightarrow ^4He + ^3H + ^1H$ and $^4He + ^4He \rightarrow ^4He + ^3He + n$ reactions were measured at the MAGNEX facility of INFN – LNS together with the elastic scattering measurement. The challenging inelastic scattering measurement was performed in a coincidence measurement by detecting the heavy ejectile ^4He in MAGNEX while the lighter ^3H and ^3He were detected by the OSCAR telescope. The high Z and mass resolutions of both MAGNEX and OSCAR guaranteed an accurate particle identification. Comprehensive kinematical simulations based on the MULTIP algorithm make it feasible to disentangle between the inelastic scattering via the first excited state of ^4He (0^+), the non-resonant breakup processes and the transfer processes leading to the same final channels. The data analysis is underway in a more quantitative basis aiming to extract differential and integrated cross sections. A theoretical analysis is also in progress.

References

1. R. Frosch, R. E. Rand, M. R. Yearian et al., Phys. Lett. **19**, 155 (1965).
2. R. F. Frosch, R. E. Rand, H. Crannell et al., Nucl. Phys. A **110**, 657 (1968)
3. Th. Walcher, Phys. Lett. **B31**, 442 (1970).
4. G. Kobshall, C. Ottermann, K. Maurer et al., Nucl. Phys. A **405**, 648 (1983).
5. E. E. Gross, E. V. Hungerford III, J. J. Malanify et al., Phys. Rev. **178**, 1584 (1969).
6. M. Baumgartner, H. P. Gubler, M. Heller et al., Nucl. Phys. A **368**, 189 (1981).
7. S. Bacca, N. Barnea, W. Leidemann, G. Orlandini, Phys. Rev. Lett. **110**, 042503 (2013).
8. Y. Kucuk, M. Karakoç, A. Vitturi, Eur. Phys. J. A **57**, 37 (2021).
9. F. Cappuzzello, C. Agodi, D. Carbone, M. Cavallaro, Eur. Phys. J. A **52**, 167 (2016).
10. M. Cavallaro, F. Cappuzzello, D. Carbone, A. Cunsolo et al., Eur. Phys. J. A **48**, 59 (2012).
11. D. Torresi, O. Sgouros, V. Soukeras et al., Nucl. Instrum. Meth. Phys. Res. A **989**, 164918 (2021).
12. D. R. Tilley, H. R. Weller, G. M. Hale, Nucl. Phys. A **541**, 1 (1992).
13. D. Dell´Aquila, I. Lombardo, G. Verde et al., Nucl. Instrum. Meth. Phys. Res. A **877**, 227 (2018).
14. F. Cappuzzello, M. Cavallaro, A. Cunsolo, A. Foti et al., Nucl. Instrum. Meth. Phys. Res. A **621**, 419 (2010).
15. F. Cappuzzello, D. Carbone, M. Cavallaro, Nucl. Instrum. Meth. Phys. Res. A **638**, 74 (2011).
16. O. Sgouros, V. Soukeras, A. Pakou, Eur. Phys. J. A **53**, 165 (2017).
17. O. Sgouros, A. Pakou, D. Pierroutsakou et al., Phys. Rev. C **94**, 044623 (2016).
18. A. Pakou, O. Sgouros, V. Soukeras et al., Phys. Rev. C **95**, 044615 (2017).
19. V. Soukeras, A. Pakou, F. Cappuzzello et al., Phys. Rev. C **95**, 054614 (2017).
20. A. Pakou, O. Sgouros, V. Soukeras et al., Phys. Rev. C **101**, 024602 (2020).
21. A. Pakou, L. Acosta, P. D. O´Malley et al., Phys. Rev. C **102**, 031601(R) (2020).
22. V. Soukeras, O. Sgouros, A. Pakou et al., Phys. Rev. C **102**, 064622 (2020).
23. A. Pakou, O. Sgouros, V. Soukeras, F. Cappuzzello, Eur. Phys. J. A **57**, 25 (2021).
24. A. Pakou, O. Sgouros, V. Soukeras et al., Nucl. Phys. A **1008**, 122155 (2021).

Neutron star structure with nuclear force mediated by hypothetical X17 boson

Vlasios Petousis[1,*], *Martin* Veselský[1], and *Jozef* Leja[2]

[1]Czech Technical University in Prague - Institute of Experimental and Applied Physics
[2]Faculty of Mechanical Engineering - Slovak University of Technology in Bratislava

Abstract. A reported 17 MeV boson, which has been proposed as an explanation to the ^{8}Be and ^{4}He anomaly, is investigated in the context of its possible influence to neutron stars structure. Implementing a m_X=17 MeV to the nuclear equation of state using different incompressibility values K_0=245 MeV and K_0=260 MeV and solving Tolman-Oppenheimer-Volkoff equations, we estimate an upper limit of $M_{TOV} \approx 2.4 M\odot$ for a non rotating neutron star with span in radius *R* between 11.5 km to 14 km. Moving away from pure - NN with admixture of 10% protons and simulating possible softening of equation of state due to hyperons, we see that our estimated limits fit quite well inside the newest reported studies, coming from neutron stars merger event, GW190814

1 Introduction

In 2016 an article of Krasznahorkay et al. appeared [1], where an anomaly in angular correlation of electron-positron decay of the 1^+ excited level of ^{8}Be nucleus at 18.15 MeV, specifically observed enhancement at folding angles close to 140 degrees, is interpreted as a signature of a decay via emission of neutral boson with mass around m_X = 17 MeV. Since initial article [1], similar effect was reported by the same group concerning a lower 1^+ excited state of ^{8}Be at 17.6 MeV [2] and more recently in the 0^- excited state of ^{4}He at 21.01 MeV [3], reporting an enhancement at the folding angles close to 115 degrees. Through a recent work [4], we came up with an idea to investigate the effect of hypothetical 17 MeV boson on nuclear matter and its influence on the structure of the compact astrophysical objects like neutron stars. The knowledge learned from the latest analysis on neutron stars, allows us to conclude that we know their mass within a good precision [5, 9], but information on their radii is not so well defined yet [10, 11]. Encouragingly, recent observations of NICER [12] neutron star radius, appeared with good accuracy and in future the mass-radius relation knowledge will be improved from the eXTP mission data [13].

In our investigation we examine the influence of equation of state assuming mediation of nuclear force by X17 boson on neutron star structure in case of pure neutron matter using two different incompressibility values K_0=245 MeV and K_0=260 MeV. Solving Tolman-Oppenheimer-Volkoff (TOV) equations for that equation of state we set up a limit in mass-radius relation for neutron stars. In our studies, we did not consider the case of a rotational neutron star which according to recent investigations done by others [14], indicate possible overestimation of radii in non-rotating case by 3 - 3.5 km. First, we worked with pure neutron matter but also investigated possible admixture of protons. We investigated also possible softening of equation of state, an expected effect of hyperons, in order to attribute to the hyperons puzzle. In particular, our calculation indicates an upper limit of $M_{TOV} \approx 2.4 M\odot$ neutron star with span in radius *R* between 11.5 km to 14 km. Concerning hyperons simulation, our findings shown a reduction in mass (down to $M_{TOV} \approx 2M\odot$), as other relevant researches also indicate [15]. All our analysis has been performed for the aforementioned incompressibility values.

Our investigation indicates that equations of state used in present case modify significantly the internal structure of neutron stars. These resulting properties such as mass-radius relation fit quite well inside the newest reported studies as presented in papers [16, 17], including also tidal deformability in neutron star radius coming from the observed neutron stars merger events. The paper is organized a s follows: in section 2 we introduce the equations of state for the assumption that nuclear force is being mediated by a 17 MeV boson and model of non-rotating neutron star using TOV equation, in section 3 we present our findings after the analysis on the mass-radius and the density pressure correlations and in section 4 we discuss the results.

*e-mail: vlasios.petousis@cern.ch

2 Tolman-Oppenheimer-Volkoff equations and the equation of state

Going from the equations of states - ϵ and *P* (the energy density and pressure at radius *r*) to a predicted mass - radius relation *R(M)*, requires integration of stellar structure equations. In general relativity, these are known as the TOV (Tolman-Oppenheimer-Volkoff) equations:

$$\frac{dP}{dr} = \frac{-G}{c^2} \frac{(P + \epsilon)(m + \frac{4\pi r^3 P}{c^2})}{r(r - \frac{2Gm}{c^2})} \tag{1}$$

$$\frac{dm}{dr} = 4\pi r^2 \frac{\epsilon}{c^2} \quad (2)$$

where $m(r)$ is the total energy contained within radius r and pressure P. The energy density - ϵ and the pressure P of pure neutron matter can be calculated using the formulae [18]:

$$\epsilon = \frac{g_v^2}{2m_v^2}\rho_N^2 + \frac{g_s^2}{2m_s^2}(m_N - m_N^*)^2 + \frac{\kappa}{6g_s^3}(m_N - m_N^*)^3 + \frac{\lambda}{24g_s^4}(m_N - m_N^*)^4 + \frac{\gamma}{(2\pi)^3}\int_0^{k_F} d^3k\,\sqrt{k^2 + (m_N^*)^2} \quad (3)$$

$$P = \frac{g_v^2}{2m_v^2}\rho_N^2 + \frac{g_s^2}{2m_s^2}(m_N - m_N^*)^2 + \frac{\kappa}{6g_s^3}(m_N - m_N^*)^3 + \frac{\lambda}{24g_s^4}(m_N - m_N^*)^4 + \frac{1}{3}\frac{\gamma}{(2\pi)^3}\int_0^{k_F} d^3k \frac{k^2}{\sqrt{k^2 + (m_N^*)^2}} \quad (4)$$

where g_s,m_s are coupling and rest mass of scalar boson, g_v,m_v are coupling and rest mass of vector boson, κ,λ are couplings of cubic and quartic self-interaction of scalar boson, m_N,m_N^* are rest mass of nucleon and its effective mass, ρ_N is the nucleonic density, k_F is Fermi momentum of nucleons at zero temperature and γ is the degeneracy (γ=2 for neutron matter). The equations of state (Eq.3 and Eq.4) used here for TOV calculations are based on assumption that nuclear force is being mediated by a 17 MeV boson possibly observed in the study of anomalous electron-positron pair production in excited states of ^{8}Be [1, 2] and ^{4}He [3].

In the simplest variant, without scalar boson self-interaction, the energy density is sensitive only on ratios g_s/m_s and g_v/m_v. The corresponding couplings g_s and g_v are constrained using experimental binding energy values of symmetric nuclear matter (E/A=-16 MeV) and nuclear saturation density (ρ_0=0.16fm^{-3}), which are deduced from the masses and radii of finite nuclei. The final constraints are [18]:

$$(\frac{m_N}{m_s})^2 g_s^2 = 357.4$$
$$(\frac{m_N}{m_v})^2 g_v^2 = 273.8 \quad (5)$$

which guarantee proper values of both properties of nuclear matter. However, such simple parameterization usually leads to unrealistically high values of incompressibility.

In order to obtain more realistic values of incompressibility, it is necessary to implement also cubic and quartic self-interaction of scalar meson. Using as a starting point the most promising values of κ from scan with cubic self-interaction term only between (-0.0032 and 0.0032 MeV), a range of λ values was examined and ultimately regions

Table 1: Constrained parameter sets for equations of state with incompressibilities K_0=245 MeV and K_0=260 MeV.

K_0	m_v	m_s	g_v	g_s	κ	λ
MeV	MeV	MeV			MeV	
245	17	25.58	0.2407	0.4666	0.0039	-0.001396
260	17	25.58	0.2417	0.4684	0.00374	-0.001204

of parameters were identified, allowing to obtain equations of state with values of incompressibility in range between K_0=240 and 260 MeV (see Table 1). Similar values of imcompressibilty were extracted recently [19] from analysis of binary neutron star merger event GW170817 [20]. The resulting parameter sets for equation of states with incompressibilities K_0=245 MeV and K_0=260 MeV, respectively, are shown in Table 1. The mass of corresponding scalar particle does not correspond to any observed particle and it can be considered as an artefact of a given model.

In our recent work [4] we further demonstrated that a hypothetical 17 MeV boson fits some theory relations of chiral symmetry and instanton theory when dynamical quark mass is replaced by values close to quarks current masses, and corresponding couplings are close to these in Table 1. Referring to our recent work, we investigated that compared to experimental pion mass around m_π=135 MeV the mass of pseudoscalar particle m_X=17 MeV means that, a ratio of dynamical quark mass around $m_{q,dyn}$=310 MeV to current quark mass of around $m_{q,curr}$=5 MeV fits the following relation almost perfectly:

$$\frac{m_X^2}{m_{q,curr}} \simeq \frac{m_\pi^2}{m_{q,dyn}} \quad (6)$$

This appears to signal that as explicit breaking of chiral symmetry is restored from dynamical mass scale down to current mass scale, and also the properties of pseudoscalar particle get closer to Goldstone boson. The relation (6) then suggests that same outcome is reached when considering sum of axial current of current quark and axial current of 17 MeV boson. The 17 MeV boson thus appears to restore the symmetry of QCD in vacuum, where current quarks are considered as a degree of freedom. Based on this, it is possible to derive an analogue of the Goldberger-Treiman relation [21]

$$g_{Xqq} = \frac{g_A m_{q,curr}}{f_\pi} \quad (7)$$

where g_{Xqq} is the coupling of X-boson to current quark, g_A is renormalization factor of axial current and f_π was adopted as proportionality factor also for axial current of X-boson. The value of coupling g_{Xqq}=0.07 is obtained, what suggests coupling with nucleon g_{XNN}=0.21 In this respect, it is apparent that also results using relativistic mean field theory of nuclear force, presented above, point practically in same direction, since the value of coupling corresponding to vector mesonic mass of 17 MeV is g_v=0.24.

This situation is depicted in the Fig.1, which compares two trends of mesonic versus quark masses. First trend com-

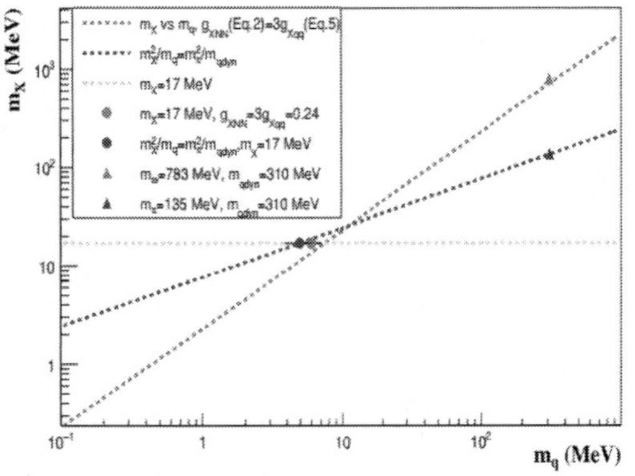

Figure 1: (Color online). Trends of mesonic versus quark masses. First trend combines formulas (2) and (5) as explained in the text. Second trend represents formula (4).

bines formulas (5) and (7) under assumption that the obtained coupling correspond to $g_{XNN} = 3g_{Xqq}$. Second trend represents formula (6). Both trends cross around the value of m_X=17 MeV and m_q=5 MeV.

Possible observation of a boson with mass m_X=17 MeV and relations (6) and (7) thus suggests the existence of a new scale in QCD, with chiral symmetry practically restored. The reason for apparent restoration of chiral symmetry suggested above is a matter of discussion. It might be related to interaction between nucleons at distance, mediated by instantons, where possibly the density of instantons mediating interaction in a region between nucleons becomes smaller and possibly a bounce into a false instanton vacuum (a quantum mechanical tunneling into a metastable vacuum and back), a mechanism based on instanton model [22], can occur.

3 Analysis results

For Tolman-Oppenheimer-Volkoff calculations [23] the equations of state $P(\rho)$ needed to be expressed in form of polytropes. Three transition densities are defined, ρ_1=2.8×10^{14} g/cm^3, ρ_2=10$^{14.7}$ g/cm^3, and ρ_3=10^{15} g/cm^3, and four parameters are calculated, three exponents of the power law polytropes $\Gamma_1, \Gamma_2, \Gamma_3$, respectively, and a value $a_0 = log(p(\rho_1)) + \Gamma_1(log(\rho_2) - log(\rho_1))$. The resulting values are shown in Table 2. The pressure as function of nuclear density for all model parameters shown in Fig.2.

The polytropes named as PNM1 and PNM2 (Fig.3) correspond to equations of state shown in Table 2, with incompressibilities K$_0$=245 MeV and K$_0$=260 MeV respectively. These equations of state represent properties of pure neutron matter. The maximum mass of neutron star reaches up to 2.4 of Solar masses, what is close to values 2.5 - 2.67 derived for the mass of secondary in the observed astrophysical event GW190814 [24]. However, the

Table 2: Parameters of polytropes used as equations of state for the Tolman-Oppenheimer-Volkoff calculations.

Models	a_0	Γ_1	Γ_2	Γ_3	K$_0$(MeV)
PNM1	34.600	3.350	3.170	2.497	245
H1 (Hyper)	34.600	3.350	2.570	1.497	245
p1 (10%)	34.553	3.470	3.307	2.546	245
PNM2	34.642	3.348	3.040	2.534	260
H2 (Hyper)	34.642	3.348	2.440	1.534	260
p2 (10%)	34.593	3.530	3.177	2.500	260

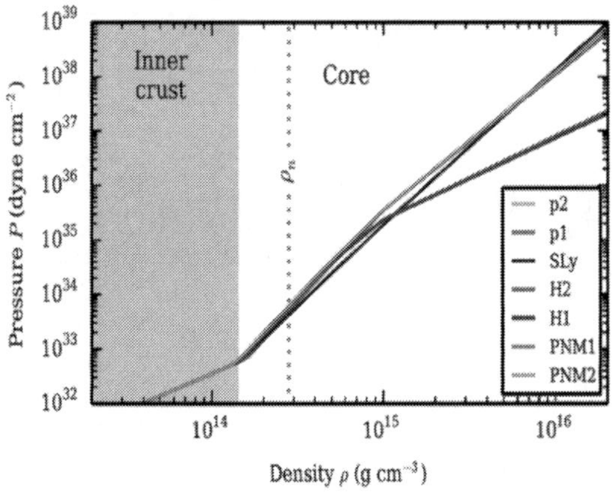

Figure 2: (Color online). The pressure as function of nuclear density for all the model parameters as they defined in the Table 2, based on assumption that nuclear force is being mediated by a 17 MeV boson.

properties of neutron stars can be influenced by admixture of protons. During formation of neutron stars most of the protons are transformed into neutrons by inverse β-process, contributing significantly to cooling of neutron star. Still, some amount of protons and electrons may remain present. The concentration of protons may vary in different layers of neutron star, depending on the local β - equilibrium, most probably it can be expected significant at outer regions with lower densities, which apparently determine the radius of neutron star.

We investigate sensitivity to proton admixture by assuming a fixed proton concentration of 10% in the whole volume of neutron star. Coulomb interaction is assumed to be counterbalanced by equal concentration of electrons. The resulting polytropes are named as p1 and p2 (Fig.3) and the admixture of protons appears to modify the radius of neutron star moderately while the maximum mass remains constant. The admixture of protons does not contradict possible observation of neutron star with mass of 2.6 times the Solar mass [24]. The radius of neutron star is reduced by a 10 % admixture of protons by approximately 3 %, which is relatively moderate change. Thus the properties of neutron star appear only moderately sensitive to the local β - equilibrium.

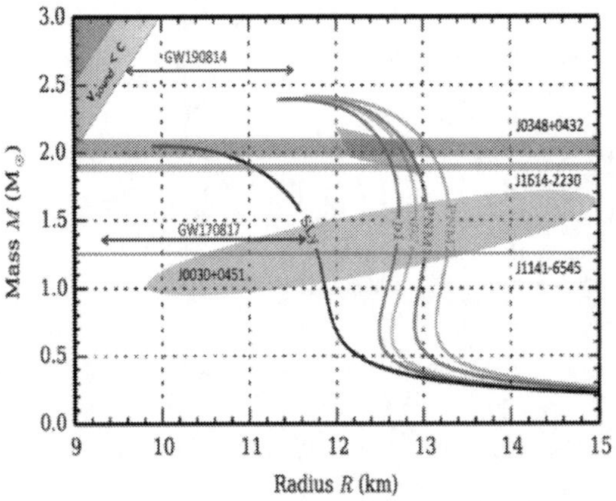

Figure 3: (Color online). The mass-radius relation for pure neutron matter (PNM1 and PNM2) and with an admixture of 10% protons (p1 and p2) for K_0=245 and 260 MeV respectively. The calculations based on assumption that nuclear force is being mediated by a 17 MeV boson.

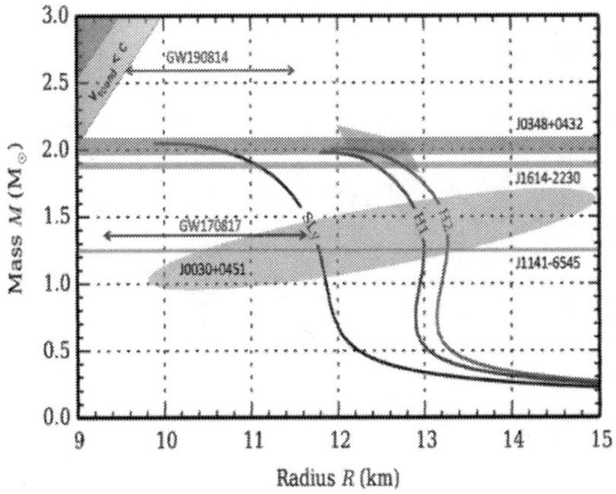

Figure 4: (Color online). The mass-radius relation for neutron matter including hyperons (H1 and H2) for K_0=245 and 260 MeV respectively. The green area shows the uncertainty on the mass and radius of stars with hyperons in the core due to the fact that the properties of these particles are barely constrained by laboratory measurements. The calculations based on assumption that nuclear force is being mediated by a 17 MeV boson.

Since it can be expected that at densities above two times saturation density, the non-nucleonic particles such as hyperons can play increasing role and soften the equation of state at these densities, such effect was simulated by polytropes under acronyms H1 and H2. The equation of state is softened significantly at high densities while the properties close to saturation density remain unchanged. Even such rough estimate may indicate the realistic behavior. The simulated effect of hyperons, represented by softening equation of state above two times the saturation density, translates mainly into decrease of maximum mass of neutron star, while the radius remains practically constant up to neutron star mass of 1.5 of Solar mass. A maximum mass around two times the Solar mass corresponds to maximum observed pulsar mass which was valid until very recently and thus the parameterizations H1 and H2 (Fig.4), represent maximum extent of such effect and no further softening of equation of state would be plausible. On the other hand, recent possible observation of neutron star with mass of 2.6 times the Solar mass [24] would apparently exclude any significant effect of hyperons at all.

4 Conclusions

In summary, the properties of a non rotating neutron stars are simulated using the equation of state, based on assumption that nuclear force is being mediated by a 17 MeV boson possibly observed in the study of anomalous electron-positron pair production in excited states of ^{8}Be [1, 2] and ^{4}He [3]. The presented analysis was performed using pure neutron matter and also an admixture of protons. We investigated also hyperons simulation in order to address the hyperons puzzle. Our investigation indicates an upper limit of $M_{TOV} \approx 2.4M\odot$ neutron stars with span in radius R between 11.5 km to 14 km for normal neutron matter and moderate reduction of radii for the admixture with 10% protons. Concerning the possible effect of hyperons, our findings shown reduction in maximum mass down to $M_{TOV} \approx 2M\odot$. We conclude that our results appear in agreement with observed properties of neutron stars. In particular, the possible existence of neutron star with mass up to 2.6 times the Solar mass, reported recently [24], appears plausible.

5 Ackwnoledgments

This work is supported by the European Regional Development Fund-Project Engineering applications of microworld physics, (Contract No. CZ.02.1.01/0.0/0.0/16_019/0000766).

References

[1] Krasznahorkay, A. J., Csatlós, M., Csige, L., Gácsi, Z., Gulyás, J., Hunyadi, M., Kuti, I., Nyakó, B. M., Stuhl, L., Timár, J., Tornyi, T. G., Vajta, Zs., Ketel, T. J., Krasznahorkay, A., Physical Review Letters 116, 042501 (2016).

[2] Krasznahorkay, A. J., Csatlós, M., Csige, L., Gulyás, J., Hunyadi, M., Ketel, T.J., Krasznahorkay, A., Kuti, I., Nagy, Á., Nyakó, B. M., Sas, N., Timár, J., & Vajda, I., EPJ Web of Conferences 137, 08010 (2017). doi: 10.1051/epjconf/201713708010

[3] Krasznahorkay, A. J., Csatlós, M., Csige, L., Gulyás, J., Koszta, M., Szihalmi, B., Timár, J., Firak, D. S., Nagy, Á., Sas, N. J., Krasznahorkay, A., (2019). https://arxiv.org/abs/1910.10459

[4] Veselský, M., Petousis, V., Leja, J., (2020). doi: 10.1088/1361-6471/ac09db, https://arxiv.org/abs/2004.09758

[5] Lattimer, J. M., Ann. Rev. Nucl. Sci. 62 (2012).

[6] Demorest, P. B., Pennucci, T., Ransom, S. M., Roberts, M.S., Hessels, J. W., Nature, 467 (2010).

[7] Fonseca, E., et al., Astrophys. J. 832, 167 (2016). doi: 10.3847/0004-637X/832/2/167

[8] Antoniadis, J., et al., Science 340, 6131, 1233232 (2013). doi: 10.1126/science.1233232

[9] Cromartie, H. Y., Fonseca, E., Ransom, S. M., Demorest, S. M., et al., Nature Astronomy, (2019). doi: 10.1038/s41550-019-0880-2

[10] Ozel, G., Freire, P., Ann. Rev. Astron. Astrophys, 54, 401-440 (2016).

[11] Guillot, S., Servillat, M., Webb, N. A., Rutledge, R. E., Astrophys. J. 772, 7 (2013).

[12] Riley, T. E., Watts, A. L., Bogdanov, S., Ray, P. S., Ludlam, R. M., Guillot, S., Arzoumanian, Z., Baker, C. L., Bilous, A. V., Chakrabarty, D., Gendreau, K. C., Harding, A. K., Ho, W. C. G., Lattimer, J. M., Morsink, S. M., & Strohmayer, T. E., Astrophys. J. Lett. 887, L21 (2019). doi: 10.3847/2041-8213/ab481c

[13] Watts, A. L., et al., Science China Physics, Mechanics and Astronomy 62, 29503 (2019). doi: 10.1007/s11433-017-9188-4

[14] Suleimanov, V. F., Poutanen, J., Werner, K., (2020). https://arxiv.org/abs/2005.09759

[15] Fortin, M., Oertel, M., & Providencia, C., New Compstar issue of PASA, 35, E044 (2018).

[16] Fattoyev, F. J., Piekarewicz, J., & Horowitz, C. J., Phys. Rev. Lett. 120, 172702 (2018).

[17] Fattoyev, F. J., Horowitz, C. J., Piekarewicz, J., & Reed, B., (2020). https://arxiv.org/abs/2007.03799

[18] Serot, B. D., Walecka, J. D., Int. Jou. Mod. Phys. E 6, 515 (1997).

[19] Perego, A., Bernuzzi, S., Radice, D., Eur. Phys. J. A, 55, 124 (2019).

[20] Abbott, R., et al., Phys. Rev. Lett., 119, 161101, (2017). doi: 10.1103/PhysRevLett.119.161101

[21] M. L. Goldberger, and S. B. Treiman, Phys. Rev. 110, 1178 (1958)

[22] Sidney Coleman, Phys. Rev. D 15, (1977) ; Erratum Phys. Rev. D 16, (1977), doi: 10.1103/PhysRevD.15.2929

[23] Nättilä, J., Columbia University, Flatiron Institute, https://github.com/natj/tov

[24] Abbott, R., et al., Astrophys J. Lett, 896,. L44 (2020). doi: 10.3847/2041-8213/ab960f

EPJ Web of Conferences **252**, 05001 (2021) https://doi.org/10.1051/epjconf/202125205001
HINPw6

Newtonian dynamics of imaginary time-dependent mean field theory

Aldo Bonasera

Cyclotron Institute, Texas A&M University, College Station, TX 77843-USA.

Laboratori Nazionali del Sud-INFN, v. Santa Sofia 64, 95123 Catania, Italy.

Abstract. A Time Dependent Hartree-Fock (TDHF) based classical model is applied to sub-barrier fusion reactions using the Feynman Path Integral Method (FPIM). The fusion cross-sections and modified astrophysical S*-factors are calculated for the $^{12}C+^{12}C$ reactions and compared to direct and indirect experimental results. Different channels cross-sections are estimated from the statistical decay of the compound nucleus. A good agreement with the direct data is found. We suggest a complementary observable given by the (imaginary) action A easily derived from theory and experiments. When properly normalized by the action in the Gamow limit it has an upper value of 1 at zero beam energies. It becomes negative at the Coulomb barrier which is V_{cb}=5.05±0.05MeV from direct data and V_{cb}=5.5MeV from model calculations.

Careful measurements of the fusion cross-sections for $^{12}C+^{12}C$ reactions are crucial to our understanding of massive stars and super-bursts from accreting neutron stars [1-4]. These measurements are especially compelling at energies close or below the Gamow peak [4], i.e., below 2 MeV. Direct data measurements [5-10] start from E_{cm}=2.1 MeV center of mass energy showing many resonances especially when expressed in terms of the modified astrophysical S* factor [13]:

$$S^*(E_{cm})=E_{cm}\sigma(E_{cm})\times\exp(87.12E_{cm}^{-1/2}+0.46E_{cm})= S(E_{cm})\times\exp(0.46E_{cm}).$$

To overcome the difficulties of vanishing fusion cross-section at very low energies, indirect methods have been proposed [14,15]. Recently, indirect measurements based on the THM have been published [11] displaying a wealth of resonances in S* at center of mass energies as low as 0.8MeV. A critical reanalysis of the same experiment to include Coulomb effects in the 3-body final channel and the Distorted Wave Born Approximation (DWBA) instead of the Plane Wave Approximation (PWA) resulted in a much lower S* up to three orders of magnitude. These contrasting results reveal a strong model dependence of the THM approach and cast doubts about its general validity. Macroscopic and microscopic models [16-19] offer a reasonable reproduction of the direct data and they are sometime considered as an upper limit of the fusion cross-section [20,21]. Predictions of these models cannot reproduce the low energy THM results of ref.[11] and are way above the Coulomb renormalization analysis of [12].

In ref. [22], we proposed a microscopic model based on the Vlasov equation and the FPIM reproducing rather well the available direct data at that time for the $^{12}C+^{12}C$ system. The approach is quite successful when applied to heavier nuclei [23] and spontaneous fission [24]. Before getting involved into complex numerical calculations, which may or may not be feasible at such low energies, it is instructive to simplify the heavy ion dynamics using a robust macroscopic model. In ref. [25], we proposed a TDHF based model, dubbed Neck Model (NM), and successfully reproduced above the barrier fusion cross section as well as deep inelastic [26] and fission [27] dynamics. In this work we generalize the NM to sub barrier energies in the FPIM framework. Newton force equations are solved assuming the collective variables given by the center of mass distance **R** of the two colliding nuclei and their relative momenta **P**. The forces acting on the nuclei are given by the Coulomb and the nuclear Bass potentials before the two nuclei touch [28]. After touching, the two nuclei are described as sections of spheres joined by a cylinder of radius r_N, the neck radius. In the rebounding phase, the nuclear geometry is given by two half spheres joined by sections of cones of radius r_N. Volume conservation is enforced which gives suitable relations between the neck radius and the relative distance of the two nuclei and it is in good agreement to TDHF calculations [25]. In such a configuration the force is given by the surface tension times the perimeter of the neck, as for a liquid drop. During this stage nucleons are transferred through the neck, giving rise to one body-dissipation described by the Randrup-window formula [29]. To describe the sub-barrier fusion dynamics, we go into imaginary times (FPIM) at the first (external) turning point [22-24] when the collective momentum P=0fm^{-1}. In imaginary times, the second derivative of R respect to (imaginary) time gets a i^2=-1 contribution and the Coulomb force becomes attractive while the nuclear part repulsive. This produces an acceleration of the two nuclei towards each other until they come to a halt (because of the short-range nuclear repulsion in imaginary times) at the second (internal) turning point. At this stage we switch from imaginary to real times again. If the nuclear force is strong enough the two nuclei fuse and this happens always for light nuclei like ^{12}C. The probability of fusion for the lth-partial wave is given by [22]:

$$T_l = (1 + \exp\{2A\})^{-1} \quad (1).$$

The action (in units of ħ) is given by

$$A = \int_1^2 PdR.$$

The cross section is given by:

$$\sigma(E_{cm}) = \frac{\pi\hbar^2}{2\mu E_{cm}} \sum_{l=0}^{\infty}(2l+1)T_l P_l \quad (2).$$

P_l gives the probability of decay of the compound nucleus into different channels, is the nuclear reduced mass. Following [22], the calculations will be performed at zero impact parameter only. In order to take into account the l-dependence of the transmission probability we shift the beam energy for each l as [30]:

$$T_l \approx T_0(E_{cm} - \frac{l(l+1)\hbar^2}{2\mu < R^2 >})$$

$\mu<R^2>$ is an effective moment of inertia of two touching spheres at the internal turning point $<R>$ and it is slowly varying with energy. Recall that for the ^{12}C+^{12}C system, because of angular momentum and parity conservation laws, only even l-values are allowed.

In figure 1, we plot, for different l-values as indicated, the collective momentum versus collective distance of the two nuclei at E_{cm}=2.5 MeV. Starting from infinite distance, the two nuclei approach each other under the influence of the Coulomb repulsion until they reach the first turning point (P=0 fm^{-1}). Clearly the first turning point is farther for higher l-values or lower beam energies. At this stage, the system enters imaginary time propagation, and the Coulomb force becomes attractive resulting in a strong acceleration. At shorter distances the (repulsive in imaginary times) nuclear force given by the Bass potential violently decelerates the two nuclei until P=0 fm^{-1} again, the second turning point. For light nuclei the imaginary dynamics is completely determined by the Coulomb and Bass potential. When returning to real times, the two nuclei accelerate again towards each other, a neck is formed, and fusion occurs. The area inside the first and second turning points gives the action A. Notice that A increases quickly with increasing l, thus the probability of fusion decreases, eq.(1), and the cross section is dominated by low l values, eq.(2). A striking difference with the microscopic Vlasov approach, see figure 2 in ref.[22], is the oscillations seen in imaginary times. These are due to collective excitation of the single particle degrees of freedom included in the microscopic model but not in the NM. As a result, oscillations are seen in the fusion cross section of figure 3 in ref.[22] and we will discuss them further in the fig.3 below.

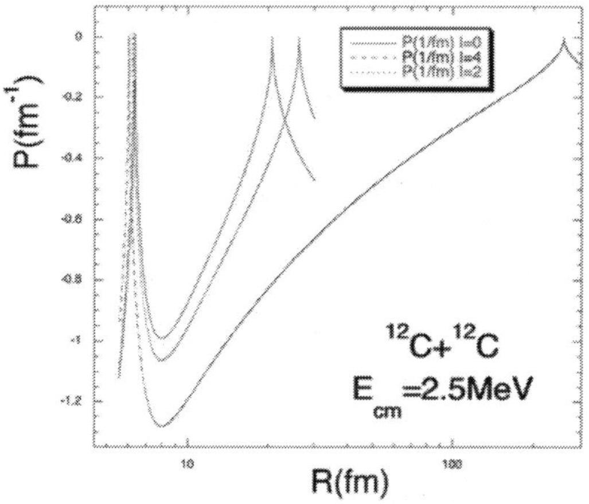

Figure 1. Collective momentum vs distance of the two nuclei for the values of E_{cm} and l indicated. The areas inside the first and second turning points give the action A.

After fusing, the excitation energy of the compound nucleus (CN) can be determined from the E_{cm}, l and the Q-values for the CN and the open decay channels. In our case the most important channels are: ^{12}C(^{12}C,γ)^{24}Mg, Q=13.9 MeV; ^{12}C(^{12}C,α) Ne Q=4.62 MeV; ^{12}C(^{12}C,p)^{23}Na, Q=2.24 MeV; ^{12}C(^{12}C,n)^{23}Mg, Q=-2.6 MeV; ^{12}C(^{12}C,^{8}Be)^{16}O, Q=-0.2MeV.

We can calculate the decay probability in certain channels within the framework of the statistical model [32]:

$$\Pi_i = g_i(T_{k_i}^2 + 2m_i T_{k_i}) \frac{\rho_i(E_{CN}^* - T_{k_i})}{\rho(E_{CN}^*)} \sigma_i^{abs} \quad (3).$$

The probability entering eq. (2) can be expressed as $P_l = \Pi_i / \sum \Pi_i$ for each l-value and excitation energy

$$E_{CN}^* = E_{cm} - \frac{l(l+1)\hbar^2}{2\mu<R^2>} + Q.$$

$$g_i = 2s_i + 1$$

and s_i is the spin of the emitted particles. The absorption cross section for the inverse process is given by: $\sigma_i^{abs} = \pi R_N^2(1 - V_C(R_N)/T_{k_i})$, where R_N is the sum of the radii of the nuclei in the exit channel, $V_C(R_N)$ the Coulomb barrier and T_{ki} the kinetic energy of the emitted particle. For gamma absorption:

$$\sigma_\gamma^{abs} = 75 A_{CN} \frac{(T_{k\gamma} - 2.226)^{\frac{3}{2}}}{T_{k\gamma}^3} (mb) \quad [32],$$

A_{CN} the mass number of the CN. The level densities contribution can be written as:

$$\frac{\rho_i(E_{CN}^* - T_{k_i})}{\rho(E_{CN}^*)} = \exp\left(-\frac{T_{k_i}}{T}\right);$$

the temperature is given by

$$T = \sqrt{E_{CN}^*/a} \text{ and } a = A_{CN}/8.0 \ (\text{MeV}^{-1}).$$

https://doi.org/10.1051/epjconf/202125205001

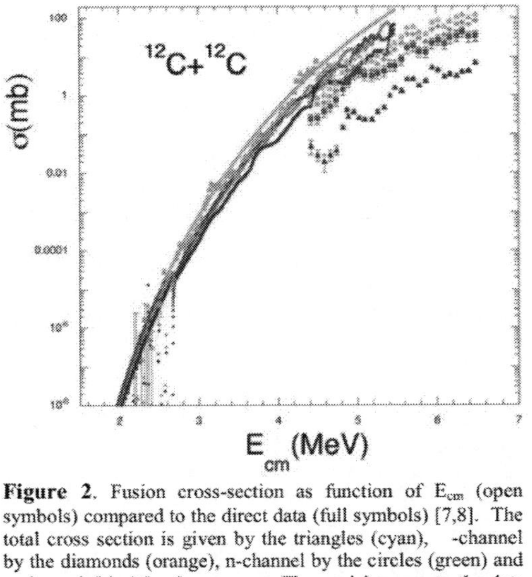

Figure 2. Fusion cross-section as function of E_{cm} (open symbols) compared to the direct data (full symbols) [7,8]. The total cross section is given by the triangles (cyan), -channel by the diamonds (orange), n-channel by the circles (green) and p-channel (blue) by the squares. The model agrees to the data from ref.[8] and it overestimate ref.[7]. Notice the faster increase of the n-channel (negative Q-value) compared to the p-channel. The model reproduces the trend but overestimates the data of ref.[7].

In figure 2, we compare the model results to the direct data [7,8] for the total cross section and the most important decay channels (α,p and n). The overall agreement is quite good especially at the highest energies. An over (under) estimate of the n (p)-channel is noticeable. At lower excitation energies the individual levels of the CN and the exit channels nuclei become important for each l-values, these levels are not included in the present calculations, and we expect them to produce oscillations as in the data if properly considered. Notice that any statistical model refinement will affect the P_l values entering eq. (1) but such a quantity is below one, thus we conclude that the calculated cross-sections represent an upper limit [20,21]. Our results are in good agreement with other model calculations [16-20].

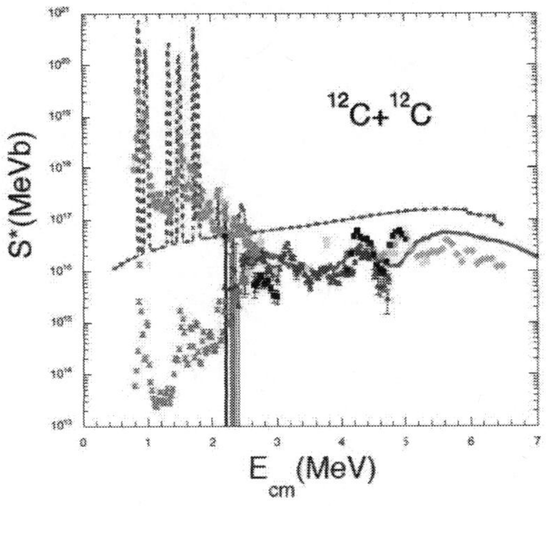

Figure 3. Modified Astrophysical S*-factor as function of E_{cm}. Direct data are from [7,8,33,34]. The THM results are given by the full (orange) circles [11] and the Coulomb renormalized one by the full (red) triangles [12]. The neck-model calculations, including ^{24}Mg resonances, are given by the dashed (green) line [35]. The Vlasov results in imaginary times are given by the full (red) line [22].

To avoid the quickly decreasing fusion cross-section at sub-barrier energies, it is customary to display the modified astrophysical factor S* as function of E_{cm}, see figure 3. The model calculations give S*=1.3 10^{16} MeV b at the lowest energy, slowly increasing over the entire energy region. Direct data [7,8] are overestimated by the calculation at higher energies and display definite 'bumps'. The indirect data are normalized to the direct data at 2.6 MeV [11]. At much lower energies, the THM diverges from the calculation up to 3 orders of magnitude higher, while the Coulomb renormalized THM analysis is up to 3 orders below the calculation. These discrepancies need to be experimentally investigated maybe within the same THM approach but for a system, which does not present the same Coulomb problem for one of the outgoing particles. A suitable reaction could be $^{13}C(^{12}C,n)$…where the neutron may act as a spectator [12]. In ref. [35] we have extended the present model to include effects of known resonances in ^{26}Mg. We found a reasonable agreement to the THM data [11]. In ref. [36], we have added e^+e^- corrections to the Coulomb field resulting in a 15% decrease in S*, closer to the direct data. The Vlasov calculations [22] are in rather good agreement to the direct data and display oscillations or 'bumps' similar to the data. A repetition of those calculations extending to lower energies and different interactions would be very interesting.

Calculated and measured values of S or S* vary widely from system to system and as a function of E_{cm} [31]. The original purpose in defining these observables was to avoid the large variation of the fusion cross-section by factoring out the Coulomb barrier penetration probability [4] thus leaving nuclear effects embedded into S or S*. Whether this goal has been reached can be judged from figure 3 and the literature [1-13,16-21,31]. In this work we would like to propose a complementary observable which can be derived from eq. (2). We have shown in figure 1 that at the lowest energies the l=0 value is dominant, thus neglecting higher l-values and solving for T_0, the action becomes:

$$A = \frac{1}{2} ln\left[\frac{\pi\hbar^2}{2E_{cm}\sigma(E_{cm})} - 1\right] \quad (4).$$

Of course, if data could be provided for a particular l-value, then eq.(4) could be modified to correct for it, see eq.(2). Notice that when dealing with the total fusion cross-section P_l=1. The value of the action has been derived analytically for the case of the Coulomb barrier only (Gamow limit) [31]

$$A_G = e^2\pi Z_T Z_P\sqrt{\frac{\mu}{2E_{CM}}}.$$

Thus, we expect $A \rightarrow A_G$ when $E_{CM} \rightarrow 0$.

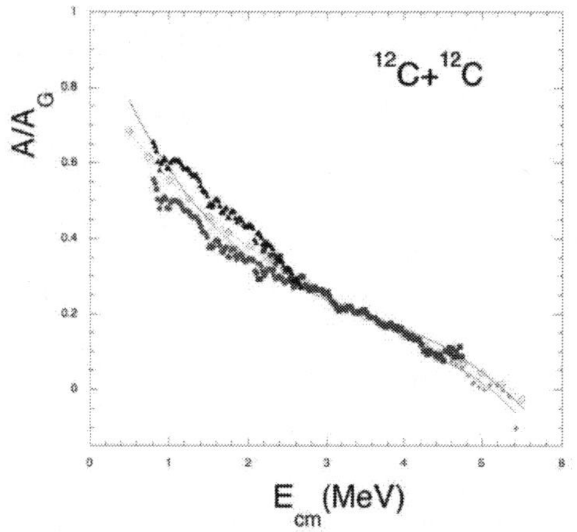

Figure 4. Normalized sub barrier action, symbols as in figure 3. The dashed line is the fit to the model using eq.(5), the full line is the fit to the direct data of ref.[7] and the dashed line to ref.[8].

In figure 4, we plot the Gamow normalized action vs E_{cm} for the model not including resonances [35] (open symbols) and the data (full symbols). We notice a reasonable agreement to the direct data, while the THM is below [11] or above the calculations for the Coulomb renormalized case [12], i.e., similar to the figure 3. Notice that the action A has an opposite behavior to S and S*, i.e., the higher is A the smaller is the cross section. Because of the clear physical interpretation of A/A_G we can parameterize its behavior as:

$$A/_{A_G} = a + be^{-cE_{cm}} \quad (5).$$

The parameters a,b,c is fitted to data or theory and the results can be seen in figure 3. The action becomes zero at the effective Coulomb barrier V_{cb}=5.05±0.05 MeV from the direct data [7,8] to compare to the model V_{cb}=5.5 MeV. The latter is the highest energy we run the simulation and its value critically depends on the Bass or other nuclear potentials. The fits to the indirect data give V_{cb}=4.3 MeV for the Coulomb renormalization case [12] and no solution for the original data [11]. Of course, this last result depends on the THM normalization [11] to the direct data [8]. There seem to be some discrepancy between the direct data of ref. [7] and [8] with the former better reproduced by the model.

In conclusion, in this paper we have discussed a TDHF inspired macroscopic model and extended it to sub-barrier energies using the Feynman Path Integral Method i.e., solving Newton dynamics below the barrier in imaginary times. The model has no free parameters and reasonably reproduces the direct data. The THM indirect data are well above or below the model depending on if the Coulomb corrections are applied [12] or not [11]. Inclusion of resonances in the model [35] supports the results of ref. [11]. Coulomb field modifications due to e+e- corrections gives results in better agreement to the direct data [36]. We have introduced a new observable, the action A, that when properly normalized by the action in the Gamow limit gives a quantity well defined and close to unity and we expect all the other systems to fall in the same region.

ACKNOWLEDGEMENTS
We thank Profs. J. Natowitz, A. Tumino and A. Zhanov for discussions and providing the THM data. In particular, we thank prof. A.Tumino for help to recover the results from the imaginary time Vlasov calculations [22] displayed in fig.3. This work was supported in part by the United States Department of Energy under Grant # DE-FG03-93ER40773 and NNSA Grant No. DENA0003841 (CENTAUR).

REFERENCES
[1] Woosley, S. E., Heger, A. and Weaver, T. A. *Rev. Mod. Phys.* **74**,1015–1071 (2002).
[2] Keek L.et al., Atron.Astrophys.,**479**,177(2008).
[3] Cooper, R. L. et al., *Astrophys.J.* **702**, 660–671 (2009).
[4] Iliadis, C. *Nuclear Physics of Stars*. (Wiley, Weinheim, 2007).
[5] H.W. Becker et al., Z. Phys. A 303, 305 (1981).
[6] K.U. Kettner, H. Lorenz-Wirzba, C. Rolfs, Z. Phys. A 298, 65(1980).
[7] E.F. Aguilera et al., Phys. Rev. C 73, 064601 (2006).
[8] T. Spillane et al., Phys. Rev. Lett. 98, 122501 (2007).
[9] J.R. Patterson, H. Winkler, C.S. Zaidins, Astrophys. J. 157, 367(1969).
[10] C.L. Jiang et al., Phys. Rev. C 97, 012801(R) (2018).
[11] A. Tumino et al., Nature 557, 687 (2018).
[12] C.Beck et al., Eur. Phys. J. A**56,** 87 (2020).
[13] Patterson, J. R., Winkler, H. & Zaidins, C. S., Astrophys. J. 157, 367–373(1969).
[14] Tribble, R. et al.,Rep. Prog.Phys. 77, 106901–106950 (2014).
[15] G. Baur, Phys. Lett. B 178, 135 (1986).
[16] H.Esbensen, X.Tang, C.L.Jiang, Phys.Rev.C **84**, 064613(2011).
[17] A. Diaz Torres, M. Wiescher, Phys.Rev.C **97**,055802(2018)12.
[18] L.R. Gasques et al., Phys.Rev.C**76**,035802(2007).
[19] Le Hoang Chien et al., Phys. Rev. C **98**, 064604 (2018).
[18] Dao T. Khoa et al., Nucl.Sci.Tech. **29**,182(2018).
[19] K. Godbey, C. Simenel, A.S. Umar, Phys. Rev. C **100**, 024619 (2019).
[20] M. Notani et al., Phys.Rev.C**85**,014607(2012).
[21] C.L. Jiang et al., Phys. Rev. Lett. **110**, 072701 (2013).
[22] A. Bonasera and V.Kondratyev, Phys. Lett. B **339**, 207(1994).
[23] V.N. Kondratyev, A. Bonasera and A.Iwamoto, PRC **61**, 044613(2000).
[24] A. Bonasera and A. Iwamoto, Phys.Rev.Lett. **78**, 187(1997).
[25] A. Bonasera, G. F. Bertsch, and E. El-Sayed, Phys. Lett. **141B**, 9 (1984).
[26] A. Bonasera, Nucl. Phys. **A439**, 353(1985).
[27] A. Bonasera, Phys. Rev. C **34**, 740(1986).
[28] R. Bass, Phys.Rev.Lett.**39**, 265(1977).

 https://doi.org/10.1051/epjconf/202125205001

[29] J. Randrup, Ann.Phys.(NY) **112**, 356(1978).
[30] A.B. Balantekin, S.E. Koonin and J.W. Negele, Phys.Rev. C **28**, 1656 (1983).
[31] S.Kimura and A.Bonasera, PRC **76**, 031602(R)(2007).
[32] J. Ahrens et al., Nucl.Phys. A **251**, 479(1975).
[33] G.Fruet et al., PRL124, 192701(2020).
[34] W.P. Tan et al., PRL124, 192702(2020).
[35] A.Bonasera and J.Natowitz, PRC **102**, 061602(**R**) (2020).
[36] T.Settlemyre et al., Nucl.Phys.A (2021) in press.

From nuclear astrophysics to fundamental nuclear physics: challenging experimental approaches at n_TOF (CERN)

Agatino Musumarra[1,2] for the n_TOF collaboration

[1]Dipartimento di Fisica e Astronomia, DFA University of Catania 95123 Catania, Italy
[2]Istituto Nazionale di Fisica Nucleare, INFN Sezione di Catania 95123 Catania, Italy

Abstract. The n_TOF installation at CERN is one of the leading neutron facilities worldwide undergoing a major update of the neutron spallation source. The update will provide improved n-TOF resolution in the experimental areas and the possibility to perform neutron cross section measurements at very high neutron flux (NEAR-Station). The renewed capabilities of the facility must be supported by smart and non-conventional experimental approaches. In this framework two examples will be reported. The first one concerns the measurement of a key reaction channel involved in Primordial Nucleosynthesis: the ^{7}Be(n,α), by using a radioactive ^{7}Be target. The second one provides a state-of-the-art scenario for the *n-n* scattering length measurement. This will be performed by neutron-deuteron (*n-d*) breakup three-body reaction. In this case, the envisaged experimental setup will provide a complete three-body kinematic reconstruction. By these important physics cases we are crossing the technological frontiers for charged particle and neutron detection.

1 The n_TOF facility

The n_TOF facility at CERN [1], operating since 2001, provides a unique opportunity for measuring neutron cross sections of interest to fundamental nuclear physics, nuclear astrophysics and neutron induced reactions for advanced nuclear technologies. The qualifying features of the facility are the wide energy range that can be covered in a single measurement and the large yield of neutrons produced by spallation of a single proton pulse provided by the CERN Proton Synchrotron (PS). More in detail, the new, high-flux experimental area (EAR2) allows us to measure neutron cross section relevant for nuclear astrophysics covering the region of interest for Big-Bang, stellar and s-process nucleosynthesis. The main advantage of the EAR2 beam line, in operation since August 2014, with respect to the first experimental area (EAR1), is the high neutron flux, of 10^8 neutrons/pulse (see fig. 1), delivered in a short time interval, which results in a convenient signal-to-background ratio when measuring small mass and/or highly radioactive targets, as well as isotopes characterized by a small reaction cross-section [2].

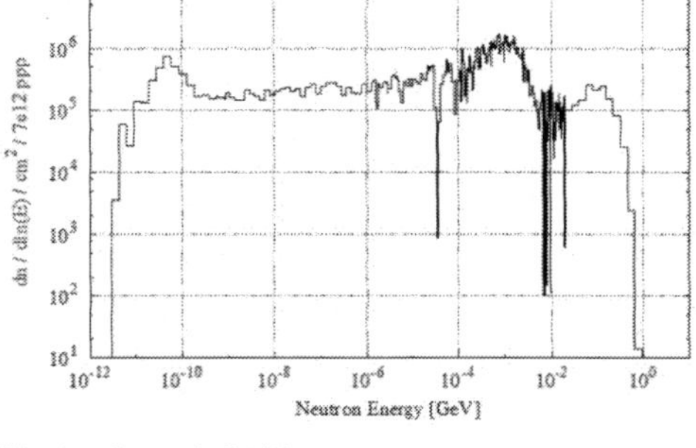

Fig. 1. *n*-fluence in EAR2.

Consequently, n_TOF is one of the few time-of-flight facilities worldwide where neutron cross sections on radioactive isotopes of short half-life can be measured. On the other hand, the facility has been complemented by a state-of-the-art detection systems and high-performance data acquisition systems. It follows that, apart for the neutron beam, another challenge in many measurements is the availability of a suitable amounts of radioactive isotopes to be used as target. For instance, the n_TOF collaboration involves a group working at Paul-Scherrer-Institute (PSI), Zurich, able to extract and separate relatively large amount of radioactive isotope for producing targets. This scientific synergy makes possible to measure on short lifetime isotopes (e.g., ^{7}Be) in a very effective way, as shown in the following chapters.

After the full exploitation of the second-generation n_TOF neutron spallation target, the facility underwent an important upgrade during the last CERN's Long Shutdown (restart in June 2021). The upgrade concerns the installation of a new spallation target. The first and second-generation lead targets were water-cooled, and they were designed for maximizing performances of the EAR1 beam line, while the new target is gas cooled, in order to avoid corrosion issues and contamination of the cooling system; the design has been also optimized for the vertical flight path of EAR2, providing an improved Time-of-Flight resolution at low beam energy.

In addition to the new spallation target, the installation of a new experimental area, close to the target and outside the shielding wall of the target-moderator, has been implemented (fig.2). The new experimental site, the NEAR station (LoI CERN-INTC-2020-073; INTC-I-2223), will take advantage of the extremely high neutron fluence expected at a position close to the spallation target. The dramatic increase in neutron fluence will make possible activation measurements on relatively small-mass samples and on radioactive isotopes. In this case the neutron beam will be transported to the NEAR Station through a small aperture in the shielding wall, traversing a suitable collimator/moderator system. As a matter of fact, the NEAR Station will operate in parallel with the two existing beam lines (EAR1 and EAR2), improving the

https://doi.org/10.1051/epjconf/202125205002

overall facility capabilities. Monte Carlo simulations show that, as a result of a correct combination of the moderator material with the filter geometry, a Maxwellian-like neutron spectrum can be produced (from a few keV to a few tens of keV). This low energy n-beam will allow measuring Maxwellian Average Cross Sections (MACS) by using activation techniques. It is worthwhile to mention that the possibility to experimentally determine MACS for short-lived unstable nuclei opens the possibility to investigate with high reliability the branching points in the s-process and weak r-process nucleosynthesis scenarios.

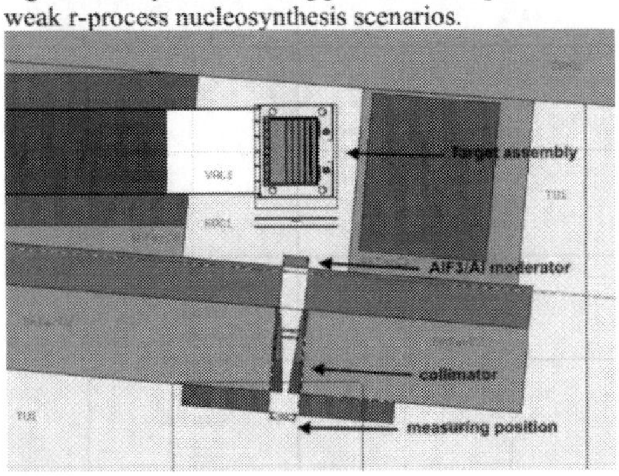

Fig. 2. n_TOF NEAR station, experimental site. From LoI CERN-INTC-2020-073.

2 The $^7Be(n,\alpha)$ reaction in the context of the Cosmological Lithium Problem

The distribution of the Big-Bang Nucleosynthesis (BBN) abundances results by a well-established cosmological model describing the production of elements in the early stage of the Universe and it represents a cornerstone of the Big-Bang model. In fact, BBN reproduces observed primordial abundances for all isotopes except ^{7}Li, for which it overestimates the relative abundance by approximately a factor three [3]. A possible explanation is related to the Nuclear Physics input of the nucleosynthesis network, describing the production and destruction rate of ^{7}Be, being primordial ^{7}Li mostly produced by the decay of this isotope. In particular, while the main production channel, the ^{3}He$(\alpha,\gamma)^7$Be, is relatively well known, the reaction cross sections responsible for its destruction remain uncertain. Several measurements were performed to solve this problem [4-7]. They have shown that neither proton- nor deuteron-induced reactions can solve the Lithium problem. In this framework, neutron-induced reactions play a key role, in particular the ^{7}Be$(n,p)^7$Li, considered one of the main reaction for ^{7}Be destruction during BBN. If ^{7}Li production is enhanced by neutron capture in the early stages of BBN ($T=50$ *keV*) it can be rapidly burned by the ^{7}Be(p,α) reaction. In 1988, a measurement of the cross-section of the ^{7}Be(n,p) reaction from thermal to 13.5 keV was performed at LANSCE (Los Alamos) [8] and more recently at n_TOF in a wider energy range [9]: the experimental results exclude a significant impact of this reaction on the ^{7}Li problem [9].

With respect to the ^{7}Be(n,p) channel, the contribution of the ^{7}Be(n,α) reaction to the depletion of ^{7}Be during BBN has been considered negligible, due to its much lower cross-section. However, this assumption was never experimentally verified. For this reason, a large uncertainty in BBN calculation has been typically assigned to this reaction and just a single (n,α) measurement at thermal energy has been reported [10], while various theoretical extrapolations in the keV neutron energy region yield completely different results [11]. Consequently, a first measurement on this reaction in the 20-100 keV neutron energy range would finally shed light on the role of the ^{7}Be(n,α) reaction in the Cosmological Lithium Problem. Due to the very important astrophysical motivations and the recent availability of new n-beam facilities (n_TOF-CERN, SARAF-Israel, ILL-France), this subject is being investigated by several research groups worldwide. The complementarity of the experimental approaches used in this scientific quest will ensure the reliability of the result.

3 The $^7Be(n,\alpha)$ reaction: the advanced experimental approach at n_TOF

The main challenge in the measurement of the ^{7}Be(n,α) reaction in the energy range of interest for BBN is strictly related to its low cross section. Thus, two major problems arise: the production and availability of a suitable amount of the short living ^{7}Be isotope for the target and the implementation of an experimental setup with a 4π coverage for ejectile detection. In order to measure the ^{7}Be(n,α) cross sections with good statistical accuracy, a target of a few micrograms of ^{7}Be is needed. In this contest the group at Paul-Scherrer-Institute (PSI), Zurich, has been able to extract and separate a relatively large amount of ^{7}Be, up to 8 μg [12], thus finally making possible this measurement. This isotope has been produced and made available within the well-established INFN-CERN-PSI collaboration. The ^{7}Be has been extracted from the water cooling of the SINQ spallation source, purified and deposited on a proper backing. Two target deposition methods have been developed so far: electrodeposition and micro-drop deposition-evaporation, both the targets are shown in fig. 3. An initial activity of 18 GBq was measured for each target, accounting for a total ^{7}Be mass of 2.8 μg.

Fig. 3. ^7Be targets (PSI): electrodeposited target (left), evaporated target by micro-drop deposition (right).

The two targets have been used in the setup, confirming a full consistency with respect to the target preparation. The main remaining challenge was to design and build an experimental setup, after suitable R&D and in-beam test, able to sustain the large count rate related to the ^{7}Be radioactivity and minimize the background induced by the neutron beam. This was one of the major task of the proposed experiment, also important for future measurements of (n,cp) reactions at n_TOF, as well as for nuclear technology, in particular for research related to fusion energy, and for Nuclear Medicine.

In the ^{7}Be$(n,\alpha)^4$He reaction channel, the two alpha-particles are emitted back-to-back with a maximum energy of 9.5 MeV (the *Q*-value of the reaction is 18.99 MeV). The design of the setup exploited a double sandwich of Si-pad detector, with a ^{7}Be layer in between. The setup, placed directly in the neutron beam, is shown in fig 4 [13].

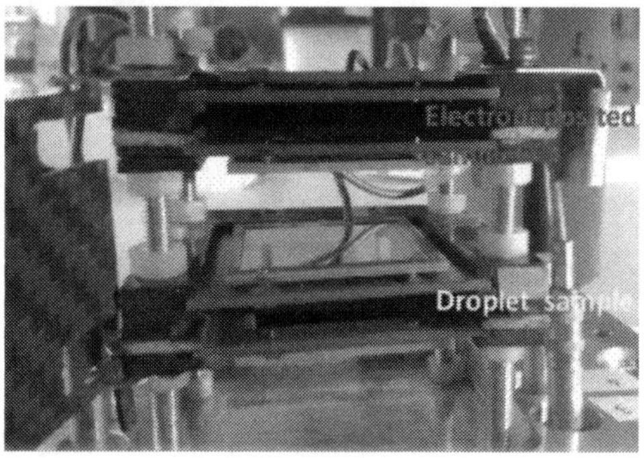

Fig. 4. Experimental setup. The two ^{7}Be targets have been inserted in between a couple of Si-pads, detecting alpha-particles in coincidence. All the system has been placed in line with the n_TOF n-beam, for further details see ref. [13].

The geometrical configuration grants a quasi-4π coverage for the emitted alpha particles. Back-to-back coincidence is a crucial requirement for background suppression, while the double target configuration provided redundancy and mutual cross-check, suitable for revealing systematic errors. In this setup the measured ^{7}Be total target activity was 36 GBq, confirming that the radiation hardness of the Si-detectors, in combination with the large *Q*-value of the reaction, leads to a very good discrimination of the coincidence events produced by the two alpha-particles. Such a performance was confirmed by the background-free alpha-alpha correlation 2D-spectrum, as shown in fig. 4.

At the same time we verified that the neutron induced background in the detectors, target contaminants, and target-backing are not affecting the measurement: in general competing reactions induced by the neutron beam on the detectors themselves, in particular *(n,p)* and (n,γ) reactions on Si (or C) and other material in the beam can pollute the data.

According to the existing predictions of the cross section, the measurement lasted several weeks. During the run, the performances of the setup were continuously monitored, confirming the excellent radiation hardness of the Si-pad, directly exposed to n-beam irradiation. Further details on the analysis and the results have been published in Physical Review Letters [14].

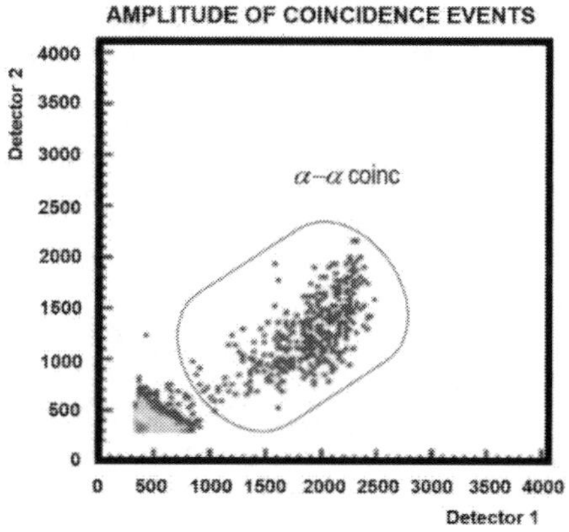

Fig. 5. Correlation plot, α–α coincidences locus shows up fully separated with respect to the γ and *n*-induced background on Silicon-pads and detector frame.

4 Towards the measurement of the n-n scattering length in EAR2

The wide energy range covered by the n-beam provided by the n_TOF facility makes the measurement of excitation function for few-body reactions involving observables relevant for fundamental Nuclear Physics possible. This is the case of the *n-n* scattering length measurement, according to a LoI recently proposed and approved by the CERN INTC-committee (LoI CERN-INTC-2020-051 / INTC-I-22021/09/202).

Due to the unavailability of free neutron targets, it has been realized, since five decades, that the neutron-neutron (*n-n*) cross section can be studied only through reactions involving two neutrons in the final state, mainly by the neutron-deuteron breakup *d(n,p)nn*.

A determination of the *n-n* scattering length in a wide energy range (5-50 MeV) joined with a contextual *n-p* scattering length measurement would represent a cornerstone for Nuclear Physics in investigating charge symmetry breaking and the nucleon-nucleon (N-N) force.

It must be mentioned that in 2006, the n_TOF collaboration submitted a proposal (INTC-P-204) for such a measurement and the experiment was recommended for approval. Due to the demanding requirements in terms of detection efficiency and beam intensity it was not possible to carry out the measurement in the EAR1 station. Now, by the overall improvements of the facility, especially in terms of neutron flux and TOF resolution in the EAR2 station, the definition of the experimental setup can be finally approached in a realistic way.

The most challenging part of the experiment concerns the detection of the three outgoing particles of the *d(n,p)nn* reaction in kinematic coincidence, thus

https://doi.org/10.1051/epjconf/202125205002

leading to a complete three-body kinematic reconstruction. The demand of large efficiency in detecting two neutrons and one proton in coincidence, requires the development of new detection techniques. In this context the collaboration is developing a new deuterated active target for proton detection and an innovative system for neutron momentum reconstruction with high efficiency. Consequently, the first n_TOF deuterated active target «*TaraT*» has been proposed and preliminarly tested at NCSR DEMOKRITOS, Athens. The target design evolved from the proof-of-concept described in fig. 6 to the first prototype presented in fig. 7.

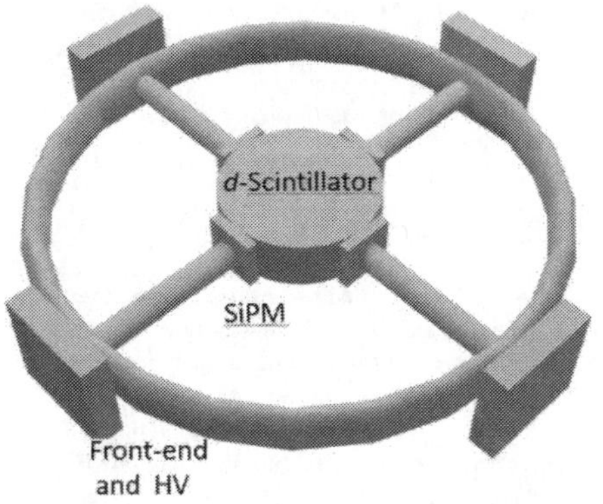

Fig. 6. The geometry of "TaraT", the Active Target.

The low-mass target assembly develops around the central deuterated crystal scintillator: in the first prototype shown in fig. 7 a *d*-stilbene crystal was kindly provided by the Lawrence Livermore National Laboratory (LLNL) [15]. The scintillation light is then collected by using four Silicon-Photomultiplier (SiPM) coupled to the side surface of the crystal. The external frame is supporting the front-end electronics (fast-amplifiers and SiPM power supply). The cylindrical symmetry of the active target allows an easy rescaling of the deuterated scintillator in terms of diameter and shape as well as a fast maintenance and replacement of the activated parts.

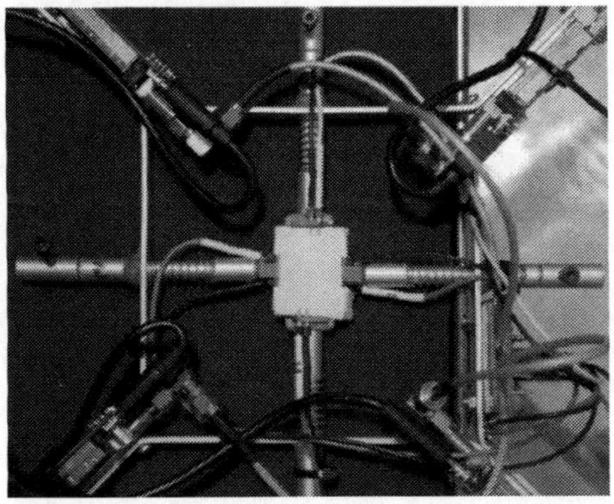

Fig. 7. The instrumented first "*TaraT*" prototype. The *d*-stilbene scintillator has been provided by LLNL.

The analysis of the n-beam test (E_n=5-20 MeV) performed at NCSR DEMOKRITOS confirmed a suitable dynamic range for the SiPM devices and a good time resolution (300 ps). The *n-γ* Pulse-Shape Discrimination (PSD) capability of the stilbene scintillator was confirmed and a new statistical approach to the PSD is being implemented.

Finally, a test of the active target is planned in the n_TOF EAR2 station at CERN, after the commissioning of the new spallation target (i.e., at the end of 2021). This test will provide a crucial information on the γ-flash response of the setup. In fact, the strong electromagnetic pulse produced by the PS proton bunch impinging on the lead spallation target can couple to the detector and the front-end electronics, producing a shortening of the TOF dynamic range.

1. N. Colonna et al., Energy Environ. Sci. **3**, 1910 (2010)
2. M. Sabaté-Gilarte et al., Eur. Phys. J. A **53**, 210 (2017)
3. R.H. Cyburt et al., Rev. Mod. Phys. **88**, 015004 (2016)
4. C. Angulo et al., Astrophys. J. **630**, L105 (2005)
5. P. O'Malley et al., Phys. Rev. C **84**, 042801(R) (2011)
6. O. S. Kirsebom and B. Davids, Phys. Rev. C **84**, 058801 (2011)
7. C. Broggini, L. Canton, G. Fiorentini, and F. L. Villante, J. Cosmol. Astropart. Phys. **06** (2012) 30
8. P.E. Koehler et al, Phys. Rev. C **37**, 917 (1988)
9. L. Damone et al., Phys. Rev. Lett. **121**, 042701 (2018)
10. P. Bassi et al., Il Nuovo Cimento **28**, 1049 (1963)
11. S. Q. Hou, J. J. He, S. Kubono, and Y. S. Chen, Phys. Rev. C **91**, 055802 (2015)
12. E. Maugeri et al., Nucl. Instrum. Methods Phys. Res., Sect. A **889**, 138 (2018)
13. L Cosentino et al., Nucl. Instrum. and Methods Phys. Res., Sect. A **830** 197 (2016)
14. M. Barbagallo et al., Phys. Rev. Lett. **117**, 152701 (2016)
15. N.P. Zaitseva et al., Nucl. Instrum. and Methods Phys. Res. A **789**, 8 (2015)

EPJ Web of Conferences **252**, 05003 (2021)
HINPw6

https://doi.org/10.1051/epjconf/202125205003

The AMS technique as an important tool for the measurement of astrophysical cross sections

Luis Acosta[1,*], *Paulina* Amador-Valenzuela[2], *Eduardo* Andrade[1], *Libertad* Barrón-Palos[1], *David* Belmont[1], *Luis E.* Charón[1], *Efraín* Chávez[1], *Hilarion S.* Cruz-Galindo[2], *Jorge* García-Ramírez[1], *David* Godos-Valencia[1], *Arcadio* Huerta[1], *Daniel J.* Marín-Lámbarri[1], *Josselin N.* Martínez[3], *Javier* Mas-Ruiz[1], *Carmen G.* Méndez[1,4], *Eliud* Moreno[2], *Ghiraldo* Murillo[2], *Santiago* Padilla[1], *Rafael* Policroniades[2], *Roberto* Raya-Arredondo[2], *Guadalupe* Reza[1], *María* Rodríguez-Ceja[1], *Ebanibaldo* Sánchez-Zúñiga[1], *Sinuhé* Sandoval-Hipólito[1], *Vijay* R. Sharma[1], *Corina* Solís[1], *Armando* Varela[2], *Pedro* Villaseñor[2], and *Amacalli B.* Zunun-Torres[1]

[1]Instituto de Física, Universidad Nacional Autónoma de México, Mexico City, Mexico
[2]Instituto Nacional de Investigaciones Nucleares, Ocoyoacac, Mexico
[3]Colegio Ciudad de México, Mexico City, Mexico.
[4]CONACYT, Instituto de Física, Universidad Nacional Autónoma de México, Mexico City, Mexico

Abstract. Accelerator Mass Spectrometry is a technique commonly used to approach low concentrations of certain long half-life radioisotopes. The most important contribution of the technique is the accurate measure of organic sample ages, by separating masses 12,13 and 14 in the case of carbon allocated in such samples. However, the reach of AMS could cover many other scientific scopes, since it can give us a precise measure of a very small concentration of a radioisotope. On this direction, AMS can be used to approach reactions of interest for astrophysics, if we spot an specific radioisotope which concentration can be measure with AMS. Starting with this, we have selected specific reactions involving ^{14}C, ^{10}Be and ^{26}Al, produced with slow neutrons from a reactor and positive ions at an accelerator. The main idea is to produce a particular reaction and later to measure the radioisotopic concentration using AMS. In this study our first results for ^{14}C and ^{10}Be nuclei produced with neutrons, and the preliminary results for ^{26}Al nuclei produced with deuterium are shown.

1 Introduction

Most of times, to approach a good measurement of an specific nuclear reaction, particularly those important for astrophysics, is a difficult task. The energy range, the specific nuclei and the low cross sections are typical problems which currently appear for this kind of experimental studies. Underground accelerators, compact accelerator-based neutron sources and indirect techniques are the main solutions to develop an important amount of interesting reactions. Several studies have been carried out using these kind of methods [1–9].

Nevertheless it is always important to find alternative solutions, considering the big importance of the understanding of certain reactions with difficult access and essential for a better description of stellar and primordial nucleosynthesis. On this direction, we started 5 years ago a campaign of measurements to test the capabilities of the Accelerator Mass Spectrometry technique (AMS) as a possible tool to approach specific reactions in a direct way. The proposal of this technique starts with the fact that it may give an accurate value of the radioisotopic concentration contained in a certain sample; if such radioisotope is a product of an interesting reaction, this concentration is then directly related to the total cross section of the reaction selected.

*e-mail: acosta@fisica.unam.mx

The main constrain of the use of AMS as a technique to find a particular reaction cross section is that it can be used only for a number of radioisotopes, which main characteristic is to have a long half-life. Some examples are ^{10}Be, ^{14}C, ^{26}Al, ^{36}Cl and ^{129}I radioisotopes, whose half-lifes are among the thousands and millions of years. Another important issue to consider when AMS is used, is to have a good radiochemical protocol to isolate the correct compound that can be studied with this technique. This important step is probably the main key to have a good and precise measurement.

An aspect not directly related to the AMS technique, but the first step to start this kind of studies, is the reaction production. We have to find the way to produce the reaction that we are planning to study. This implies to choose a target, a projectile and an energy range. The projectile may be produced, as usual, using some kind of particle accelerator. A good alternative studied in this work, is to use thermal neutrons from a research reactor. On the other hand, the targets must be carefully chosen, considering they have to be suitable to be radiochemically processed to achieve the isolation of the final compound which can be studied with AMS.

In the following sections the different reactions studied, the facilities used and the results achieved will be described.

2 Candidate reactions

To select the possible reactions feasible to be studied with AMS, we have to base them in the radioisotopes that can be measured in our AMS facility. ^{10}Be, ^{14}C and ^{26}Al are radioisotopes well characterized and previously measured in our Mass Spectrometer. On that way, the ^{14}N(n,p)^{14}C reaction is a feasible one, if we consider it can be produced with thermal neutrons. Its cross section is large and well known, hence it is a good candidate reaction to test the technique proposed [10, 11].
Another reaction involving thermal neutrons which is just known by indirect measurements in the ^{9}Be(n,γ)^{10}Be reaction. This later reaction is important to complete for instance the knowledge regarding the primordial and star synthesis of light elements [12, 13].
In the case of possible reactions produced with beam particles we found two interesting ones involving ^{26}Al: the ^{28}Si(d,α)^{26}Al reaction at low energies could be a possible answer to the anomalous amount of ^{26}Mg found in meteorites [14]; on the other hand the ^{25}Mg(p,γ)^{26}Al is one of the important reactions in the stellar evolution, and the production of ^{26}Al found with the existing models still disagree the ^{26}Al observed by its gamma decay [15, 16].

3 Facilities

There is a tradition of 40 years in the Ion Beam Analysis studies by using the different Mexican accelerators. In that sense, for our present studies it was easy to find low energy beams with high intensities. That is the case of the 5.5 MV Van de Graaff Accelerator [17] (Fig 1 left) of the Carlos Graef Fernández Laboratory (Physics Institute, UNAM) and the 6 MV Tandem Van de Graaff Accelerator [18] (Fig 1 center) of the National Institute of Nuclear Research (ININ). Both facilities may produces high intensity deuterium beams at energies from 0.8 till 4.5 MeV. These beams were used to produce our ^{26}Al reactions.
On the other hand, we started to explore the possibilities to use the TRIGA Mark III (Fig 1 right) research reactor [19] also at ININ, which is currently used just for chemical studies and analytical techniques using neutrons. The low energy neutrons ($\sim$ 25 meV) produced by this facility (fluxes of thermal neutrons of up to 3.3×10^{13} n/cm^{2}s) are suitable to approach astrophysical neutron capture reactions. Starting with this, we tested the ^{10}Be and ^{14}C reactions chosen for our studies using the facility described.

For the AMS measurement we used the 1 MV Tandetron accelerator of the Accelerator Mass Spectrometry National Laboratory (LEMA, for their meaning in Spanish) of the Physics Institute, UNAM. This is the most recent accelerator in Mexico (it was inaugurated on 2013) and is well calibrated to discriminate radionuclei from its simile stable ones in a ratio of 1×10^{-15} [20]. This high precision makes to LEMA an ideal tool to explore reactions with small cross sections.

4 Targets

The target selection is probably the most important thing for this kind of studies, considering they have to be used as cathodes (to use AMS technique), once the reaction was produced. For our first attempts to produce ^{26}Al with deuterium beams we use a combination of silicon and aluminum powder (50-50), inserted directly in one of the cylinders which are used as part of a cathode for the AMS system. Unfortunately, as it has shown in the RBS spectrum of the Fig. 3a, the area facing the beam is small enough to loss most of the beam (which is mostly hitting on copper, the material of the cylindrical container) [21, 22]. This problem was solved a couple of years later, by using a silicon wafer target with a size 10 times larger than the LEMA cylinder used as cathode (see spectrum of Fig. 3b). It is important to mention that this solution implies a radiochemical separation, which was not consider in the first test. Nevertheless, the radiochemical process to disintegrate the silicon wafer was previously tested and proved, finding totally acceptable results.

On the other hand, for thermal neutron reactions the following target were used: For the ^{14}C reaction we used uracil ($C_4H_4N_2O_2$), a well-known material with a good amount of nitrogen, enough for our experimental proposal; in the case of ^{10}Be reaction, BeO samples were chemically prepared separating this compound from 8 ml of Be standard solution (Beryllium ICP Standard Solution, $Be_4O(C_2H_3O_2)_6$, 1000 mg/l Be Merck), which is currently used either as carrier or blank in AMS studies.

5 Experimental results.

5.1 ^{28}Si(d,α)^{26}Al reaction.

In our first attempt to produce ^{26}Al by means of ^{28}Si(d,α)^{26}Al reaction, we used deuterium beams at different energies starting from 1 MeV till 2 MeV. Considering we used a copper cylinder with a small area as target, we lost around 80% of the beam for most of the samples. This situation taken us to measure extremely small total cross sections (of the order of μb), making questionable our methodology (a more detailed description can be consulted in [21, 22]).
For such reason we decided to make a second test, but this time using a silicon wafer as target, starting from 1 MeV till 4.8 MeV deuterium energies. The results of this new measurement were very promising, considering we achieved much larger ^{26}Al currents, once AMS was employed. The plot of the Fig. 4 shows a very stable ^{26}Al/^{27}Al ratio of around 5.9×10^{-10}, at least three orders of magnitud larger that in the first measurement. The same figure shows one of the ΔE-E spectra obtained with these samples, in which a clean and well-defined spot of ^{26}Al counts can be observed. A deep description of this preliminar study can be found in [23].

In spite the good results achieved in this second attempt, some modifications to our setup are necessary, considering the measure of the beam current impinging the

Figure 1. Mexican facilities used for the irradiation stage of the present studies. Left: 5.5 MV Van de Graaff Accelerator, Carlos Graef Fernández Laboratory, IFUNAM, Mexico. Center: 6 MV Tandem Van de Graaff Accelerator and Right: TRIGA Mark III Research reactor, both from the National Institute of Nuclear Research, Mexico.

Figure 2. 1 MV Tandetron of the Accelerator Mass Spectrometry National Laboratory, LEMA IFUNAM, Mexico, which was used for the AMS stage of the present studies.

target has to be achieved with a better precision, since we observed instabilities during the data taking. A most precise measurement is programmed in the coming months.

5.2 ^{14}N(n,p)^{14}C reaction.

In order to test a good protocol for thermal neutron capture reaction measurements using the technique proposed, the ^{14}N(n,p)^{14}C reaction was produced in the TRIGA III research reactor before mentioned, using the highest flux of thermal neutrons for 20 seconds in two different samples (targets) of uracil. An example of the ΔE-E spectra achieved with AMS is shown in Fig. 5, where the ^{14}C events can be well separated from other beams produced from the sample irradiated. After the analysis of the ^{14}C concentrations obtained, our data gave a total cross section for ^{14}N(n,p)^{14}C of 2.07 ± 0.37 b, which is in good agreement with other results reported in the literature [10, 11], demonstrating the effectiveness of the method used. A more detailed description of this measurement can be consulted in [24].

5.3 ^{9}Be(n,γ)^{10}Be reaction.

Regarding ^{10}Be reaction, it was also produced by using thermal neutrons from the TRIGA III research reactor. In this case a thermal neutron flux of 2.3×10^{13} n/cm^{2}s in BeO samples was used for periods of 30 minutes and 2 hours. Fig. 6 shows a typical ΔE-E spectrum obtained with the AMS system for the BeO samples irradiated. In the spectrum is easy to observe by eye, the separation among ^{10}Be and ^{10}B, its isobar. Once the analysis of the data achieved was performed, we obtained a value for the total cross section of the ^{9}Be(n,γ)^{10}Be reaction of 9.7 ± 0.53 mb. This value is in acceptable agreement with previous measurements found in the literature [11, 25], nevertheless it is important to remark that ours is the first direct measurement reported. This last aspect makes even more valuable the technique used, considering the possibility to obtain direct measurements for complicated reactions. More details of this study can be consulted in [24].

6 Conclusions and perspectives.

Taking advantages of low energy accelerators, a research reactor and a AMS system, all of them Mexican facilities, we could develop a campaign of measurements of different nuclear reactions, most of them important in astrophysics and not previously measured in a direct way. The method has three stages: sample irradiation, radiochemical preparation and AMS analysis. The final result of this combination may decant in the measure of a total cross section of the initial reaction selected.

We presented in this work a review of our first measurements regarding ^{28}Si(d,α)^{26}Al, ^{14}N(n,p)^{14}C and ^{9}Be(n,γ)^{10}Be reactions. The first case is still in progress

https://doi.org/10.1051/epjconf/202125205003

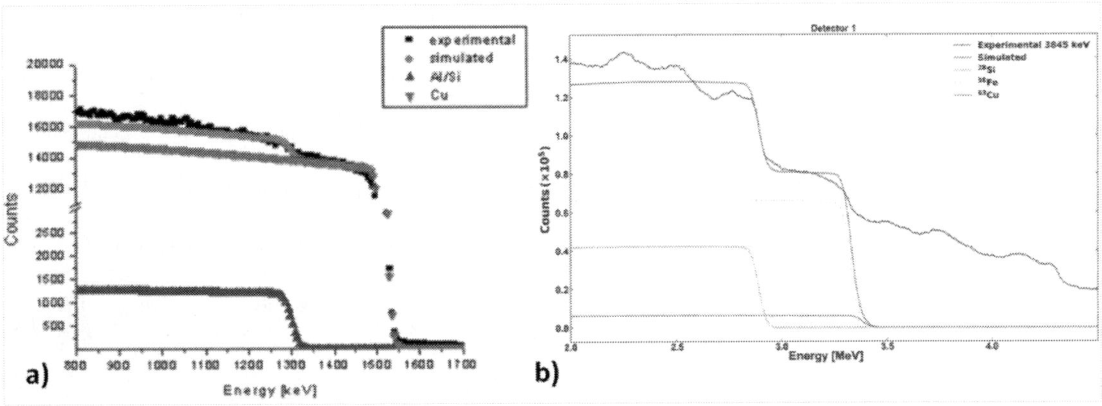

Figure 3. Rutherford Back Scattering (RBS) spectra for two different approaches to the ^{26}Al reaction. The spectra show the different components where the deuterium beam is hitting. In (a) we use a small size cylinder facing the beam, then most of the beam was scattered by copper, which explains the large step of Cu in the spectrum (taken from [22]). A better situation is presented in (b) where a larger Si target was used. In this last, the beam scattered by silicon is much more notorious, indicating that most of the beam is been used to produce the reaction of interest.

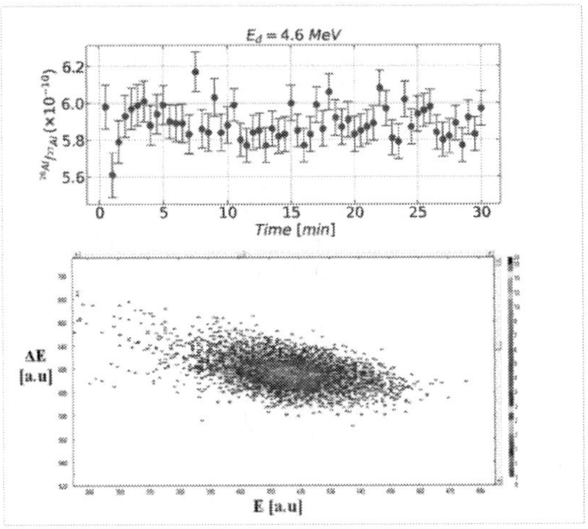

Figure 4. Some results achieved for the counting of ^{26}Al using AMS. On top the ^{26}Al/^{27}Al ratio for 4.6 MeV. The ΔE-E spectrum on bottom figure shows exclusive ^{26}Al events produced from one of the targets irradiated with deuterium.

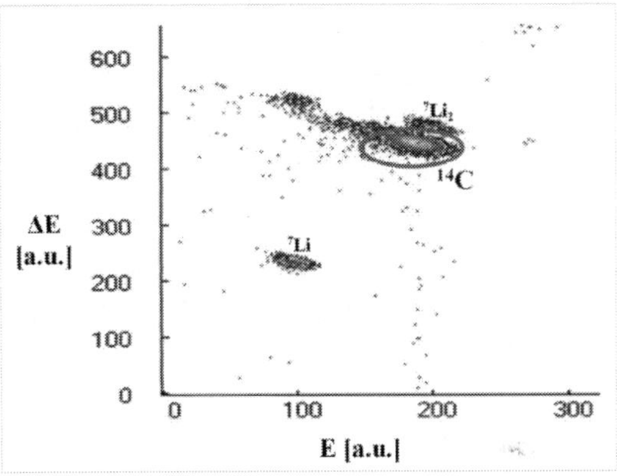

Figure 5. ΔE-E spectrum coming from uracil target after the neutron irradiation to produce ^{14}C. Enclosed in red the ^{14}C events, which are well separated from ^{7}Li$_2$ molecule.

showing promising results till this moment; for the second and third cases we obtained total reaction cross sections which are in a good agreement with other data found in the literature. The case of the ^{10}Be reaction is special, considering the reaction studied was not directly measured in previous studies, showing thus, the importance of the experimental method used in this work.
For further studies, and once we proved the effectiveness of the method, we plan to obtained the total cross sections for different energies of the ^{28}Si(d,α)^{26}Al reaction, as well as to revisit the ^{25}Mg(p,γ)^{26}Al, which was previously studied with the same technique [16] and some others, showing some discrepancies around 400 keV, according to that found in the literature. This last reaction is now possible at LEMA facility thanks to its recent new beam line [26], where intense proton beams at low energies can be produced.

7 Acknowledgements.

The authors want to thank the technical support from ININ, Carlos Graef Laboratory and LEMA staffs. This work was partially supported for DGAPA-UNAM IN107820, AG101120 and CONACyT 314857 projects.

References

[1] L. Csedreki, et. al., Nucl. Inst. Meth. Phys. Res. A **994**, 165081 (2021)

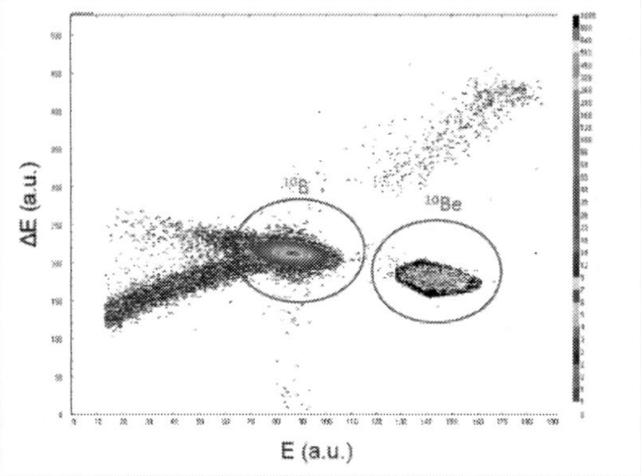

Figure 6. ΔE-E spectrum from BeO samples irradiated with neutrons to produce ^{10}Be. The circles show the well separation between ^{10}Be events and those of its isobar, the ^{10}B.

[2] G.G. Rapisarda, et. al., Front. Astron. Space Sci. **7** 589240 (2021)

[3] T. Aumann and C. Bertulani, Progr. Part. Nuclear Physics **112**, 103753 (2020)

[4] C. Spitaleri et al., Eur. Phys. J. A **52**, 77 (2016)

[5] I. Mardor et. al., Eur. Phys. J A **54**, 91 (2018)

[6] I.S. Anderson et. al., Phys. Reports **654**, 1 (2016)

[7] I. Dillmann and R. Reifarth JINST **7** C04014 (2012)

[8] R. E. Tribble et. al., Rep. Prog. Phys. **77**, 106901 (2014)

[9] A. Best et al., Eur. Phys. J. A **52**, 72 (2016)

[10] J. Wagemans, C. Wagemans, G. Goeminne, and P. Geltenbort, Phys. Rev. C **61**, 064601 (2000)

[11] S. Mughabghab, *Atlas of Neutron Resonances*, 5th ed. (Elsevier, Amsterdam, 2006)

[12] C. M. Conneely, W. V. Prestwich, and T. J. Kennett, Nucl. Instrum. Methods Phys. Res. A **248**, 416 (1986)

[13] D. D. Clayton, S. A. Colgate, and G. J. Fishman, Astrophys. J. **155**, 75 (1969)

[14] T. Lee, D. a. Papanastassiou, and G. J. Wasserburg, Geophys. Res. Lett. **3**, 1 (1976)

[15] J. Knödlseder et. al., Astron. and Astrophys. **344** (1999)

[16] A. Arazi et. al., Phys. Rev. C. **74** 025802 (2006)

[17] E. Andrade, et. al., Nucl. Inst. Meth. Phys. Res. B **332** (2014)

[18] G. Murillo, et. al., Nucl. Inst. Meth. Phys. Res. B **136–138** 888–892 (1998)

[19] H. S. Cruz-Galindo and R. Raya-Arredondo, AIP Conf. Proc. **1544**, 53 (2013)

[20] C. Solís, E. Chávez, M.E. Ortiz, A. Huerta, E. Andrade, E. Barrios. Nucl. Inst. Meth.Phys. B **331** (2014)

[21] V. Araujo-Escalona, et. al., Phys. Procedia **90**, 421–428 (2017)

[22] L. Acosta, et. al., Eur. Phys. J. Web Conf. **165**, 01001 (2017)

[23] G. Reza, et. al., Eur. Phys. J. Plus **135**, 899 (2020)

[24] D.J. Marín-Lámbarri et. al., Phys. Rev. C **102**, 044601 (2020)

[25] K. Nishiizumi, M. Imamura, M.W. Caffee, J. R. Southon, R. C. Finkel, and J. McAninch, Nucl. Instrum. Methods Phys. Res. B **258**, 403 (2007)

[26] L. Acosta, et. al., J. Phys. Conf. Ser. **1643**, 012029 (2020)

Thermal properties of hot and dense matter: Influence of rapid rotation on protoneutron stars, hot neutron stars, and neutron star merger remnants

Polychronis Koliogiannis[1,*] and *Charalampos* Moustakidis[1]

[1]Department of Theoretical Physics, Aristotle University of Thessaloniki, Greece

Abstract. The knowledge of the equation of state is a key ingredient for many dynamical phenomena that depend sensitively on the hot and dense nuclear matter, such as the formation of protoneutron stars and hot neutron stars. In order to accurately describe them, we construct equations of state at finite temperature and entropy per baryon for matter with varying proton fractions. This procedure is based on the momentum dependent interaction model and state-of-the-art microscopic data. In addition, we investigate the role of thermal and rotation effects on microscopic and macroscopic properties of neutron stars, including the mass and radius, the frequency, the Kerr parameter, the central baryon density, etc. The latter is also connected to the hot and rapidly rotating remnant after neutron star merger. The interplay between these quantities and data from late observations of neutron stars, both isolated and in matter of merging, could provide useful insight and robust constraints on the equation of state of nuclear matter.

1 Introduction

Neutron stars are born in the gravitational collapse of massive, degenerate stellar cores. Newly born stars are quite different than the ordinary ones, as they are hot and lepton rich [1–4]. In view of these, a study of hot neutron star matter is of increasing interest in relation to astrophysical phenomena, such as the neutron star mergers, the formation of protoneutron stars, and the formation of black holes. Their study requires the knowledge of the equation of state (EoS), which at several times the normal nuclear matter density remains unknown.

Due to the limit of the experimental data concerning the dense nuclear matter, constraints on the EoS are possible though the observation of non/slow rotating neutron stars. The most robust ones are related to the maximum mass of a neutron star, including the resent PSR J0740+6620 with $M = 2.14^{+0.10}_{-0.09}\ M_{\odot}$ [5]. Nonetheless, the recent observations of binary neutron star mergers, GW170817 [6] and GW190425 [7], improved our knowledge of the EoS, especially in the intermediate region of densities. Furthermore, after the merger, the remnant and its maximum stable mass, spin period, and lifetime strongly depend on the dense matter properties at high temperatures and entropy. The latter is a valid candidate for constraints in the high density region.

In the present work we construct and apply in astrophysical processes, a set of thermodynamically consistent isothermal and isentropic EoSs, based on the parameterized cold one. Specifically, we dedicate a part for analyzing the free energy and the effect of the proton fraction related to dynamical processes. We explore the significant meaning of the constant baryon mass sequences and connect them with the microscopic properties of neutron stars. In addition, we focus on the extensive study on some of the bulk properties of both nonrotating and rotating at the mass-shedding limit hot neutron stars, including the moment of inertia, Kerr parameter, ratio of rotational kinetic to gravitational binding energy T/W, etc. Finally, we study the hot, rapidly rotating remnant of a binary neutron star merger in accordance with the derived EoSs.

The paper is structured as follows. In Section 2 we present the framework of the nuclear model. Section 3 provides the discussion of the results while in Section 4, the scientific remarks are pointed out. Finally, Appendix A includes the instability criteria.

2 The Nuclear Model

For the description of neutron stars, the EoSs have been constructed using the momentum-dependent interaction (MDI) model [8–10] and the data provided by Akmal et al. [11] for the APR-1 EoS (hereafter MDI+APR1). The schematic presentation of the cold, catalyzed matter of the MDI+APR1 EoS, for the pure neutron matter (PNM) and symmetric nuclear matter (SNM), is presented in Figure 1.

This parameterization is used for the construction of one cold EoS, 10 hot EoSs based on temperatures in the range $[1, 60]$ MeV, and nine hot EoSs based on lepton fractions and entropies per baryon in the ranges $[0.2, 0.4]$ and $[1, 3]\ k_B$, respectively.

For the solid crust region of the cold case, we applied the EoS of Feynman et al. [12] and also Baym et al. [13], while for the finite temperature cases and the low-density region ($n_b \leq 0.08\ \mathrm{fm}^{-3}$), as well as the finite entropies per baryon and lepton fractions, the EoSs of Lattimer and Swesty [14] and the specific model corresponding to the incomprehensibility modulus at the saturation density of

*e-mail: pkoliogi@physics.auth.gr

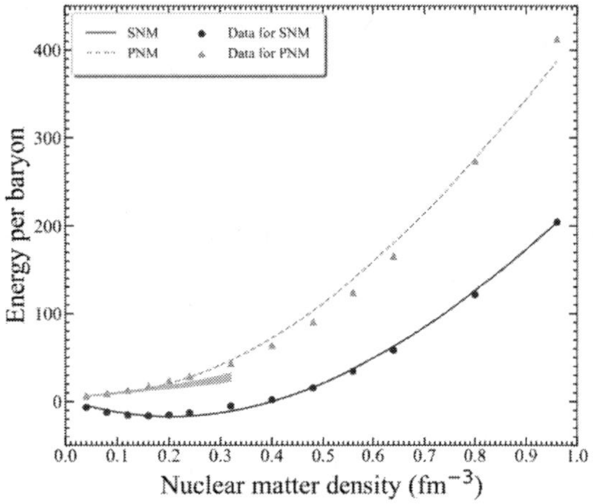

Figure 1. The fits of SNM and PNM of MDI+APR1 cold EoS. The SNM is presented by the circles and solid line, while the PNM is presented by the triangles and dashed line. The shaded region corresponds to benchmark calculations of the energy per particle of PNM extracted from Piarulli et al. [16].

SNM K_s = 220 MeV are used (https://www.stellarcollapse.org).

For an extensive study and information about the construction of the EoSs see Ref. [15].

3 Results and discussion

3.1 Free energy and proton fraction

Figure 2 displays the free energy per particle as a function of the baryon density for (a) PNM and (b) SNM and temperatures in the range [0, 60] MeV. The free energy per particle is an important quantity directly related to the calculation of the proton fraction. The effect of the temperature in free energy is more pronounced at the low density region, while at the high density region, due to the quantum character of hadronic matter, the system tends to be more symmetric.

Figure 3 displays the proton fraction as a function of the baryon density for temperatures in the range [0, 60] MeV. As the proton fraction is related to the structure of neutron stars, similar sensitivity from temperature with the free energy is presented. More specific, while in the low density region there is a high dependence from temperature, in the high density region, the effects are very mild.

3.2 Baryon mass sequences

The baryon mass sequences are a way to study the evolution of a neutron star. In particular, we construct sequences related to the cooling of a neutron star, based on isothermal EoSs, and study the thermal evolution at the mass-shedding limit. In Figure 4 we display the Kepler frequency as a function of (a) the temperature and (b) the central baryon density. Specifically, while Figure 4(a) presents the reduction of the Kepler frequency along the temperature, Figure 4(b) presents a twofold relation between the Kepler frequency and the central baryon density. For relatively low temperatures, $T < 30$ MeV, there is a linear dependence between the Kepler frequency and the central baryon density, assuming constant temperature in each case. In addition, for higher temperatures, a linear relation described as

$$f(n_b^c) = -473.144 + 2057.271 n_b^c \quad \text{(Hz)}, \qquad (1)$$

holds on, independently of the selected baryon mass. This relation defines the allowed region for rotation with the Kepler frequency of a hot neutron star for a specific value of the central baryon density, and vice versa.

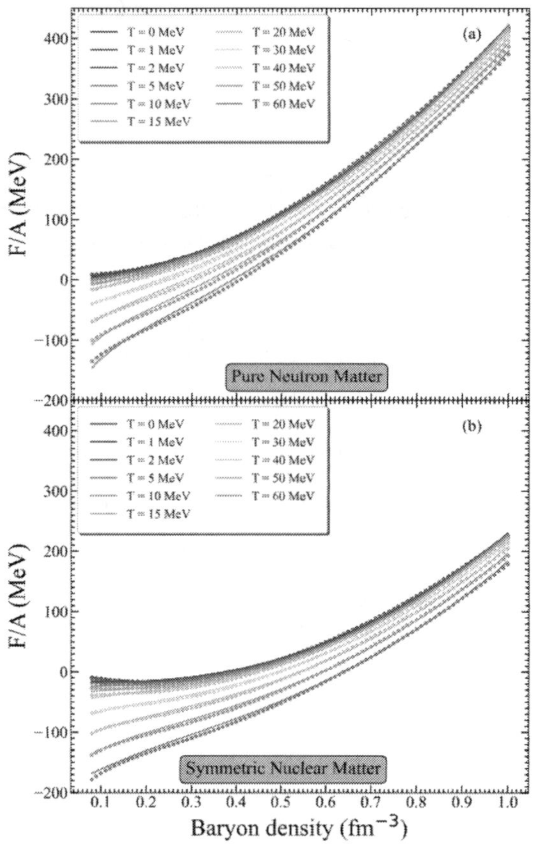

Figure 2. The free energy per particle as a function of the baryon density for (a) PNM and (b) SNM and temperatures in the range [0, 60] MeV. Fits are presented with the solid lines and data with circles (see also Ref. [15]).

3.3 Moment of inertia, Kerr parameter, and T/W ratio

Moment of inertia is an important parameter that provides information about the structure of neutron stars. In Figure 5(a) we display the dimensionless moment of inertia as

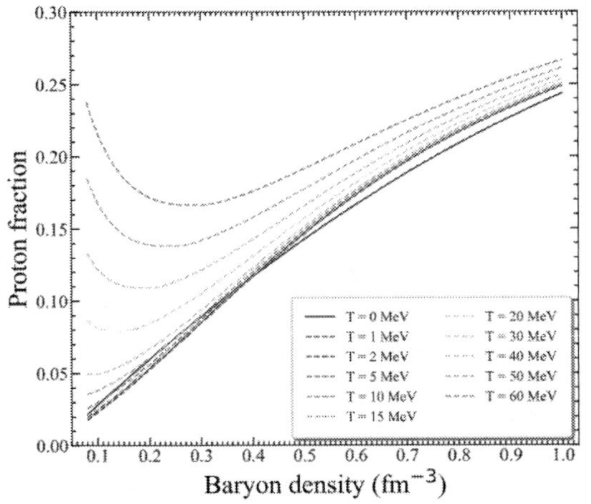

Figure 3. The proton fraction as a function of the baryon density for temperatures in the range [0, 60] MeV. The cold configuration is presented with the solid line and the hot ones with the dashed lines.

a function of the compactness parameter, defined as

$$\beta = \frac{G}{c^2}\frac{M}{R}, \tag{2}$$

for isothermal neutron stars at the mass-shedding limit. The outcome is that in general, the increase of temperature leads to lesser compact stars with lower values of dimensionless moment of inertia than the cold neutron star.

Figure 5(b) displays the dimensionless moment of inertia as a function of the compactness parameter for isentropic neutron stars at the mass-shedding limit. In this case too, for a constant lepton fraction, the increase of the entropy per baryon leads to lesser compact stars with lower values of dimensionless moment of inertia than the cold neutron star.

Kerr parameter is a quantity directly related to neutron stars and black holes, defined as

$$\mathcal{K} \equiv \frac{c}{G}\frac{J}{M^2}. \tag{3}$$

We study the mass-shedding limit which is connected to the final fate of the collapse of a rotating compact star. Following the Refs. [17, 18], and the relations for angular velocity and moment of inertia, we found that the Kerr parameter, at the mass-shedding limit for a cold, catalyzed neutron star, is given by the simple universal expression

$$\mathcal{K}_k \simeq 1.34\sqrt{\beta_{max}}. \tag{4}$$

Assuming that, for the majority of realistic cold EoSs, the relation $0.24 \leq \beta_{max} \leq 0.32$ holds, we concluded that $0.66 \leq \mathcal{K}_k \leq 0.76$ [15].

Figures 5(c) and 5(d) display the Kerr parameter as a function of the gravitational mass for isothermal and isentropic neutron stars, respectively. As the temperature or the entropy per baryon increases, assuming a constant lepton

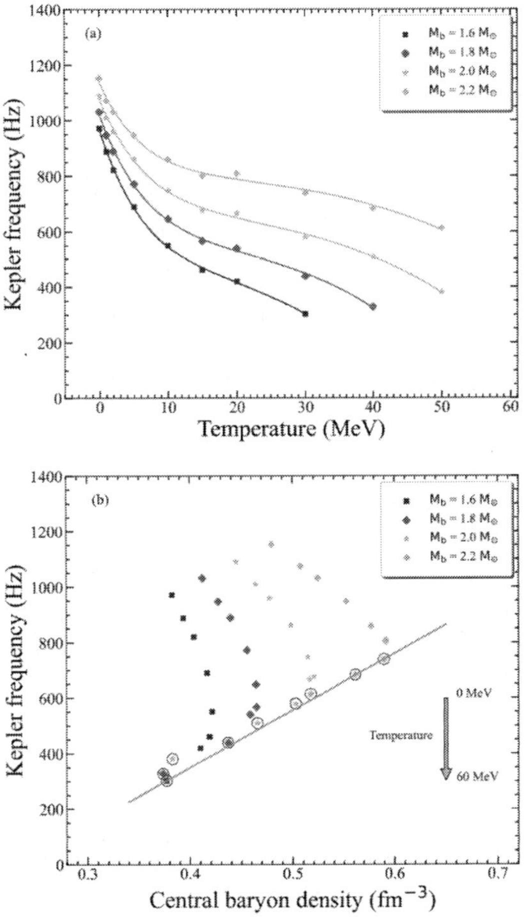

Figure 4. The Kepler frequency as a function of (a) the temperature and (b) the central baryon density for constant baryon mass sequences. (a) Solid lines correspond to fits (for more details see Ref. [15]). (b) The solid line corresponds to Equation (1) and open circles mark the high-temperature region ($T \geq 30$ MeV).

fraction in the second case, the Kerr parameter decreases. The introduction of the temperature leads to the conclusion that the existed limit for Kerr black holes [19] and the one for neutron stars, through Equation (4), cannot be violated. Henceforth, the gravitational collapse of a hot, uniformly rotating neutron star, constrained to mass-energy and angular momentum conservation, cannot lead to a maximally rotating Kerr black hole.

Figures 5(e) and 5(f) display the angular velocity as a function of the ratio T/W for isothermal and isentropic neutron stars, respectively. In neutron stars, the point that locates the nonaxisymmetric instability, which is responsible for the emission of gravitational waves, is defined via the ratio of rotational kinetic to gravitational binding energy T/W. In particular, instabilities driven by gravitational radiation for models with $M_{gr} = 1.4\ M_\odot$ [20] would set in at $T/W \sim 0.08$. Figures 5(e) and 5(f) show that

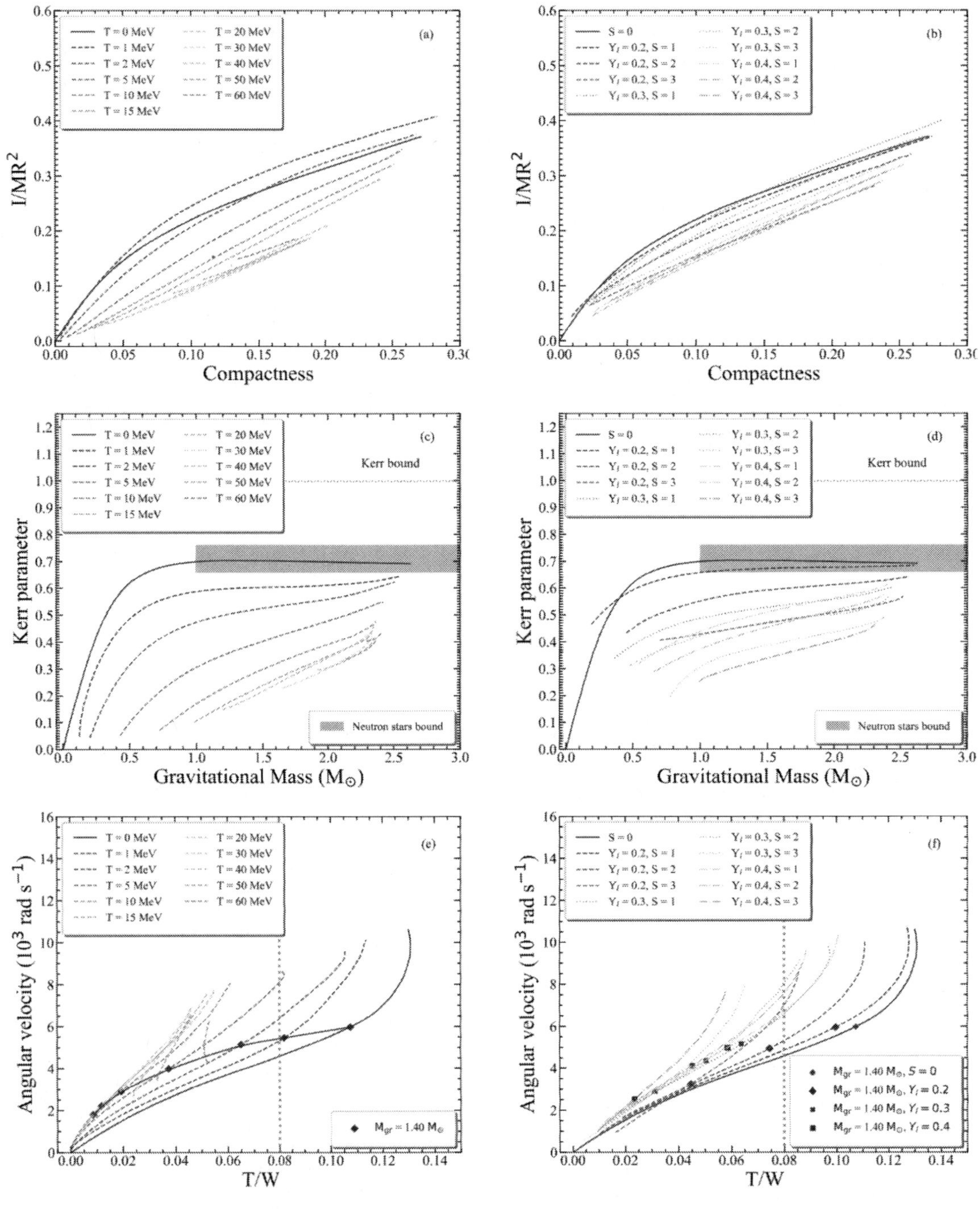

Figure 5. (a,b) The dimensionless moment of inertia as a function of the compactness parameter for (a) isothermal and (b) isentropic neutron stars at the mass-shedding limit. (c,d) The Kerr parameter as a function of the gravitational mass for (c) isothermal and (d) isentropic neutron stars at the mass-shedding limit. The horizontal dotted line marks the Kerr bound for astrophysical Kerr black holes, $\mathcal{K}_{\rm B.H.} = 0.998$ [19]. The shaded region represents the limits for neutron stars from Equation (4). (e,f) The angular velocity as a function of the ratio of rotational kinetic to gravitational binding energy for (e) isothermal and (f) isentropic neutron stars at the mass-shedding limit. The vertical dotted line marks the critical value, $T/W = 0.08$ [20], for gravitational radiation instabilities. Black plus sign, diamonds, squares, and crosses represent the $M_{\rm gr} = 1.4\,M_\odot$ configuration. For all Figures, the cold configuration is presented with the solid line while hot configurations, with the dashed ones.

for sufficiently compact neutron stars, the nonaxisymmetric instability will set in before the mass-shedding limit is reached. The information that can be gained is that the maximum gravitational mass and the angular velocity will be lowered.

According to the analyzed quantities, useful insight can be gained for the hot, rapidly rotating remnant (at least $T \geq 30$ MeV for isothermal EoSs, $S = 1$ and $Y_l = 0.2$ for isentropic ones) after the neutron star merger, which is a compact object with neutron star matter. Assuming that the postmerger remnant is rotating at its mass-shedding limit, we study the maximum mass configuration. In this case, (a) the compactness parameter is constraint to $\beta_{\rm rem}^{\rm iso} \leq 0.19$ and $\beta_{\rm rem}^{\rm ise} \leq 0.27$, (b) the Kerr parameter is constraint to $\mathcal{K}_{\rm rem}^{\rm iso} \leq 0.42$ and $\mathcal{K}_{\rm rem}^{\rm ise} \leq 0.68$, and (c) the ratio T/W is constraint to $(T/W)_{\rm rem}^{\rm iso} \leq 0.05$ and $(T/W)_{\rm rem}^{\rm ise} \leq 0.127$ (the superscripts "iso" and "ise", correspond to isothermal and isentropic profiles). Two hypotheses can be made for the aftermath of a neutron star merger. In particular, in the isothermal case, it creates a lesser compact star than the cold neutron star with lower values of maximum gravitational mass and frequency, and stable toward dynamical instabilities, where for the isentropic aftermath, the object is comparable to the cold one [15].

4 Remarks

Thermal properties of hot and dense matter in the isolated neutron stars and in the matter of merging, remain an open problem. In order to gain insight into the role of thermal support, we constructed and applied in astrophysical phenomena a set of thermodynamically self-consistent EoSs (isothermal and isentropic). Specifically, we constructed nonrotating and uniformly rotating axisymmetric equilibrium sequences. Such equilibrium models may be acceptable for a first-order study of protoneutron stars, as well as of the postmerger remnant that has survived the initial highly dynamical and strongly nonaxisymmetric postmerger evolution and settled down into a stable quasi-equilibrium state [21].

The nuclear model which applied for the construction of the EoSs, presents merits compared to other models: (a) the thermal effects (both in isothermal and isentropic profiles) have been included in a self-consistent way, (b) the model is flexible enough to produce a wide range of EoSs by properly modifying the density dependence of the symmetry energy, (c) reproduces microscopic calculations concerning both the SNM and the PNM, (d) the momentum dependence of the potential interaction (which is absent in the majority of the proposed models) is in accordance with the terrestrial studies and experiments of heavy-ion reactions for both low and high densities and temperatures, and (e) the model ensures the causal behavior of the EoS at high densities.

For the numerical integration of the equilibrium equations we used the publicly available numerical code *nrotstar* from the C++ Lorene/Nrotstar library [22] (for more details, see Ref. [15]).

In the postmerger phase, the baryon mass of the merger remnant is approximately conserved, an evidence that causes its dependence on temperature significant. Specifically, as the temperature rises, neutron stars lead to lower baryon masses. As in rapidly rotating configuration supramassive neutron stars exist, we study the mass-shedding limit in connection to the remnant. In particular, in the cold case, the baryon mass is 3.085 $M_\odot$, while for a hot one at $T = 30$ MeV is 2.427 $M_\odot$ and one at $S = 1$ is 3.05 $M_\odot$. By assuming equal masses of merger components, these limits correspond to merger components of $\sim$ 1.5425, $\sim$ 1.2135, and $\sim$ 1.525 $M_\odot$ baryon masses, respectively. Additionally, the immediate aftermaths of GW170817 [6] and GW190425 [7] had created hot, rapidly rotating remnants. Although the merger remnants are born with differential rotation, the uniform rotation approach can provide us with useful insight about the EoS. In the case of GW170817, a remnant with a total mass of $\sim$ 2.7 $M_\odot$ has been created. In correlation with the MDI+APR1 EoS, with respect to baryon mass, only cold and isentropic neutron stars with uniform rotation can support the remnant. As for the GW190425 event, the existence of the $\sim$ 3.7 $M_\odot$ remnant cannot be described only by uniform rotation.

Moment of inertia is a quantity that informs us about the distribution of matter in the star as it continuously changes its angular velocity and loses angular momentum due to radiation. The introduction of thermal effects in neutron stars lead to the conclusion that moment of inertia has lower values than the cold neutron star. This particular behavior is influenced by the unique interplay between the gravitational mass and the equatorial radius.

In our study, the maximum Kerr parameter existed in the cold neutron star. In particular, thermal support lead to lower values of the Kerr parameter. The outcome is that thermal support cannot lead a star to collapse into a maximally rotating Kerr black hole. In addition, it is observed that after $\sim$ 1 $M_\odot$, while in the cold configuration the Kerr parameter is approximately constant, in hot configurations the Kerr parameter is an increasing function of the gravitational mass, leading to a unique maximum value.

The gravitational collapse to a black hole and the existence of stable supramassive neutron stars is related to the ratio T/W. In the specific case of the gravitational collapse due to gravitational radiation, the critical value of this ratio is $\sim$ 0.08 for the $M_{\rm gr} = 1.4$ $M_\odot$ configuration [20]. As thermal support leads to lower values for the ratio T/W, instabilities occurred by gravitational radiation never occur in a postmerger remnant or a hot, rapidly rotating neutron star. In some cases for low values of temperature or entropy per baryon, the ratio exceeds the limit. In this case, the critical value of T/W set the limit for the maximum gravitational mass and angular velocity.

The effects of temperature on the EoS can be efficiently manifest though the evolutionary sequences of constant baryon mass. In particular, the attention is focused on the dependence of the Kepler frequency on the central baryon density. For temperatures $T \geq 30$ MeV, the existence of a linear relation between these quantities, defines the allowed region of the pair of the central baryon density and corresponding Kepler frequency for a rotating hot neutron star at its mass-shedding limit.

Future work should be focused on differential rotation, with emphasis on the various differential laws, and consider rotating configurations constructed with a broader set of hot microphysical EoSs. Finally, the postmerger remnant and its dynamical processes should be thoroughly investigated, as the LIGO and Virgo collaboration will provide us with more events of neutron star mergers.

The authors thank Prof. K. Kokkotas for his constructive comments on the preparation of the manuscript and Prof. L. Rezzolla for his useful and helpful considerations and clarifications.

A Instabilities in Neutron Stars

A.1 Cold Neutron Stars

The stability of cold rotating neutron stars is acquired via the turning-point criterion. It has to be noted that it is only a sufficient and not a necessary one. To be more precise, the neutral stability line is positioned to the left of the turning-point line in (M, ρ_c) space. Hence, the latter indicates that the star will collapse before reaching the turning-point line [15, 23, 24].

A.2 Hot Neutron Stars

The stability of hot neutron stars is achieved with a specific version of the secular instability criterion of Friedman et al. [25], which follows the Theorem I of Sorkin [26]. We choose a continuous sequence of equilibria to be at a fixed baryon number $N_{\rm bar}$ and total entropy of the neutron star $S_{\rm t}^{\rm ns}$, and the extremal point of the stability loss is [2]

$$\left.\frac{\partial J}{\partial n_b^c}\right|_{N_{\rm bar},S_{\rm t}^{\rm ns}} = 0, \qquad (5)$$

where J and n_b^c are the angular momentum and central baryon density of the star, respectively.

In addition, a turning point in the sequence occurs where three out of four derivatives, $\partial M_{\rm gr}/\partial n_b^c$, $\partial M_{\rm b}/\partial n_b^c$, $\partial J/\partial n_b^c$, and $\partial S_{\rm t}^{\rm ns}/\partial n_b^c$, where $M_{\rm gr}$ and $M_{\rm b}$ denote the gravitational and baryon mass, vanish [15, 21, 27]. At this point, the turning-point theorem shows that the fourth derivative also vanishes, and the sequence has transitioned from stable to unstable.

References

[1] M. Prakash, I. Bombaci, M. Prakash, P.J. Ellis, J.M. Lattimer, R. Knorren, Phys. Rep. **280**, 1 (1997)

[2] J.O. Goussard, P. Haensel, J.L. Zdunik, Astron. Astrophys. **321**, 822 (1997)

[3] N. Glendenning, *Compact Stars: Nuclear Physics, Particle Physics, and General Relativity* (Springer, Berlin, 2000)

[4] F. Weber, *Pulsars as astrophysical laboratories for nuclear and particle physics* (UK: Institute of Physics, Bristol, 1996)

[5] H. Cromartie, E. Fonseca, S. Ransom, P.B. Demorest, Z. Arzoumanian et al., Nat. Astron. **4**, 72 (2020)

[6] B.P. Abbott, R. Abbott, T.D. Abbott et al., Phys. Rev. Lett. **119**, 161101 (2017)

[7] B.P. Abbott, R. Abbott, T.D. Abbott et al., Astrophys. J. **892**, L3 (2020)

[8] C.C. Moustakidis, Phys. Rev. C **78**, 054323 (2008)

[9] C.C. Moustakidis, Inter. J. Mod. Phys. D **18**, 1205 (2009)

[10] C.C. Moustakidis, C.P. Panos, Phys. Rev. C **79**, 045806 (2009)

[11] A. Akmal, V.R. Pandharipande, D.G. Ravenhall, Phys. Rev. C **58**, 1804 (1998)

[12] R.P. Feynman, N. Metropolis, E. Teller, Phys. Rev. **75**, 1561 (1949)

[13] G. Baym, C. Pethick, P. Sutherland, Astrophys. J. **170**, 299 (1971)

[14] J.M. Lattimer, F.D. Swesty, Nucl. Phys. A **535**, 331 (1991)

[15] P.S. Koliogiannis, C.C. Moustakidis, Astrophys. J. **912**, 69 (2021)

[16] M. Piarulli, I. Bombaci, D. Logoteta, A. Lovato, R.B. Wiringa, Phys. Rev. C **101**, 045801 (2020)

[17] P.S. Koliogiannis, C.C. Moustakidis, Phys. Rev. C **101**, 015805 (2020)

[18] D.S. Shao, S.P. Tang, X. Sheng, J.L. Jiang, Y.Z. Wang, Z.P. Jin, Y.Z. Fan, D.M. Wei, Phys. Rev. D **101**, 063029 (2020)

[19] K.S. Thorne, Astrophys. J. **191**, 507 (1974)

[20] S.M. Morsink, N. Stergioulas, S.R. Blattnig, Astrophys. J. **510**, 854 (1999)

[21] J.D. Kaplan, C.D. Ott, E.P. O'Connor, K. Kiuchi, L. Roberts, M. Duez, Astrophys. J. **790**, 19 (2014)

[22] LORENE, *Lorene: Langage objet pour la relativité numérique* (1998), `http://lorene.obspm.fr/`

[23] K. Takami, L. Rezzolla, S. Yoshida, Mon. Not. Roy. Astron. Soc. Lett. **416**, L1 (2011)

[24] L.R. Weih, E.R. Most, L. Rezzolla, Mon. Not. Roy. Astron. Soc. Lett. **473**, L126 (2017)

[25] J.L. Friedman, J.R. Ipser, R.D. Sorkin, Astrophys. J. **325**, 722 (1988)

[26] R.D. Sorkin, Astrophys. J. **257**, 847 (1982)

[27] M. Marques, M. Oertel, M. Hempel, J. Novak, Phys. Rev. C **96**, 045806 (2017)

https://doi.org/10.1051/epjconf/202125205005

Constraints on the speed of sound of dense nuclear matter through the tidal deformability of neutron stars

Alkiviadis Kanakis-Pegios[1,*], *Polychronis* Koliogiannis[1], and *Charalampos* Moustakidis[1]

[1]Department of Theoretical Physics, Aristotle University of Thessaloniki, Thessaloniki, Greece

Abstract. One of the greatest interest and open problems in nuclear physics is the upper limit of the speed of sound in dense nuclear matter. Neutron stars, both in isolated and binary system cases, constitute a very promising natural laboratory for studying this kind of problem. This present work is based on one of our recent study, regarding the speed of sound and possible constraints that we can obtain from neutron stars. To be more specific, in the core of our study lies the examination of the speed of sound through the measured tidal deformability of a binary neutron star system (during the inspiral phase). The relation between the maximum neutron star mass scenario and the possible upper bound on the speed of sound is investigated. The approach that we used follows the contradiction between the recent observations of binary neutron star systems, in which the effective tidal deformability favors softer equations of state, while the high measured masses of isolated neutron stars favor stiffer equations of state. In our approach, we parametrized the stiffness of the equation of state by using the speed of sound. Moreover, we used the two recent observations of binary neutron star mergers from LIGO/VIRGO, so that we can impose robust constraints on the speed of sound. Furthermore, we postulate the kind of future measurements that could be helpful by imposing more stringent constraints on the equation of state.

1 Introduction

One of the current unsolved problems in Nuclear Physics is the properties of dense nuclear matter. Neutron stars are among the most promising sources in universe for studying the dense nuclear matter [1–4]. The speed of sound consists a microscopic quantity of great interest, especially the possible upper bounds that can get, since it affects the equation of state (EoS).

The main assumption about the upper bound of the speed of sound is that fullfills the causality, i.e. the speed of sound cannot exceed the speed of light, which is the main restriction in the speed of sound imposed from general principles according to Zel'dovich [5, 6]. Various studies have been made about the upper bound on the speed of sound (see Refs. [4, 7, 8]). Furthermore, in Ref. [8] the authors found also that the existence of neutron stars with masses about two solar masses, combined with the knowledge of the EoS of hadronic matter at low densities, is not consistent with the bound $c/\sqrt{3}$. The speed of sound and its effects on the tidal deformability has been studied also in Refs. [9–12].

We notice that there are two main considerations for the possible upper bound in the speed of sound of dense nuclear matter. Firstly, the causality which imposes an upper bound on the speed of sound $v_c = c$, and secondly a conformal upper limit with $v_c = c/\sqrt{3}$. Additionally, the limit $v_s = c/\sqrt{3}$ is related to the existence of quark matter in neutron star interior, consisting a very important reason to study this case of speed of sound bound (for more details see Refs. [13, 14]).

The main idea that motivated our study is the examination of possible constraints on the tidal deformability of neutron stars through the speed of sound and its upper bounds. We notice that there is a contradiction that arises regarding the stifness of the EoS; the recent high measured values of neutron star masses favor stiff EoS, while the upper limit on the effective tidal deformability (from gravitational-wave detections) favors softer EoSs. Our study aims to shed more light on these antagonized points of view, by parametrizing the stifness (softness) of the EoS via the speed of sound upper bounds and the transition density. The predictions are combined with the recent observations of two events, GW170817 and GW190425 [15–18], as well as the current observed maximum neutron star masses ($1.908 \pm 0.016\ M_\odot$ [19], $2.01 \pm 0.04\ M_\odot$ [20], $2.14^{+0.10}_{-0.09}\ M_\odot$ [21], $2.27^{+0.17}_{-0.15}\ M_\odot$ [22]) (see also Ref. [23]).

The demand of a) a soft EoS for low densities (in order to be in accordance with the upper limit of the tidal deformability) and b) a stiff EoS for high densities (in order to predict the high neutron star measurements) leads to robust constraints on the EoS. In addition, we postulate what kind of future measurements could help us to more stringent constraints. For more details, we refer to the original study Ref. [24].

Moreover, we mention the recent detection of GW190814 event [25, 26], in which the lighter component compact object might be a neutron star. Following

*e-mail: alkanaki@auth.gr

this hypothesis, in a recent work we studied the role of the speed of sound in such a case [27].

2 Theoretical background

2.1 Speed of sound and EoS

First of all, we constructed the necessary maximum mass configuration, by parametrizing the neutron star EoS as follows [28–36]

$$P(\mathcal{E}) = \begin{cases} P_{\rm crust}(\mathcal{E}), & \mathcal{E} \leq \mathcal{E}_{\rm c-edge} \\ P_{\rm NM}(\mathcal{E}), & \mathcal{E}_{\rm c-edge} \leq \mathcal{E} \leq \mathcal{E}_{\rm tr} \\ \left(\frac{v_s}{c}\right)^2 (\mathcal{E} - \mathcal{E}_{\rm tr}) + P_{\rm NM}(\mathcal{E}_{\rm tr}), & \mathcal{E}_{\rm tr} \leq \mathcal{E}. \end{cases} \quad (1)$$

where P and $\mathcal{E}$ are the pressure and energy density, respectively, and $\mathcal{E}_{\rm tr}$ is the transition energy density. In region $\mathcal{E} \leq \mathcal{E}_{\rm c-edge}$, we used the equation of Feynman *et al.* [37] and also of Baym *et al.* [38] for the crust and low densities of neutron star. In the intermediate region, $\mathcal{E}_{\rm c-edge} \leq \mathcal{E} \leq \mathcal{E}_{\rm tr}$, we employed a specific EoS based on the MDI model and data from Akmal *et al.* [39], while for $\mathcal{E}_{\rm tr} \leq \mathcal{E}$, the EoS is maximally stiff with the speed of sound, defined as $v_s = c\sqrt{(\partial P/\partial \mathcal{E})_S}$ (where S is the entropy). In the framework of the present work, the speed of sound is fixed on the two values, $c/\sqrt{3}$ and c. The cases which took effect in this study were the ones where the fiducial baryon transition density is $n_{\rm tr} = pn_0$, where n_0 is the saturation density of symmetric nuclear matter ($n_0 = 0.16$ fm^{-3}) and p takes the values $1, 1.5, 2, 3$. Therefore, the predicted EoSs are functional of $n_{\rm tr}$ and v_s and implemented to study their effects on the bulk neutron star properties including $M_{\rm max}$ (maximum mass of a neutron star), and Λ (tidal deformability; for definition see below Eq. (10)).

While in approach (1) the continuity on the EoS is well ensured, its artificial character does not ensure the continuity in the speed of sound at the transition density. We underline that the speed of sound must be treated carefully, especially in regions of discontinuity [40]. Therefore, in order to ensure the continuity and a smooth phase transition, a method is needed, in which we considered that above the transition density,the speed of sound is parametrized as follows (for more details see Ref. [41])

$$\frac{v_s}{c} = \left(a - c_1 \exp\left[-\frac{(n-c_2)^2}{w^2}\right]\right)^{1/2}, \quad a = 1, 1/3 \quad (2)$$

where the parameters c_1, c_2, and w are fit to the speed of sound and its derivative at $n_{\rm tr}$, and also to the demands $v_s(n_{\rm tr}) = [c, c/\sqrt{3}]$ [28]. Using Eq. (2), the EoS for $n \geq n_{\rm tr}$ can be constructed with the help of the following recipe [41]

$$\mathcal{E}_{i+1} = \mathcal{E}_i + \Delta\mathcal{E}, \quad P_{i+1} = P_i + \left(\frac{v_s}{c}(n_i)\right)^2 \Delta\mathcal{E}, \quad (3)$$

$$\Delta\mathcal{E} = \Delta n \left(\frac{\mathcal{E}_i + P_i}{n_i}\right), \quad (4)$$

$$\Delta n = n_{i+1} - n_i. \quad (5)$$

2.2 Tidal deformability of neutron stars

One of the most important sources for ground-based gravitational wave detectors are the gravitational waves from the final stages of inspiraling binary neutron stars [40, 42–49]. Especially, Flanagan and Hinderer [43] articulated that tidal effects are also potentially measurable during the early part of the evolution when the waveform is relatively clean. The response of a component neutron star in a binary neutron star system to the tidal field is described by the tidal Love number k_2, which relates the quadrupole moment Q_{ij} to the applied tidal field E_{ij} [43, 50]

$$Q_{ij} = -\frac{2}{3}k_2\frac{R^5}{G}E_{ij} \equiv -\lambda E_{ij}, \quad (6)$$

where R is the neutron star radius and $\lambda = 2R^5k_2/3G$ is the tidal deformability. The tidal Love number k_2 is given by [43, 44]

$$\begin{aligned} k_2 = & \frac{8\beta^5}{5}(1-2\beta)^2[2 - y_R + (y_R - 1)2\beta] \\ & \times \Big[2\beta(6 - 3y_R + 3\beta(5y_R - 8)) \\ & + 4\beta^3\left(13 - 11y_R + \beta(3y_R - 2) + 2\beta^2(1 + y_R)\right) \\ & + 3(1-2\beta)^2 \\ & + [2 - y_R + 2\beta(y_R - 1)]\ln(1-2\beta)\Big]^{-1} \quad (7) \end{aligned}$$

where $\beta = GM/Rc^2$ is the compactness parameter. The parameter y_R is determined by solving the relevant differential equation of $y(r)$ simultaneously with the Tolman-Oppenheimer-Volkoff equations [40].

Moving on to the binary neutron star systems, one of the binary parameters that is well constrained by the gravitational wave detectors is the chirp mass $\mathcal{M}_c$, which is a combination of the component masses [15, 17]

$$\mathcal{M}_c = \frac{(m_1m_2)^{3/5}}{(m_1+m_2)^{1/5}} = m_1\frac{q^{3/5}}{(1+q)^{1/5}}, \quad (8)$$

where m_1 is the mass of the heavier component star and m_2 is the lighter's one. Hence, the binary mass ratio $q = m_2/m_1$ is within $0 \leq q \leq 1$.

The information regarding the tidal effects of the system is transfered via the gravitational waves. This information is characterized by the effective tidal deformability [15, 17]

$$\tilde{\Lambda} = \frac{16}{13}\frac{(12q+1)\Lambda_1 + (12+q)q^4\Lambda_2}{(1+q)^5}, \quad (9)$$

where the key quantity q characterizes the mass asymmetry. Moreover, Λ_i is the dimensionless deformability defined as [15, 17]

$$\Lambda_i = \frac{2}{3}k_2\left(\frac{R_ic^2}{M_iG}\right)^5 \equiv \frac{2}{3}k_2\beta_i^{-5}, \quad i = 1, 2. \quad (10)$$

Now, by replacing in Eq. (10) the value of k_2 from Eq. (7), we found that Λ_i depends both on the compactness of the star as well as on the value of $y(R)$.

https://doi.org/10.1051/epjconf/202125205005

3 Results and Discussion

In our study we used two scenarios for the value of speed of sound, the lower bound of $(v_s/c)^2 = 1/3$ and the upper one of $(v_s/c)^2 = 1$, and four transition densities $n_{tr} = \{1, 1.5, 2, 3\}n_0$ [28]. In Fig. 1 the corresponding mass-radius diagram is displayed. The red colored lines correspond to the $(v_s/c)^2 = 1/3$ limit, while the blue ones correspond to the $(v_s/c)^2 = 1$ limit. One can observe that each taken value of transition density leads to bifurcations in M-R diagram. Between the same kind of linestyle, the upper (lower) bound $v_s = c$ ($v_c = c/\sqrt{3}$) of speed of sound corresponds to higher (lower) masses. As shown in Fig. 1, the higher the transition density, the softer the EoS, with the lower limit of $(v_s/c)^2 = 1/3$ leading to a more soft EoS, compare to the $(v_s/c)^2 = 1$ branch. In addition, the estimation of the GW170817 event and the NICER's data are also displayed [16, 51]. We notice that the specific gravitational-wave's origin (GW170817) information is more suitable for our study due to the fact that while the NICER's contour estimation covers almost all the cases, the neutron stars merger that was detected (GW170817), restricts the cases, excluding at least those with transition density $n_{tr} = n_0$, for both cases of speed of sound.

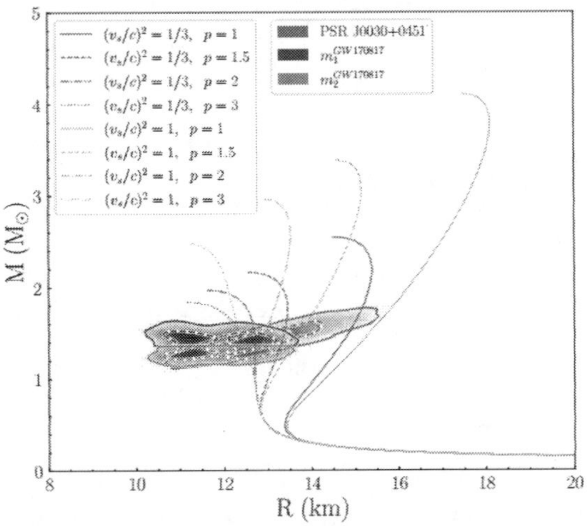

Figure 1: Mass vs radius for an isolated neutron star, and for the two cases of speed of sound bounds. The blue (red) lines correspond to the upper (lower) bound. The green diagonal shaded region corresponds to NICER's observation (data taken from Ref. [51]), while the purple upper (orange lower) shaded region corresponds to the higher (smaller) component of GW170817 event (data taken from Ref. [16]). The solid (dashed) contour lines describe the 90% (50%) confidence interval.

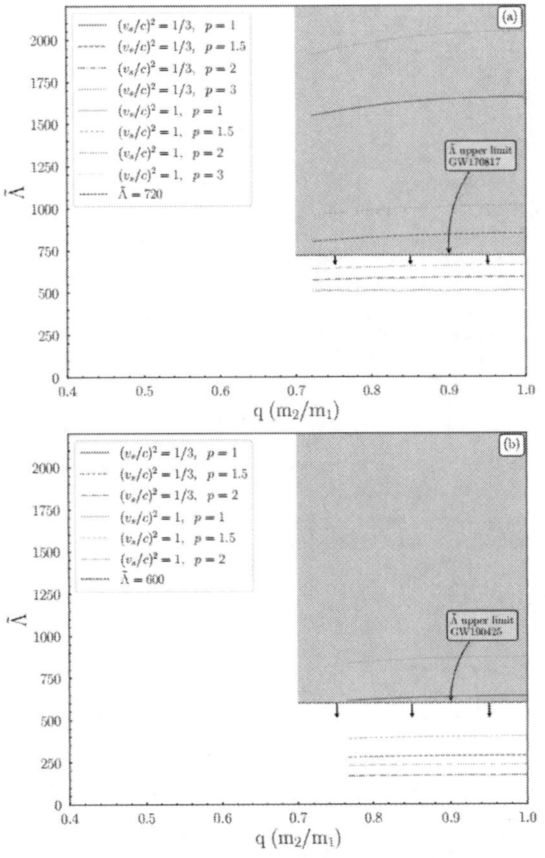

Figure 2: The tidal deformability $\tilde{\Lambda}$ as a function of the binary mass ratio q for the event (a) GW170817 and (b) GW190425. The corresponding upper observation limits for $\tilde{\Lambda}$ are also indicated, with the grey shaded region marking the excluded area. The red (blue) lines correspond to the $(v_s/c)^2 = 1/3$ ($(v_s/c)^2 = 1$) limit.

Our study takes into consideration the information from the gravitational wave emission of binary neutron stars coalescences. Hence, we used the constraints on the effective tidal deformability $\tilde{\Lambda}$, provided by the events GW170817 and GW190425 [16–18]. The chirp masses for the two events are $\mathcal{M}_c = 1.186\ M_\odot$ [15] and $\mathcal{M}_c = 1.44\ M_\odot$ [18], respectively. The component masses vary in the ranges $m_1 \in (1.36, 1.60)\ M_\odot$ and $m_2 \in (1.16, 1.36)\ M_\odot$ [17] (GW170817) and $m_1 \in (1.654, 1.894)\ M_\odot$ and $m_2 \in (1.45, 1.654)\ M_\odot$ (GW190425). We notice that we modified the range of the component masses (especially in the second event) to have an equal-mass boundary, hence $q \leq 1$.

In Fig. 2 we display the effective tidal deformability $\tilde{\Lambda}$ as a function of q for both events. The upper limit on $\tilde{\Lambda}$ in Fig. 2(a), provided by GW170817, leads to the exclusion of both cases of speed of sound for transition densities $n_{tr} = 1, 1.5\ n_0$. By comparing to Fig. 1, the constraints on the upper limit of $\tilde{\Lambda}$ in Fig. 2 distinguish in a more efficient way the cases that must be excluded. Furthermore, for the second event in Fig. 2(b), we observed that in general all EoSs are shifted downwards. This behavior is because of the higher value of chirp mass and the component masses. Contrary to the GW170817 event, the upper limit on $\tilde{\Lambda}$, provided by GW190425 event, ex-

https://doi.org/10.1051/epjconf/202125205005

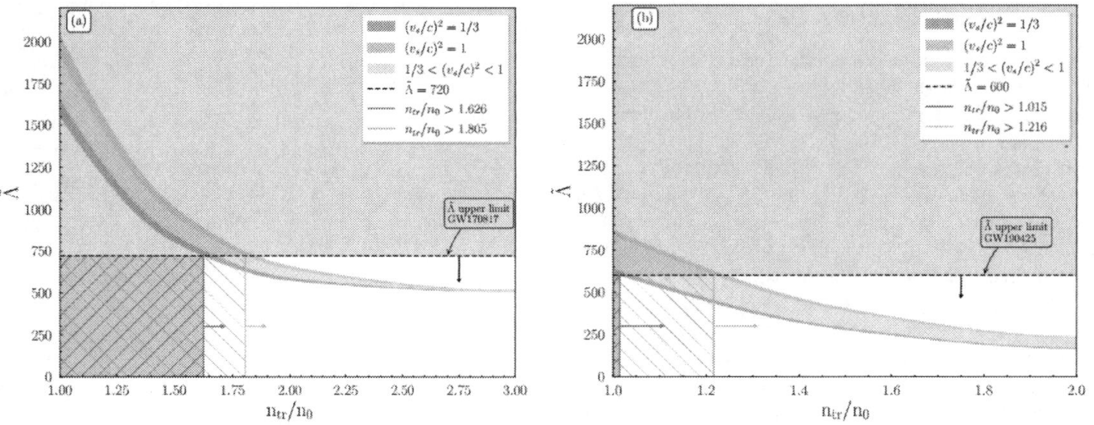

Figure 3: $\tilde{\Lambda}$ as a function of the transition density n_{tr} (in units of saturation density n_0) at the maximum mass configuration for the two speed of sound bounds $v_s = c/\sqrt{3}$ and $v_s = c$ and for the event (a) GW170817 (left panel) and (b) GW190425 (right panel). The corresponding upper observation limits for $\tilde{\Lambda}$ [17, 18] as well as the compatible lower transition density values are also indicated for both events. The red (blue) arrow marks the accepted region of transition density for the $v_s = c/\sqrt{3}$ ($v_s = c$) case. The red lower (blue upper) curved shaded region corresponds to the $v_s = c/\sqrt{3}$ ($v_s = c$) limit. The intermediate region between the two cases of speed of sound bounds is displayed with yellow color.

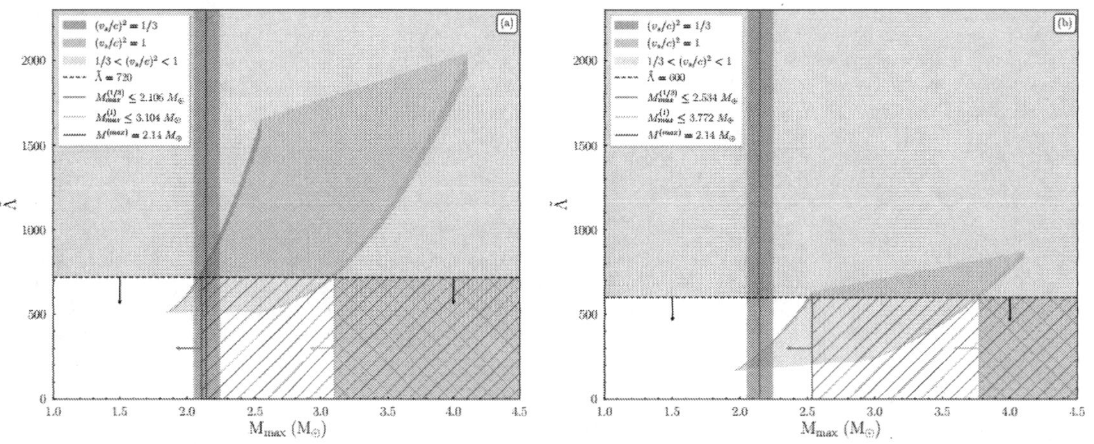

Figure 4: $\tilde{\Lambda}$ as a function of the maximum mass for the two speed of sound bounds $v_s = c/\sqrt{3}$ and $v_s = c$ and for the event (a) GW170817 (left panel) and (b) GW190425 (right panel). The corresponding upper observation limits for $\tilde{\Lambda}$ (black dashed lines with arrows, see Refs. [17, 18]), the compatible maximum mass shaded regions, for $v_s = c/\sqrt{3}$ (left red) case, $v_s = c$ case (right blue) and the middle cases (yellow), as well the current observed maximum neutron star mass $M = 2.14^{+0.10}_{-0.09}\ M_\odot$ (purple shaded vertical region, see Ref. [21]) are also indicated. The red left (blue right) arrow marks the accepted region of maximum mass M_{max} for $v_s = c/\sqrt{3}$ ($v_s = c$) case.

cludes only the EoS with transistion density $n_{tr} = n_0$, for both speed of sound cases. In general for both events, the EoSs with higher values of transition density n_{tr} have smaller values of $\tilde{\Lambda}$. Therefore, the constraints on $\tilde{\Lambda}$ derived from gravitational-waves events favor softer EoSs. At this point we have to underline that for the GW190425 event, we removed from our study the cases with transition density $n_{tr} = 3\ n_0$ because of the fact that the EoS with $(v_s/c)^2 = 1/3$ and $n_{tr} = 3n_0$ can not reproduce the masses of this event.

In Fig. 3 we display the dependence of the effective tidal deformability $\tilde{\Lambda}$ on the transition density n_{tr} at the maximum mass configuration for the two speed of sound bounds, $v_s = c/\sqrt{3}$ and $v_s = c$, and the two events GW170817 (Fig. 3(a)) and GW190425 (Fig. 3(b)). The corresponding upper observational limits for $\tilde{\Lambda}$, as well as the compatible lower transition density values, are also indicated in each case. The middle region corresponds to bounds on the speed of sound between the two limit cases that we studied. In the case of the GW170817 event the lower limit for the transition density is $1.626n_0$ for $v_s = c/\sqrt{3}$ and $1.805n_0$ for $v_s = c$. Correspondingly, for the second event GW190425 the limits are $1.015n_0$ for $v_s = c/\sqrt{3}$ and $1.216n_0$ for $v_s = c$. Therefore, the first event imposes more stringent constraints on the EoS. Especially, the result for the case of $v_s = c/\sqrt{3}$ is in agreement with other studies (see Ref. [52]). For the second event GW190425, the higher masses that characterize it lead to smaller values on $\tilde{\Lambda}$, hence the upper limit on $\tilde{\Lambda}$ cannot provide any further constraints. In Ref. [24] there is a more detailed description of the results.

In Fig. 4 we display the effective tidal deformability $\tilde{\Lambda}$ as a function of the maximum mass for the two speed of sound bounds and for both events. The corresponding upper observational limit for $\tilde{\Lambda}$ (black dashed horizontal line), the compatible maximum mass in each case (horizontal arrows), as well as the current observed maximum neutron star mass $M = 2.14^{+0.10}_{-0.09}\ M_\odot$ [21] (shaded purple vertical region) are also indicated. For the first event the upper bound of $\tilde{\Lambda}$ is compatible with a maximum mass value $2.106\ M_\odot$ for $v_s = c/\sqrt{3}$ and $3.104\ M_\odot$ for $v_s = c$. The upper limit on $M_{\rm max}$ for the case of $(v_s/c)^2 = 1/3$ lays roughly inside the estimation of the observed maximum mass. There are two different points of view that antagonize one another. To be more specific, the constraints derived by the upper limit on $\tilde{\Lambda}$ lead to more soft EoSs, contrary to the observational estimations of the maximum mass of neutron stars which lead to more stiff EoSs. This difference becomes less contradictory as the speed of sound takes larger values, with the causal scenario of $v_s = c$ leading to a very wide area for the maximum mass. Moving on to the second event, the correspondig value for the maximum mass is $M_{\rm max} \leq 2.534\ M_\odot$ for $v_s = c/\sqrt{3}$ and $M_{\rm max} \leq 3.772\ M_\odot$ for $v_s = c$. As one can observe, the second event (which is in general less informative than the first one) cannot provide any further constraints.

4 Concluding remarks

In this work we studied possible constraints on the speed of sound (which affects the stiffness of the EoS) and transition density $n_{\rm tr}$ which are based on the recent observations of the events GW170817 and GW190425. The method which was implemented involved mainly the upper limits of the effective tidal deformability $\tilde{\Lambda}$ (estimated from the mentioned events), combined with measurements and estimations of the maximum neutron star mass. As basis of our study, we used the APR1-MDI EoS, for two boundary cases of speed of sound [24]; the lower bound of $v_s = c/\sqrt{3}$ and the upper one of $v_s = c$.

Firstly, we examined the behavior of EoSs for each case, by using the information on the component masses of GW170817 event in combination with the recent NICER's data. The M-R diagram was able to impose some robust constraints, by excluding the more stiff EoSs with $n_{\rm tr} = n_0$. Moreover, the upper limit on $\tilde{\Lambda}$, from $\tilde{\Lambda} - q$ diagram for both events, imposed more stringent constraints on the speed of sound bounds. Especially, the constraints from the first event (GW170817), which is more informative than the second one (GW190425), led to the additionally exclusion of EoSs with $n_{\rm tr} = 1.5n_0$.

Subsequently, we studied the $\tilde{\Lambda}^{(1/3,1)}_{\rm min,max}$ as a function of $n_{\rm tr}$, where the parenthesis (min,max) denotes the minimum and maximum case respectively, for both speed of sound bounds (indicated as (1,1/3)). Therefore we were able to impose more stringent constraints, by using the upper limit on $\tilde{\Lambda}$ (derived from gravitational-wave observations) and taking advantage of the variety on $\tilde{\Lambda}$ for each EoS. From the first event (GW170817) we found that the speed of sound must be lower than the value $v_s = c/\sqrt{3}$ at least up to densities $n_{\rm tr} \approx 1.6n_0$ and lower than $v_s = c$ up to densities $n_{\rm tr} \approx 1.8n_0$. The respective values derived from the GW190425 event are $n_{\rm tr} \approx n_0$ for the lower speed of sound bound and $n_{\rm tr} \approx 1.2n_0$ for the upper one. We notice that the constraints from the event GW170817 are more stringent compared to the GW190425 event.

Moreover, we extended our approach by treating the effective tidal deformability $\tilde{\Lambda}$ as a function of the maximum mass $M_{\rm max}$ for both cases of speed of sound. From the first event (GW170817) we obtained that the maximum mass should be $M_{\rm max} \leq 2.106\ M_\odot$ for the $v_s = c/\sqrt{3}$ bound and $M_{\rm max} \leq 3.104\ M_\odot$ for the upper bound $v_s = c$. The limit of $M_{\rm max} \approx 2.11\ M_\odot$ corresponds to a transition density equal to $n_{\rm tr} \approx 1.5n_0$. According to this finding, the limit $v_s = c/\sqrt{3}$ is in contradiction with the observational evidence on the $M_{\rm max}$ of neutron stars and must be violated in order to be able to simultaneously describe small values of the effective tidal deformability and high values for neutron star mass. This contradiction lays into the antagonized points of view; the upper limit on $\tilde{\Lambda}$ favors softer EoSs leading to higher values of $n_{\rm tr}$, while the observational information regarding to the maximum mass $M_{\rm max}$ of a neutron star requires stiffer EoSs, leading to smaller values of $n_{\rm tr}$. We notice that the second event GW190425, was not able to offer further constraints.

We postulate that future observations may offer even more rigorous constraints on the bound of the speed of sound. To be more specific, the detection of future events could provide further information on the upper limit of $\tilde{\Lambda}$, leading to more stringent constraints on $n_{\rm tr}$ and the sound speed bounds. According to our approach, the more informative events, for the lower limit of $n_{\rm tr}$, would be those with lighter masses. Moreover, further detection of neutron stars mergers will assist both on the neutron stars maximum mass determination and its link to the speed of sound, shedding more light on the open problem of the speed of sound of dense nuclear matter and EoS.

References

[1] S. L. Shapiro and S. A. Teukolsky, *Black Holes, White Dwarfs, and Neutron Stars* (John Wiley and Sons, New York, 1983)

[2] N. K. Glendenning, *Compact Stars: Nuclear Physics, Particle Physics, and General Relativity* (Springer, Berlin, 2000)

[3] P. Haensel, A. Y. Potekhin, and D. G. Yakovlev, *Neutron Stars 1: Equation of State and Structure* (Springer-Verlag, New York, 2007)
[4] S. Weinberg, *Gravitational and Cosmology: Principle and Applications of the General Theory of Relativity* (Wiley, New York, 1972) 47–52
[5] Ya. B. Zel'dovich and I. D. Novikov, *Stars and Relativity* (Dover Publications, INC, Mineapolis New York, 1971)
[6] Ya. B. Zel'dovich, Sov. Phys. JETP **14**, 1143 (1962)
[7] J. B. Hartle, Phys. Rep. **46**, 201 (1978)
[8] P. Bedaque and A.W. Steiner, Phys. Rev. Lett. **114**, 031103 (2015)
[9] B. Reed and C.J. Horowitz, Phys. Rev. C **101** (4), 045803 (2020)
[10] C. C. Moustakidis, T. Gaitanos, C. Margaritis, and G. A. Lalazissis, Phys. Rev. C **95**, 045801 (2017)
[11] E. D. Van Oeveren and J. L. Friedman, Phys. Rev. D **95**, 083014 (2017)
[12] Y.-L. Ma and M. Rho, Phys. Rev. D **100**, 114003 (2019)
[13] G. Baym *et al.*, Rept. Prog. Phys. **81** (5), 056902 (2018)
[14] E. Annala, T. Gorda, A. Kurkela, J. Nattila, and A. Vuorinen, Nat. Phys. **16**, 907 (2020)
[15] B.P. Abbott *et al.*, Phys. Rev. Lett. **119**, 161101 (2017)
[16] B. P. Abbott *et al.*, Phys. Rev. Lett. **121**, 161101 (2018)
[17] B. P. Abbott *et al.*, Phys. Rev. X **9**, 011001 (2019)
[18] B. P. Abbott *et al.*, Astrophys. J. Lett. **892**, L3 (2020)
[19] Z. Arzoumanian *et al.*, Astrophys. J. Suppl. Ser. **235**, 37 (2018)
[20] J. Antoniadis *et al.*, Science **340**, 1233232 (2013)
[21] H. Cromartie *et al.*, Nat. Astron. **4**, 72 (2020)
[22] M. Linares, T. Shahbaz, and J. Casares, Astrophys. J. **859**, 54 (2018)
[23] P.S. Koliogiannis and C.C. Moustakidis, Phys. Rev. C **101**, 015805 (2020)
[24] A. Kanakis-Pegios, P. S. Koliogiannis, and C. C. Moustakidis, Phys. Rev. C **102**, 055801 (2020)
[25] R. Abbott *et al.*, Astrophys. J. **896**, L44 (2020).
[26] H. Tan, J. Noronha-Hostler, and N. Yunes, (2020) [arXiv:2006.16296 [astro-ph.HE]].
[27] A. Kanakis-Pegios, P. S. Koliogiannis, and C. C. Moustakidis, Symmetry **13**, 183 (2021)
[28] Ch. Margaritis, P. S. Koliogiannis, and Ch. C. Moustakidis, Phys. Rev. D **101**, 043023 (2020)
[29] C. E. Rhoades and R. Ruffini, Phys. Rev. Lett. **32**, 324 (1974)
[30] V. Kalogera and G. Baym, Astrophys. J. **470**, L61 (1996)
[31] S. Koranda, N. Stergioulas, and J. L. Friedman, Astrophys. J. **488**, 799 (1997)
[32] N. Chamel, P. Haensel, J. L. Zdunik, and A. F. Fantina, Int. J. Mod. Phys. E **22**, 1330018 (2013)
[33] C. Breu and L. Rezzolla, Mon. Not. R. Astron. Soc. **459**, 646 (2016)
[34] J. Alsing, O. H. Silva, and E. Berti, Mon. Not. R. Astron. Soc. **478**, 1377 (2018)
[35] D. M. Podkowka, R. F. P. Mendes, and E. Poisson, Phys. Rev. D **98**, 064057 (2018)
[36] C. Xia, Z. Zhu, X. Zhou, and A. Li, (2019) [arXiv: 1906.00826 [nucl-th]]
[37] R. Feynman, N. Metropolis, and E. Teller, Phys. Rev. **75**, 1561 (1949)
[38] G. Baym, C. Pethik, and P. Sutherland, Astrophys. J. **170**, 299 (1971)
[39] A. Akmal, V.R. Pandharipande, and D.G. Ravenhall, Phys. Rev. C **58**, 1804 (1998)
[40] S. Postnikov, M. Prakash, and J. M. Lattimer, Phys. Rev. D **82**, 024016 (2010)
[41] I. Tews, J. Carlson, S. Gandolfi, and S. Reddy, Astrophys. J. **860**, 149 (2018)
[42] L. Baiotti, Prog. Part. Nucl. Phys. **109**, 103714 (2019)
[43] E.E. Flanagan and T. Hinderer, Phys. Rev. D **77**, 021502(R) (2008)
[44] T. Hinderer, Astrophys. J. **677**, 1216 (2008)
[45] T. Damour and A. Nagar, Phys. Rev. D **80**, 084035 (2009)
[46] T. Hinderer, B. D. Lackey, R. N. Lang, and J. S. Read, Phys.Rev. D **81**, 123016 (2010)
[47] F. J. Fattoyev, J. Carvajal, W. G. Newton, and B. A. Li, Phys. Rev. C **87**, 015806 (2013)
[48] B. D. Lackey and L. Wade, Phys. Rev. D **91**, 043002 (2015)
[49] J. Takátsy and P. Kovács, Phys. Rev. D **102** (2), 028501 (2020)
[50] Kip S. Thorne, Phys. Rev. D **58**, 124031 (1998)
[51] M. C. Miller *et al.*, Astrophys. J. Lett. **887**, L24 (2019)
[52] S.K. Greif, G. Raaijmakers, K. Hebeler, A. Schwenk, and A.L. Watts, Mon. Not. Roy. Astron. Soc. **485** (4), 5363 (2019)

Momentum dependent mean-fields of hyperons & antihyperons

Arsenia Chorozidou[1,*] and *Theodoros* Gaitanos[1,**]

[1]Department of Physics, Aristotle University of Thessaloniki

Abstract. The in-medium properties of hyperons and antihyperons are studied with the Non-Linear Derivative (NLD) model and focus is made on the momentum dependence of strangeness optical potentials. The NLD model is based on the Relativistic Mean Field (RMF) approximation to Relativistic Hadrodynamics (RHD) approach of nuclear systems, but it incorporates an explicit momentum dependence of mean-fields. The extension of the NLD model to the baryon and antibaryon octet is based on SU(6) and G-parity arguments. It is demonstrated that with a proper choice of momentum cut-offs, the Λ and Σ optical potentials are consistent with recent studies of the chiral effective field theory (χ -EFT) and optical potentials are consistent with Lattice-QCD calculations, over a wide momentum region. We also present NLD predictions for the in-medium momentum dependence of $\bar{\Lambda}$, $\bar{\Sigma}$ and $\bar{\Xi}$ hyperons. This work is important for future experimental studies, like CBM, PANDA at FAIR and is relevant to nuclear astrophysics as well.

1 Introduction

In an attempt to explain both astrophysical observations ($2M_{\odot}$ Neutron Stars, that seems to exclude soft hadronic Equation of State (EoS) at high baryon densities) and results from Heavy-Ion Collisions (they lead to a softening of the high-density EoS) with a proper nuclear matter model, an inclusion of hyperons degrees of freedom is performed in the NLD model. The NLD model is based on a basic RMF Lagrangian formulation, but it incorporates higher order derivatives in the nucleon-nucleon (NN) interaction Lagrangians. In previous work [1], it has been demonstrated that this Ansatz corrects the high-momentum behaviour of the interaction, makes the EoS softer at densities just above saturation, but at the same time it reproduces the $2M_{\odot}$ pulsars at densities far beyond saturation. Given that the NN interaction is fairly well known in contrast to the hyperon-nucleon (YN) interaction, in order to approach YN interaction, we take the tested for nucleons NLD model and extend it to the strangeness degrees of freedom, in the spirit of SU(6) and G-parity arguments. Having done that, we discuss the momentum dependence of the in-medium YN-potentials.

2 The NLD model for the baryon octet

The NLD Lagrangian is based on the conventional RHD and for the baryon octet it reads as

$$
\begin{aligned}
\mathcal{L} =& \frac{1}{2}\sum_B \left[\overline{\Psi}_B\gamma_\mu i\overrightarrow{\partial}^\mu\Psi_B - \overline{\Psi}_B i\overleftarrow{\partial}^\mu\gamma_\mu\Psi_B\right] \\
& - \sum_B m_B\overline{\Psi}_B\Psi_B - \frac{1}{2}m_\sigma^2\sigma^2 + \frac{1}{2}\partial_\mu\sigma\partial^\mu\sigma - U(\sigma) \\
& + \frac{1}{2}m_\omega^2\omega_\mu\omega^\mu - \frac{1}{4}F_{\mu\nu}F^{\mu\nu} \\
& + \frac{1}{2}m_\rho^2\vec{\rho}_\mu\vec{\rho}^\mu - \frac{1}{4}\vec{G}_{\mu\nu}\vec{G}^{\mu\nu} - \frac{1}{2}m_\delta^2\vec{\delta}^2 + \frac{1}{2}\partial_\mu\vec{\delta}\,\partial^\mu\vec{\delta} \\
& + \mathcal{L}_{int}^\sigma + \mathcal{L}_{int}^\omega + \mathcal{L}_{int}^\rho + \mathcal{L}_{int}^\delta \,. \qquad (1)
\end{aligned}
$$

The sum over B runs over the baryonic octet

$$\Psi_B = (\Psi_N, \Psi_\Lambda, \Psi_\Sigma, \Psi_\Xi)^T \qquad (2)$$

with

$$\Psi_N = (\psi_p, \psi_n)^T, \quad \Psi_\Lambda = \psi_\Lambda \qquad (3)$$

$$\Psi_\Sigma = (\psi_{\Sigma^+}, \psi_{\Sigma^0}, \psi_{\Sigma^-})^T, \quad \Psi_\Xi = (\psi_{\Xi^0}, \psi_{\Xi^-})^T \qquad (4)$$

*e-mail: achoro@auth.gr
**e-mail: tgaitano@auth.gr

https://doi.org/10.1051/epjconf/202125205006

for the isospin-doublets Ψ_N and Ψ_Ξ, isospin-triplet Ψ_Σ and the neutral Ψ_Λ. The interactions between the nucleon fields are described by the exchange of meson fields. These are the scalar σ and vector ω^μ mesons in the isoscalar channel, as well as the scalar $\vec{\delta}$ and vector $\vec{\rho}^\mu$ mesons in the isovector channel. Their corresponding Lagrangian densities are of the Klein-Gordon and Proca types, respectively Eq. (5, 6 and 7). The term $U(\sigma) = \frac{1}{3}b\sigma^3 + \frac{1}{4}c\sigma^4$ contains the usual selfinteractions of the σ meson. The notations for the masses of fields in Eq. (1) are obvious. The field strength tensors are defined as $F^{\mu\nu} = \partial^\mu\omega^\nu - \partial^\nu\omega^\mu$, $\vec{G}^{\mu\nu} = \partial^\mu\vec{\rho}^{\,\nu} - \partial^\nu\vec{\rho}^{\,\mu}$ for the isoscalar and isovector fields, respectively. In the following we restrict to a minimal set of interaction degrees of freedom. In the iso-scalar sector, the σ- and ω-fields are obviously considered. In the iso-vector channel, we keep the vector, isovector ρ-meson field and neglect the δ-field.

The NLD interaction Lagrangians contain the conventional and simple combinations between the bilinear baryon-and linear meson-fields, however, they are extended by the inclusion of non-linear derivative operators $\overrightarrow{\mathcal{D}}, \overleftarrow{\mathcal{D}}$ for each baryon species B:

$$\mathcal{L}^{\sigma}_{int} = \sum_B \frac{g_{\sigma B}}{2}\left[\overline{\Psi}_B \overleftarrow{\mathcal{D}}_B \Psi_B \sigma + \sigma \overline{\Psi}_B \overrightarrow{\mathcal{D}}_B \Psi_B\right] \quad (5)$$

$$\mathcal{L}^{\omega}_{int} = -\sum_B \frac{g_{\omega B}}{2}\left[\overline{\Psi}_B \overleftarrow{\mathcal{D}}_B \gamma^\mu \Psi_B \omega_\mu + \omega_\mu \overline{\Psi}_B \gamma^\mu \overrightarrow{\mathcal{D}}_B \Psi_B\right] \quad (6)$$

$$\mathcal{L}^{\rho}_{int} = -\sum_B \frac{g_{\omega\rho}}{2}\left[\overline{\Psi}_B \overleftarrow{\mathcal{D}}_B \gamma^\mu \vec{\tau} \Psi_B \vec{\rho}_\mu + \vec{\rho}_\mu \overline{\Psi}_B \vec{\tau} \gamma^\mu \overrightarrow{\mathcal{D}}_B \Psi_B\right] \quad (7)$$

for the isoscalar-scalar, isoscalar-vector and isovector-vector vertices, respectively. The arrows on the non-linear operator $\mathcal{D}$ indicate the direction of their action. The only difference with respect to the conventional RHD interaction Lagrangian is the presence of additional operator functions $\overrightarrow{\mathcal{D}}_B$, $\overleftarrow{\mathcal{D}}_B$. These operator functions will regulate the high momentum component of hyperons. These operator functions (or regulators) $\overrightarrow{\mathcal{D}}_B$, $\overleftarrow{\mathcal{D}}_B$ are hermitician and generic functions of partial derivative operator. That is, $\overrightarrow{\mathcal{D}}_B := \mathcal{D}\left(\overrightarrow{\xi}_B\right)$ and $\overleftarrow{\mathcal{D}}_B := \mathcal{D}\left(\overleftarrow{\xi}_B\right)$ with the operator arguments $\overrightarrow{\xi}_B = -\zeta^\alpha_B i\overrightarrow{\partial}_\alpha$, $\overleftarrow{\xi}_B = i\overleftarrow{\partial}_\alpha \zeta^\alpha_B$. The four vector $\zeta^\mu_B = v^\mu/\Lambda_B$ contains the cut-off Λ_B and v^μ is an auxiliary vector. They are assumed to act on the baryon spinors Ψ_B and $\overline{\Psi}_B$ by a formal Taylor expansion with respect to the operator argument. The functional form of the regulators is constructed such that in the limit $\Lambda_B \to \infty$ the original RHD Lagrangians are recovered, that is, $\overrightarrow{\mathcal{D}}_B = \overleftarrow{\mathcal{D}}^\dagger_B \to 1$.

From the generalised Euler-Lagrange formalism, we obtain the equations of motion for the degrees of freedom in the NLD model. The meson field equations of motion read

$$\partial_\alpha\partial^\alpha\sigma + m^2_\sigma\sigma + \frac{\partial U}{\partial\sigma} = = \frac{1}{2}\sum_B g_{\sigma B}\left[\overline{\Psi}_B \overleftarrow{\mathcal{D}}_B \Psi_B + \overline{\Psi}_B \overrightarrow{\mathcal{D}}_B \Psi_B\right], \quad (8)$$

$$\partial_\mu F^{\mu\nu} + m^2_\omega\omega^\nu = = \frac{1}{2}\sum_B g_{\omega B}\left[\overline{\Psi}_B \overleftarrow{\mathcal{D}}_B \gamma^\nu \Psi_B + \overline{\Psi}_B \gamma^\nu \overrightarrow{\mathcal{D}}_B \Psi_B\right], \quad (9)$$

$$\partial_\mu G^{\mu\nu} + m^2_\rho\vec{\rho}^{\,\nu} = = \frac{1}{2}\sum_B g_{\rho B}\left[\overline{\Psi}_B \overleftarrow{\mathcal{D}}_B \gamma^\nu \vec{\tau} \Psi_B + \overline{\Psi}_B \vec{\tau} \gamma^\nu \overrightarrow{\mathcal{D}}_B \Psi_B\right], \quad (10)$$

for the isoscalar-scalar, isoscalar-vector and isovector-vector exchange mesons, respectively. Each baryon-field obeys a Dirac-equation of the following type

$$\left[\gamma_\mu(i\partial^\mu - \Sigma^\mu_B) - (m_B - \Sigma_{sB})\right]\psi_B = 0\,, \quad (11)$$

with the selfenergies Σ^μ_B and Σ_{sB} defined as

$$\Sigma^\mu_B = g_{\omega_B}\omega^\mu \overrightarrow{\mathcal{D}}_B + g_{\rho B}\vec{\tau}_B \cdot \vec{\rho}^{\,\mu} \overrightarrow{\mathcal{D}}_B\,, \quad (12)$$

$$\Sigma_{sB} = g_{\sigma B}\sigma \overrightarrow{\mathcal{D}}_B\,. \quad (13)$$

Both Lorentz-components of the selfenergy, Σ^μ and Σ_s, show an explicit linear behavior with respect to the meson fields σ, ω^μ and $\vec{\rho}^{\,\mu}$ as in the standard RHD. However, they contain an additional dependence on the regulators. If we apply the RMF approximation to static hadronic matter (the spatial components of the meson fields in Minkowski- and isospin-spaces vanish, $\omega^\mu \to (\omega^0, \vec{0})$ and $\vec{\rho}^{\,\mu} \to (\rho^0_3, \vec{0}_3)$), we end up to the following forms

https://doi.org/10.1051/epjconf/202125205006

for the NLD meson-field equations

$$m_\sigma^2\sigma + \frac{\partial U}{\partial\sigma} = \sum_B g_{\sigma B}\left\langle\overline{\psi}_B \mathcal{D}_B \psi_B\right\rangle = \sum_B g_{\sigma B}\,\rho_{sB}\,, \tag{14}$$

$$m_\omega^2\omega = \sum_B g_{\omega B}\left\langle\overline{\Psi}_B\gamma^0\mathcal{D}_B\Psi_B\right\rangle = \sum_B g_{\omega B}\,\rho_{0B} \tag{15}$$

with the scalar and vector density sources

$$\rho_{sB} = \int\limits_{|\vec{p}|\leq p_{F_B}} d^3p\,\frac{m_B^*}{\Pi_B^0}\,\mathcal{D}_B(p)\,, \tag{16}$$

$$\rho_{0B} = \int\limits_{|\vec{p}|\leq p_{F_B}} d^3p\,\frac{E_B^*}{\Pi_B^0}\,\mathcal{D}_B(p)\,. \tag{17}$$

The isovector densities are calculated through the standard isospin relations. For a hyperon with a given momentum relative to nuclear matter at rest (at a given nucleon density and isospin asymmetry) the mesonic sources contain only nucleons, $B = p, n$.

The meson-field equations of motion show a similar structure as those of the standard RMF approximation. However, the substantial difference between NLD and other conventional RMF models appears in the source terms which now contain in addition the momentum-dependent regulators $\mathcal{D}_B$. This is an important feature of the NLD model. The cut-off leads naturally to a particular suppression of the vector field at high densities or high Fermi-momenta in agreement with phenomenology, as discussed in detail in the previous work [1]. This feature is absent in conventional RHD approaches, except if one introduces by hand additional and complicated scalar/vector self-interactions.

The key observable for general discussions related to momentum or energy dependencies of in-medium hadronic potentials is the Schrödinger-equivalent optical potential U_{opt}, which is a complex quantity. The imaginary part describes the scattering processes of a given particle, e.g. a hyperon, with a nucleon of nuclear matter. The real part of the optical potential is related to the mean-field that a particle, e.g. a hyperon with a given momentum, experiences in the nuclear medium at a given density and isospin-asymmetry. The imaginary part of U_{opt} cannot be calculated within a conventional RMF prescription. However in the NLD model (given that it contains an explicit momentum dependence of the mean-fields and thus of the optical potential) we can give, at least, estimations for the imaginary part. We do this in the case of anti-hyperons. The real part of the Schrödinger-equivalent optical potential for hyperons is obtained from a non-relativistic reduction of the Dirac-equation and reads

$$U_{opt}^B = -S_B + \frac{E_B}{m_B}V_B + \frac{1}{2m_B}\left(S_B^2 - V_B^2\right)\,. \tag{18}$$

It describes the in-medium interaction of a baryon species B with a momentum p, or single-particle energy $E_B = E_B(p)$, relative to nuclear matter at rest at a given density and isospin asymmetry. We will compare the NLD results with recent microscopic calculations from χ-EFT [2, 3] and Lattice-QCD [4] for the hyperon in-medium potentials.

3 Results

We consider nuclear matter at rest, at a given density, isospin asymmetry and at zero temperature, in which hyperons are situated at a given momentum relative to the nuclear matter at rest. The quantity of interest is the optical potential U_{opt} of the in-medium hyperons. There is no experimental information on the momentum dependence of the in-medium hyperonic potentials, so, as it has been mentioned before, we use for comparison some recent microscopic calculations. For Λ and Σ hyperons we compare with calculations based on the chiral-Effective Field Theory [2, 3] and for the Ξ hyperon we compare with calculations based on Lattice-QCD [4]. In the NLD calculations we take as a parameter only the strangeness cut-off of the hyperons and it has a monopole form

$$\mathcal{D}_Y(p) = \frac{{\Lambda_{\gamma_1}}^2}{{\Lambda_{\gamma_2}}^2 + \vec{p}^{\,2}}\,, \tag{19}$$

with $\gamma = \sigma, \omega, \rho$ indicating the cut-off values for the hyperon-nucleon σ, ω and ρ vertices, respectively, and $Y = \Lambda, \Sigma, \Xi$ denotes the hyperon type.

At first, the cut-offs of the hyperons have to be determined. The strangeness S=1 cut-offs are adjusted to the corresponding hyperonic optical potentials at saturation density of symmetric and cold nuclear matter from χ-EFT calculations. This is shown in Fig. 1 for the optical potential of Λ-hyperons and in in Fig. 2 for the optical potential of Σ-hyperons. The NLD calculations are able to describe the microscopic χ-EFT results in NLO very well. The corresponding values for the strangeness cut-off's are tabulated in Table 1.

Λ cut-off				
Λ_σ	Λ_{ω_1}	Λ_{ω_2}	Λ_{ρ_1}	Λ_{ρ_2}
0.7	0.85	0.79	–	–
Σ cut-off				
Λ_σ	Λ_{ω_1}	Λ_{ω_2}	Λ_{ρ_1}	Λ_{ρ_2}
0.67	0.95	0.79	0.47	0.47
			0.63	0.5
Ξ cut-off				
Λ_σ	Λ_{ω_1}	Λ_{ω_2}	Λ_{ρ_1}	Λ_{ρ_2}
0.6	0.8	0.71	1.3	1.2

Table 1. Λ-, Σ- and Ξ-regulators for σ- (Λ_σ), ω-($\Lambda_{\omega1,2}$) & ρ-hyperon ($\Lambda_{\rho_{1,2}}$) vertices in units of GeV. In the cases for Σ and Ξ the isospin cut-offs ($\Lambda_{\rho_{1,2}}$) are relevant for the charged particles only. For the Σ-hyperon different cut-offs values are used for Σ^- (upper line) and for Σ^+ (bottom line).

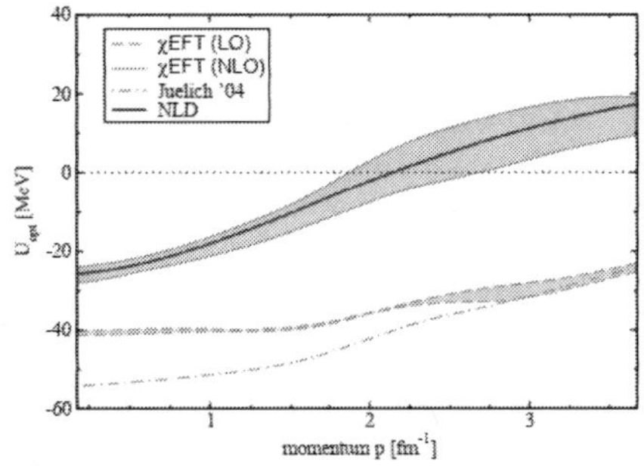

Figure 1. Optical potential of Λ-hyperons as function of their momentum p in symmetric nuclear matter at saturation density. The NLD-results (thick-solid curve) are compared with χ-EFT microscopic calculations at different orders LO (band with closed dashed borders) and NLO (band with closed solid borders). Further microscopic calculations of the Jülich group (dot-dashed curve) are shown too.

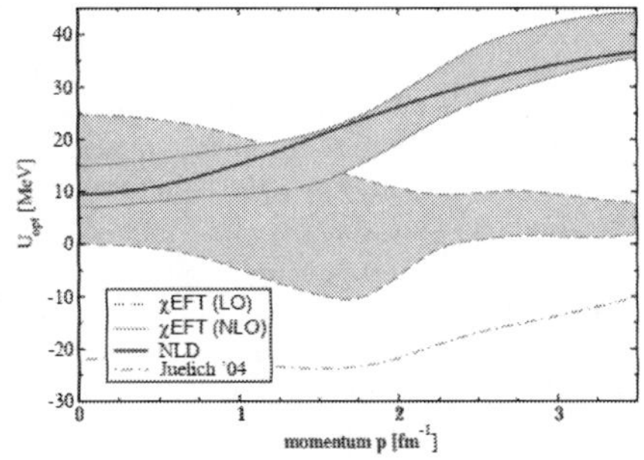

Figure 2. Same as in Fig. 1, but for Σ-hyperons.

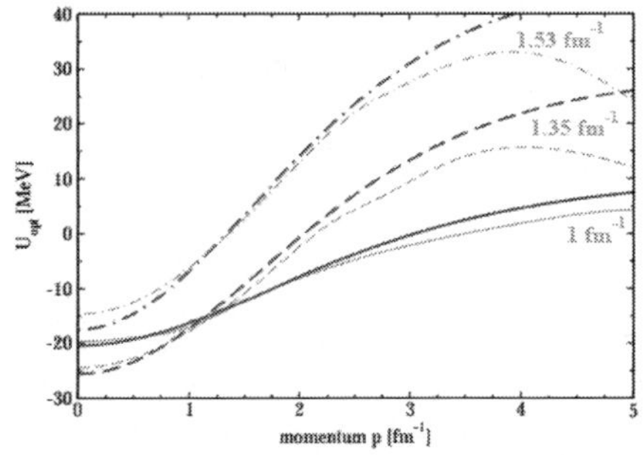

Figure 3. Optical potential of Λ-hyperons versus their momentum at various densities of symmetric nuclear matter. The NLD calculations at three Fermi-momenta (thick-solid, thick-dashed and thick-dot-dashed curves) as indicated are compared to the χ-EFT calculations in NLO.

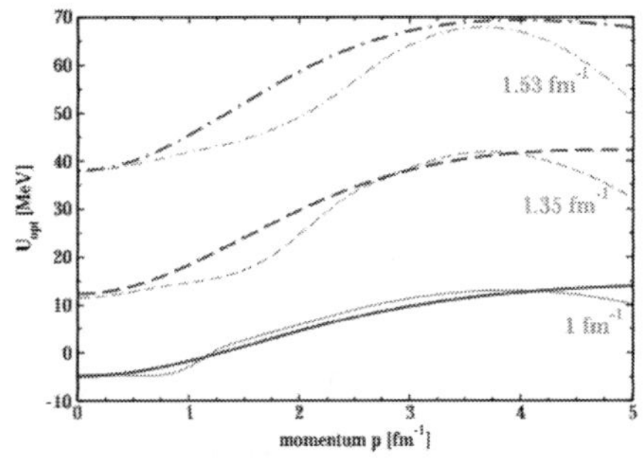

Figure 4. Same as in Fig. 3, but for the Σ-hyperons.

In Figs. 3 and 4 we show the predictive power of the NLD approach for Λ and Σ hyperons respectively, by comparing in more detail the density and momentum dependence of the NLD formalism with the χ-EFT calculations. We do this at various densities of symmetric nuclear matter.

In Figure 5 there are results for pure neutron matter at three different baryon densities. The NLD model does predict the general microscopic trends.

https://doi.org/10.1051/epjconf/202125205006

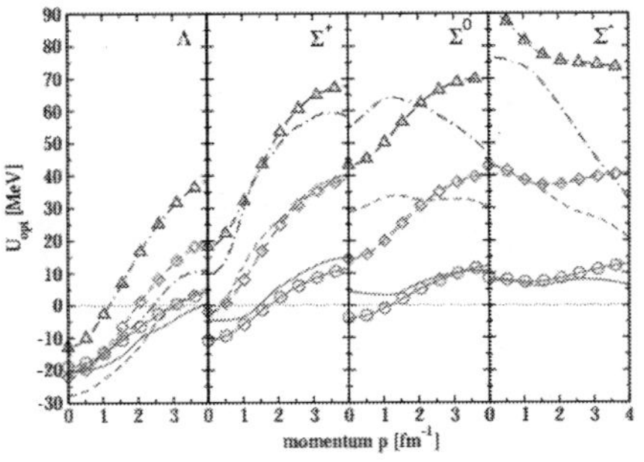

Figure 5. Optical potentials for hyperons (as indicated) versus their momentum for pure neutron matter. Solid curves with symbols indicate the NLD calculations while pure curves without symbols are the microscopic χ-EFT results from [2]. Green pairs (circles-solid for NLD and solid for χ-EFT) refer to low density of $p_F = 1$ fm^{-1}, red pairs (diamonds-solid for NLD and dashed for χ-EFT) refer to saturation density of $p_F = 1.35$ fm^{-1} and blue pairs (triangles-solid for NLD and dashed-dotted for χ-EFT) refer to a density of $p_F = 1.53$ fm^{-1}.

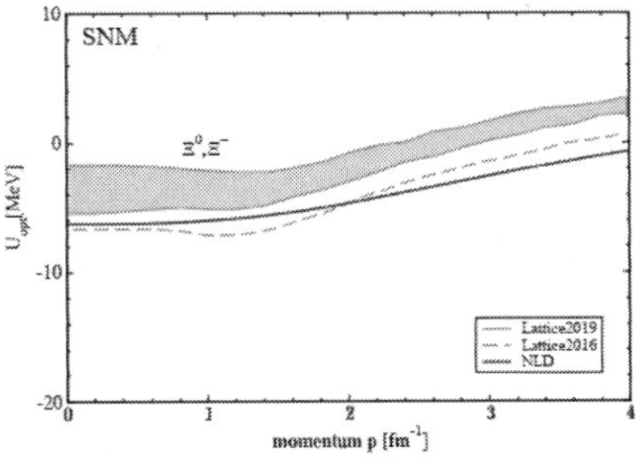

Figure 6. Optical potential of cascade hyperons versus their momentum for symmetric nuclear matter (SNM) at saturation density. The solid curve indicates the NLD predictions while the dashed curve and the gray band refer to recent Lattice calculations from Refs. [4].

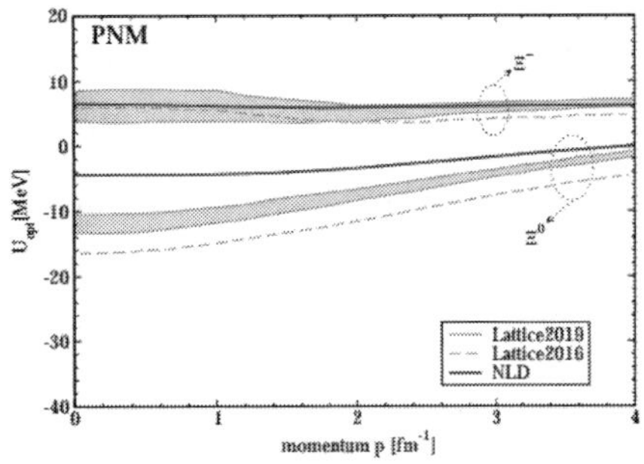

Figure 7. Same as in Fig. 6, but for pure neutron matter (PNM). The curves and bands belonging to Ξ^- and Ξ^0 are indicated in this figure.

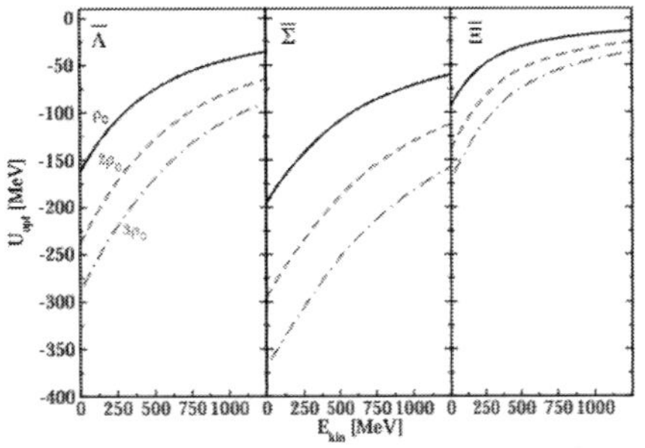

Figure 8. Optical potentials for anti-hyperons versus their kinetic energy for symmetric nuclear matter (SNM) at various densities, as indicated. The NLD predictions for saturation density ρ_0 (solid) and higher densities of $2\rho_0$ (dashed) and $3\rho_0$ (dashed-dot) are shown.

Finally, in Figs. 6 and 7 we have the comparison of NLD model with recent microscopic calculations from Lattice-QCD for the cascade-hyperons, for both symmetric nuclear matter (SNM) and pure neutron matter (PNM).

In Figure 8 the predictions of NLD for anti-hyperon in-medium potentials are presented as well. The motivation for this study of anti-hyperons is the future experiments, such as those at FAIR, where the in-medium properties of anti-hadrons will be investigated too. In general for anti-particles the imaginary part of U_{opt} may be also important, so it is showed in the same figure as well. In the NLD approach the Im(U_{opt}) has been estimated from dispersion relations [1] and it is noticeable that it is comparable with its corresponding real part, particularly at very low energies. Possible physical explanations for that is that in antinucleon-nucleon scattering annihilation can occur through the production of light pions or the interaction of anti-hyperons with nucleons can happen via the production of the heavier kaons due to strangeness conservation (it may affect the Im(U_{opt}) at low energies).

References

[1] T. Gaitanos, M. Kaskulov, Nucl. Phys. **A 899**, 133 (2013) .
T. Gaitanos, M. Kaskulov, Nucl. Phys. **A 940**, 181 (2015).

[2] S. Petschauer et al. , Eur. Phys. J. **A 52**, 15 (2016).

[3] J. Haidenbauer, U.G. Meißner, A. Nogga, Eur. Phys. J. **A 56 (3)**, 91 (2020).

[4] T. Inoue, LATTICE-HAL QCD PoS INPC2016, 277 (2016).
T. Inoue, HAL QCD AIP Conf. Proc. 2130(1), 020002 (2019).

EPJ Web of Conferences **252**, 06001 (2021)
HINPw6
https://doi.org/10.1051/epjconf/202125206001

Twin stars: probe of phase transition from hadronic to quark matter

Themistoklis Deloudis[1,*], *Polychronis* Koliogiannis[1,**], and *Charalampos* Moustakidis[1,***]

[1]Department of Theoretical Physics, Aristotle University of Thessaloniki, 54124 Thessaloniki, Greece

Abstract. In agreement with the gravitational-wave events which are constantly increasing, new aspects of the internal structure of compact stars have come to light. A scenario in which a first order transition takes place inside these stars is of particular interest as it can lead, under conditions, to a third gravitationally stable branch (besides white dwarfs and neutron stars). This is known as the twin star scenario. The new branch yields stars with the same mass as normal compact stars but quite different radii. In the current work, we focus on hybrid stars undergone a hadron to quark phase transition near their core and how this new stable configuration arises. Emphasis is to be given especially in the aspects of the phase transition and its parametrization in two different ways, namely with Maxwell construction and with Gibbs construction. Qualitative findings of mass-radius relations of these stars will also be presented.

1 Introduction

Compact stars yield the most prominent natural laboratories for the study of exotic forms of matter [1, 2]. Recently discovered pulsars alongside with gravitational waves detection, such as GW170817, have revealed new aspects of the internal structure of these stars, mainly in terms of their composition [3, 4]. Whilst the equation of state of nuclear matter is well established up to nuclear saturation density, one encounters the challenge of describing matter in fairly higher densities realized in the interior of these stars. At these densities the type of matter is yet to be determined and in turn, the construction of stellar models to agree with the aforementioned observations is still an open issue. Possible candidates are strange quark stars composed of deconfined quarks, pure neutron stars composed by hadrons, and hybrid stars composed by hadronic outer shells and cores of deconfined quarks. In the present article the latter case is claimed.

The idea of a third family of compact stars and in particular, the connection with the possibility to be a signature of a strong phase transition in the interior of the star, was introduced first by Gerlach [5]. Later on, Kämpfer worked also on this issue [6, 7]. Glendenning and Kettner introduced the term "twins" in their paper [8], while at the same time Schertler et al. [9] worked out that idea in detail. However, in all the previous studies, the maximum mass was approximately at the canonical binary pulsar mass 1.4 $M_{\odot}$. The revival of the idea of the twin stars started a few years later by Blaschke et al. [10, 11]. Specifically, in the mentioned papers is suggested that high-mass twin stars, once detected by simultaneous mass and radius measurements, could provide the evidence for a strong first-order phase transition in cold matter, which then would imply the existence of at least one critical endpoint in the quantum chromodynamics phase diagram. Moreover, in the same work is also presented, for the first time, examples of equations of state that would not only provide twin solutions, but also fulfill the constraint on the maximum mass from the existence of pulsars as heavy as 2 $M_{\odot}$ [10, 11]. The previous idea was elaborated by Benic et al. [12] (see also Ref. [13]). A systematic Bayesian analysis of the new twin star equation of state with observational constraints was presented in Ref. [14]. Finally, the analysis of the robustness of twin solutions against changing the Maxwell to a mixed phase construction, and the formation of structures in the mixed phase due to the interplay of the surface tension and Coulomb interaction effects, have been also considered respectively in Refs [15, 16].

Stars of this branch are expected to have masses in the same range as normal neutron stars, yet fairly smaller radii. The existence of such stars is a strong indication that a HQPT is a physical reality, a result of utmost importance especially in the study of dense matter physics. A study of the HQPT is presented here, examining the conditions under which the twin star configuration arises. The compatibility with the mass and radius constraints as they are formed through up to date observations is also considered.

2 Theoretical Framework

Examining the configuration of a compact star is efficiently achieved by separating the interior into regions. Generally, it is suitable to claim a hadronic phase composing the outer shells, a core of deconfined quarks and a connecting region embodying the phase transition. As it will become clear in the following, this last region can be omitted yielding a sharp phase transition or, as it is

*e-mail: tdeloudi@physics.auth.gr
**e-mail: pkoliogi@physics.auth.gr
***e-mail: moustaki@auth.gr

commonly called, a Maxwell construction (MC). Oppositely, Gibbs construction (GC) assumes the existence of this transitional region, which is commonly referred as the mixed phase. In each case, the phase transition is assumed to be of first order [3, 4, 8, 17–24].

2.1 Hadronic phase

In the outer layers of a compact star, matter is expected to be mainly hadronic. It can either consist solely of nucleons or other hadrons, usually hyperons. In our study the APR model introduced in Ref. [25] combined with the momentum dependent interactions (MDI) parametrization in the way presented in Refs. [26, 27] is applied. Hence, no hadrons other than protons and neutrons are assumed. The selection of this particular hadronic model is due to the numerous advantages provided, namely the reproduction of symmetric nuclear matter properties, value and slope of symmetry energy at the saturation density and the agreement with limits provided by experimental predictions. Furthermore, state-of-the-art calculations considering high densities, as well as calculations of the Chiral model considering pure neutron mater are successfully reproduced. Lastly, as far as the results are concerned, this model can reproduce maximum masses of neutron stars even higher than those observed [26–30].

2.2 Quark phase

Deconfined quark matter composing the stellar core is in general difficult to formulate [31]. One needs sophisticated models to study it in detail. Nevertheless, one possibility is to mimic these models by assuming constant value of the speed of sound throughout the quark phase. This is a robust feature on top of being practical, as it simplifies the parametrization of the energy density expression. Apparently, the speed of sound $c_s^2 = \partial p/\partial\mathcal{E}$ is subjected to the constraint $c_s \leq 1$ in all possible cases. Its exact value plays a decisive role on the final configuration that the star will achieve.

2.3 Phase transition

As the density of stellar matter soars above the nuclear saturation density, a phase transition must take place in order for the quark core to occur. A common approach to describe the former is with a polytropic equation of state of the form [3]

$$p = K\rho^{\Gamma}, \tag{1}$$

where p stands for pressure, ρ for baryon density and K, Γ are constants. Through Eq. (1) it is easy to distinguish between a sharp transition (MC), by imposing $\Gamma = 0$, and a non-sharp transition (GC), assuming $\Gamma \neq 0$.

A phase transition under constant pressure, as in MC, leads to a specific structure of the star. Due to the monotonic reduction of the pressure relative to the radius, points of equal pressure must be mapped onto the same radial coordinate inside the star [1]. That is, the hadronic phase of the outer shells and the quark phase of the core are essentially in direct contact with each other at the point of transition, as a result of the absence of a finite transitional region. This is not the case in GC where the configuration includes this region and a regular behaviour of the pressure inside the star. In the following, these two cases are examined in detail.

2.3.1 Maxwell construction

An abrupt phase change yields discontinuity in at least one physical quantity. In this study the energy density is of primary concern, claiming the form [3, 4]

$$\mathcal{E}(p) = \begin{cases} \mathcal{E}_{\mathrm{APR+MDI}}(p), & p < p_{\mathrm{tr}} \\ \mathcal{E}(p_{\mathrm{tr}}) + \Delta\mathcal{E} + (p - p_{\mathrm{tr}})\, c_s^{-2}, & p \geq p_{\mathrm{tr}} \end{cases} \tag{2}$$

where $\mathcal{E}(p)$ denotes the energy density, p the pressure, c_s the speed of sound, and $\Delta\mathcal{E}$ is the magnitude of the energy density jump at the transition point. Subscript "tr" denotes the corresponding quantity at this point. It should be clear that the first line in Eq. (2) refers to the hadronic phase while the second one refers to the quark phase.

By definition, MC is characterized by a precondition which, when satisfied, leads to an unstable regime essential for the third branch of stable hybrid stars to arise [3, 4, 17]. This constraint, first shown by Seidov [32], is expressed as

$$\Delta\mathcal{E} \geq \frac{3p_{\mathrm{tr}} + \mathcal{E}(p_{\mathrm{tr}})}{2}. \tag{3}$$

The star will immediately become unstable after its central pressure reaches the value p_{tr}. Otherwise, a stable state occurs. Therefore, when seeking twin stars with MC one should always ensure that Eq. (3) is satisfied.

2.3.2 Gibbs construction

In a case of a non-sharp HQPT (GC), a finite region to embody the transition is implied. The mixed phase of this transition is composed, as its name suggests, by intermittent domains of pure hadronic and quark phases [19, 20]. Contrary to MC, no discontinuities in energy density appear, giving rise to the profile

$$\mathcal{E}(p) = \begin{cases} \mathcal{E}_{\mathrm{APR+MDI}}(p), & p \leq p_{\mathrm{tr}} \\ \Lambda p^{1/\Gamma} + p\,(\Gamma - 1)^{-1}, & p_{\mathrm{tr}} \leq p \leq p_{\mathrm{css}} \\ \mathcal{E}(p_{\mathrm{css}}) + (p - p_{\mathrm{css}})\, c_s^{-2}, & p \geq p_{\mathrm{css}}. \end{cases} \tag{4}$$

Again, the energy density is denoted by $\mathcal{E}(p)$, the pressure by p, the speed of sound by c_s, while Λ and the polytropic index Γ are constants. Subscript "css" denotes the corresponding quantity at the start of the quark phase. In addition, the continuity at the two transition points imposes

$$\Lambda = \frac{\mathcal{E}(p_{\mathrm{tr}}) - p_{\mathrm{tr}}\,(\Gamma - 1)^{-1}}{p_{\mathrm{tr}}^{1/\Gamma}}, \tag{5}$$

https://doi.org/10.1051/epjconf/202125206001

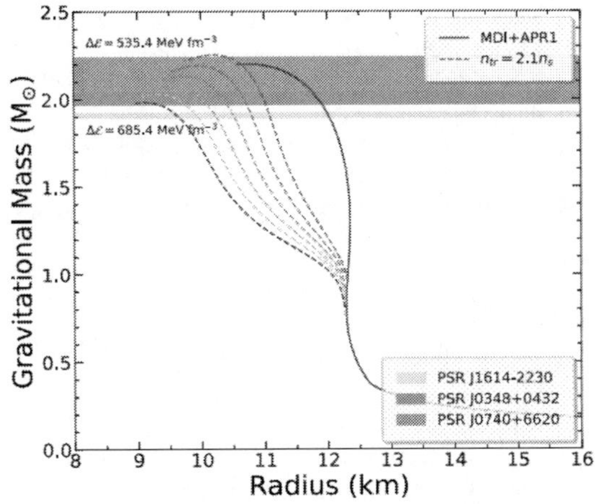

Figure 1. Mass-Radius diagram in the case of a sharp transition at $n_{tr} = 2.1\ n_s$ (dashed lines) for various values of the energy density jump $\Delta\mathcal{E}$. The solid line refers to the absence of a phase transition and is added for comparison. The coloured strips near the top of the figure denotes the range (uncertainty) around the maximum observed mass values of three pulsars [28–30].

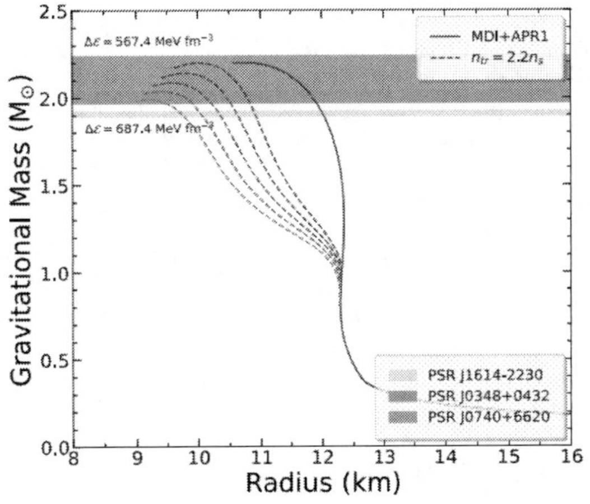

Figure 2. Mass-Radius diagram in the case of a sharp transition at $n_{tr} = 2.2\ n_s$ (dashed lines) for various values of the energy density jump $\Delta\mathcal{E}$. The solid line refers to the absence of a phase transition and is added for comparison. The coloured strips near the top of the figure denotes the range (uncertainty) around the maximum observed mass values of three pulsars [28–30].

and

$$\mathcal{E}(p_{css}) = \left(\frac{p_{css}}{p_{tr}}\right)^{1/\Gamma}\left(\mathcal{E}(p_{tr}) - \frac{p_{tr}}{\Gamma - 1}\right) + \frac{p_{css}}{\Gamma - 1}, \quad (6)$$

Apparently, the first line in Eq. (4) corresponds to the hadronic outer shells, the second one to the mixed phase, embodying the HQPT, and the third one to the quark core. An immediate consequence of GC is that an energy density jump is not defined as in MC. Therefore, in comparing the energy density change due to the transition in the two constructions one needs to use the rise of the energy density throughout the mixed phase in GC as a counterpart of $\Delta\mathcal{E}$ of MC. It is immediately seen that

$$\begin{aligned}\Delta\mathcal{E}_{GC} &= \mathcal{E}(p_{tr})\left[\left(\frac{p_{css}}{p_{tr}}\right)^{1/\Gamma} - 1\right] \\ &+ \frac{1}{\Gamma - 1}\left[p_{css} - \left(\frac{p_{css}}{p_{tr}}\right)^{1/\Gamma} p_{tr}\right]. \quad (7)\end{aligned}$$

Furthermore, in GC no constraint similar to Eq. (3) is present. In order to achieve twin star configuration only the appropriate choice of the free parameters is needed.

3 Results

Applying our aforementioned assertions on a non-rotating stellar model a first approach of the phenomenon is achieved and the basis is set for its study in rotating models. Remarkably enough, a HQPT introduces only two free parameters in each construction. Specifically, these correspond to the critical pressure p_{tr} of the hadronic phase and the magnitude of the discontinuity $\Delta\mathcal{E}$ in the case of MC, or the critical pressure p_{tr} and the mixed phase interval, commonly described by the pressure difference of its boundaries, Δp, in the case of GC. Other free parameters may exist overall, though not introduced by the implementation of a HQPT. For example, the value of the energy density at the transition point depends on the hadronic equation of state that is used.

Figure 1 is a M-R diagram visualising the case of a sharp transition taking place at $n_{tr} = 2.1 n_s$, where n is the baryon density and the subscripts "tr" and "s" denote its value at the transition and the saturation point, respectively. In examining Figure 1, the branching due to the existence of a HQPT is immediately evident. The change of $\Delta\mathcal{E}$ leads to a different position of the second minimum but the shape of the second branch remains almost identical. Moreover, it is worth to notice that higher values of $\Delta\mathcal{E}$ lead to smaller radii for the corresponding maximum mass configurations. A possible explanation is that larger $\Delta\mathcal{E}$ yield heavier stellar cores with an accompanying greater gravitational pull on the outer shells, as suggested in Ref. [4]. These conclusions are in agreement with Ref. [3]. As predicted by theory, twin star branches yield fairly smaller radii compared with the normal neutron star branch for each value of stellar mass. It is also clear that the higher the energy density jump at the transition point, the lower the resulting stellar mass. Another important remark is that the parametrization used for the HQPT reproduces the maximum masses imposed by observations [3, 4, 17, 28–30]. To explicitly show that, the masses of three pulsars belonging to different binary systems are included in Figure 1.

The last point one needs to remark in that figure is the fact that the transition occurs at baryon density equal to 2.1 times the saturation density. We display also the Figure 2, in order to study the effect of the location of the transition density on the various twin stars branches.

We observed that all of the fundamental properties mentioned above are still valid for $n_{tr} = 2.2n_s$. The only difference is that in this case the point of branching is slightly pushed towards higher masses, a fact that obviously does not affect the agreement with the observations. Lastly, it should be noted that the results for the MC were derived by assuming $c_s = 1$.

It is worth noting that the case of a non-sharp phase transition (GC) should not be omitted, at least in a form of reference. Gibbs construction yields a more complicated configuration of the compact star. This difficulty has resulted in not many publications reviewing this case in detail. We are currently working on the twin star scenario with GC, with the results yet in progress. We are confident that the extraction of these results will allow not only for a comparison and discrimination between the two constructions, but also to the conclusion on which of the two is the most physically plausible.

4 Conclusions and outlook

Analyzing the twin star scenario, it is primarily seen that two ways to formulate the HQPT, MC and GC, are dominant. These seem to cover every plausible scenario for the transition. It has to be noted that configurations achieved with MC are simpler than GC. However, in MC case the condition Eq. (3) must also be fulfilled. Results posed in Figures 1 and 2 with MC show clearly the differentiation of cases in a compact star with and without a phase transition. The formulation achieved for the former case reproduces masses and radius in agreement with observations in a satisfying interval of values of the baryon density at the transition point. For a more complete study, the same results are expected to be deducted for GC.

Future considerations may include: (a) designating the conditions under which the twin star branch and the normal neutron star branch are separated; (b) achieving plausible configurations with $c_s^2 = 1/3$, a value which is closer to reality than causality; and (c) examining the phenomenon in the frame of rotating stellar models.

The authors would like to thank Prof. D. Blaschke for his useful insight and comments.

References

[1] N. K. Glendenning, Compact stars - Nuclear physics, Particle physics and General Relativity, Springer (1997)

[2] P. Haensel, A. Y. Potekhin, and D. G. Yakovlev, Neutron Stars 1: Equation of State and Structure (Springer-Verlag, New York, 2007)

[3] J.E. Christian, A. Zacchi, and J. Schaffner-Bielich, Eur. Phys. J. A **54**, 28 (2018)

[4] G. Montana, L. Tolos, M. Hanauske, L. Rezzolla, Phys. Rev. D **99**, 103009 (2019)

[5] U.H. Gerlach, Phys. Rev. **172**, 1325 (1968)

[6] Kämpfer, J. Phys. A: Math. Gen. **14**, L471, (1981)

[7] Kämpfer, Phys. Lett. B 101, 366 (1981)

[8] N.K. Glendenning, C. Kettner, Astron. Astrophys. **353**, L9–L12 (2000)

[9] K. Schertler, C. Greiner, J. Schaffner-Bielich, and M.H. Thoma, Nucl. Phys. **A** 677, 463 (2000)

[10] D. Blaschke, D.E. Alvarez-Castillo, and S. Benic, PoS CPOD2013 063 (2013); [arXiv:1310.3803]

[11] D.E. Alvarez-Castillo and D. Blaschke, arXiv:1304.7758

[12] S. Benic, D. Blaschke, D.E. Alvarez-Castillo, T. Fischer, and S. Typel, A&A 577, A40 (2015)

[13] D.E. Alvarez-Castillo and D. Blaschke, Phys. Part. Nucl. **46**, 846 (2015)

[14] D.E. Alvarez-Castillo, A. Ayriyan, S. Benic, D. Blaschke. H. Grigorian, and S. Typel, Eur. Phys. J. **A** 52, 69(2016)

[15] A. Ayrinan, N.-U. Vastian, D. Blaschke, H. Grigorian, K. Maslov, and D.V. Voskresensky, Phys. Rev. C **97**, 45802 (2018)

[16] K. Maslov, N. Yasutake, D. Blaschke, A. Ayriyan, H. Grigorian, and T. Maruyama, Phys. Rev. C **100**, 025802 (2019)

[17] D.E. Alvarez Castillo and D.B. Blaschke , Phys. Rev. C **96**, 045809, (2017)

[18] M. Bejger, D. Blaschke, P. Haensel, J.L. Zdunik, and M. Fortin, A&A **600**, A39 (2017)

[19] A. Bhattacharyya, I. N. Mishustin, W. Greiner, J. Phys. G **37**, 025201 (2010)

[20] T. Endo, T. Maruyama, S. Chiba, T. Tatsumi, arXiv: hep-ph/0502216

[21] J.E. Christian, A. Zacchi, and J. Schaffner-Bielich, Phys. Rev. D **99**, 023009 (2019)

[22] J.E. Christian, and J. Schaffner-Bielich, Phys. Rev. D **103**, 063042 (2021)

[23] M.G. Alford, S. Han, Phys. Rev. D **88**, 083013 (2013)

[24] S. Han, M.A.A. Mamun, S. Lalit, C. Constantinou, M. Prakash, Phys. Rev. D **100**, 103022 (2019)

[25] A. Akmal, V.R. Pandharipande, and D.G. Ravenhall, Phys. Rev. C **58**, 1804 (1998)

[26] P. S. Koliogiannis, Ch. C. Moustakidis, Phys. Rev. C **101**, 015805 (2020)

[27] P.S. Koliogiannis and Ch.C. Moustakidis, Astroph.J. **912**, 69 (2021)

[28] Z. Arzoumanian, A. Brazier, S. Burke-Spolaor, et al., Astrophys. J. Suppl. Ser. **235**, 37 (2018)

[29] J. Antoniadis, P. Freire, N. Wex, et al., Sci. **340**, 448 (2013)

[30] H. Cromartie, E. Fonseca, S. Ransom, et al., Nat. Astron. **4**, 72 (2020)

[31] G. Baym, T. Hatsuda, Toru Kojo, P. D. Powell, Y. Song, and T. Takatsuka, Rep. Progr. Phys. **81**, 056902 (2018)

[32] Z.F. Seidov, Soviet Astronomy, **15**, 347 (1971)

Low energy proton induced reactions for CANS applications

V. Soukeras[1*], *O.* Sgouros[1], *A.* Pakou[2,3]

[1]INFN – Laboratori Nazionali del Sud, Catania, Italy
[2]Department of Physics, University of Ioannina, Ioannina, Greece
[3]Hellenic Institute of Nuclear Physics (HINP), Ioannina, Greece

Abstract. The significance of comprehensive studies in the field of low energy proton induced reactions for applications related to Compact Accelerator – driven Neutron Sources was underlined in our recent work (O. Sgouros et al., Eur. Phys. J. A 57, 125 (2021)). In the present short article, the scientific concept together with the relevant simulations are discussed.

1 Introduction

The low – energy Compact Accelerator – driven Neutron Sources (CANS) have a significant role in research and development in science, as well as in engineering and education [1-4]. Their applications pertain to several scientific fields with high economical interest. Among them one may mention the materials characterization, applications in environmental studies, medical applications and non – destructive techniques useful in cultural heritage and engineering. The economical competitiveness and easy operation of CANS make them a good solution for long experimental runs needed in nuclear astrophysics or nuclear data campaigns [1]. Nowadays several CANS are currently operating or under construction worldwide from high to very low energies [3-11].

Two of the most widespread reactions for high neutron rate production at low – energies are the ^{7}Li(p,n)^{7}Be and ^{9}Be(p,n)^{9}B reactions due to their large cross section. Therefore, a broad knowledge of the p + ^{7}Li and p + ^{9}Be reactions is essential for low – energy CANS projects using a proton beam as underlined in Ref. [12].

Recent global experimental and theoretical studies in the Continuum Discretized Coupled Channels (CDCC) approach, for p + ^{6,7}Li and p + ^{9}Be [13-23], provide an important input to this direction. They also underline the role of the fundamental research in the field of applications. The reactions of interest for the visualization of CANS are presented in Table 1. The relevant neutron rate for a 1 μg/cm^2 target was calculated in Ref. [12]. In the same work, the neutron rates for the proton energies of 3 and 5 MeV were provided for a thick ^{7}Li or ^{9}Be target. Following the same methodology, the energy dependence of the neutron rates for thick ^{7}Li and ^{9}Be targets is further investigated here.

Table 1. The reactions under study in the global investigation reviewed in [12-13]. The reactions associated to the neutron production are highlighted with bold fonts.

Input channel	Output channel
^{7}Li + p	→ ^{7}Li + p
	→ ^{7}Li* + p → ^{4}He + ^{3}H + p
	→ **^{7}Be + n**
	→ **^{7}Be* + n**
	→ ^{4}He + ^{4}He
^{9}Be + p	→ ^{9}Be + p
	→ **^{9}Be*+p→ ^{4}He + ^{4}He + n + p**
	→ **^{9}Be*+p→ ^{8}Be*+p→^{4}He + ^{4}He + n + p**
	→ **^{9}Be*+p→ ^{5}He +^{4}He→^{4}He + ^{4}He + n + p**
	→ **^{9}B + n → ^{4}He +^{4}He + p + n**
	→ ^{8}Be + ^{2}H → ^{4}He + ^{4}He + ^{2}H
	→ ^{6}Li + ^{4}He
	→ ^{6}Li* + ^{4}He

2 Calculation of the neutron rates

The neutron production for a 1 μg/cm^2 lithium or beryllium target is presented and discussed in details in Ref. [12]. As it was shown, the ENDF/B-VIII.0 library [24] can be safely used for CANS applications at least in respect with (p,n) reactions. However, one should take into account the breakup for the beryllium target since it accounts for at least 30% of the total neutron rate. Slowing the protons down to the threshold energy of (p,n) and breakup reactions and taking into account energy loss algorithms according to Andersen – Ziegler [25], we can calculate the neutron rate for a thick lithium or beryllium target where the proton beam stops inside. The relevant rates are presented in Table 2, while a comparison among themselves is illustrated in Fig. 1. It should be noted that an energy dependence interpolation of the neutron production was assumed paying attention and taking into account the existing resonances.

* Corresponding author: vasileios.soukeras@lns.infn.it

 https://doi.org/10.1051/epjconf/202125206002

Table 2. Neutron rates for ^{7}Li and ^{9}Be thick target. The rates are calculated taking into account (p,n) cross sections from the ENDF/B-VIII.0 library [24]. For the ^{9}Be target, the neutron production due to breakup was also included, taking into account our recent measurements [13,19-20]. For a smaller energy step see Fig. 1.

Proton energy (MeV)	Neutron production rate with ^{7}Li target (s^{-1} mA^{-1} sr^{-1})	Neutron production rate with ^{9}Be target (s^{-1} mA^{-1} sr^{-1})
2.2	3.1 E+10	1.6 E+08
2.5	7.9 E+10	2.1 E+09
3.0	1.4 E+11	1.8 E+10
3.5	2.1 E+11	3.8 E+10
4.0	2.8 E+11	7.0 E+10
4.5	3.8 E+11	1.3 E+11
5.0	5.0 E+11	2.0 E+11
5.2	5.5 E+11	2.3 E+11

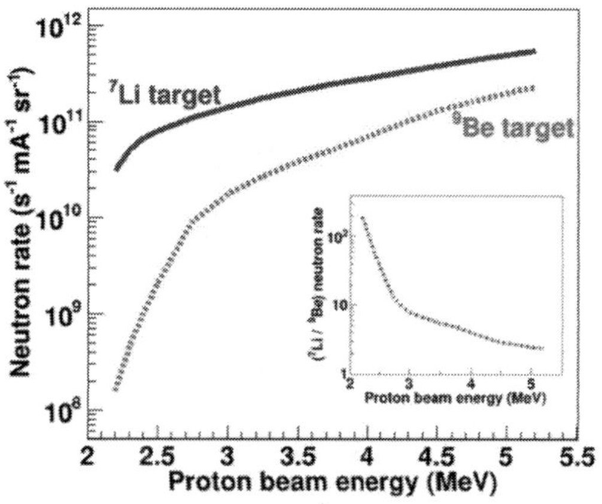

Fig. 1. Neutron rates for ^{7}Li and ^{9}Be thick target as a function of proton beam energy. See text for more details. The energy evolution of the ratio of the neutron rates for the two targets is illustrated in the inset.

3 Summary

Promoting the importance of comprehensive studies in the field of low energy proton induced reactions and their connection to societal applications of broad interest, we have presented here neutron rate calculations for thick ^{7}Li and ^{9}Be targets at the proton beam energy range between 2.2 and 5.2 MeV. It should be pointed out that breakup reaction is a substantial source of neutrons for the ^{9}Be target however no breakup data are existing in nuclear data libraries. Finally, the neutron production for a ^{7}Li target is higher however special care should be given at the unnecessary production of radioactive ^{3}H and ^{7}Be in the case of ^{7}Li target [12].

References

1. I. S. Anderson et al., Phys. Reports **654**, 1 (2016).
2. F. Ott, *Compact Neutron Sources for Neutron Scattering* [Technical Report cea-01873010] CEA Paris Saclay (2018). https://hal-cea.archives-ouvertes.fr/cea-01873010/document
3. N. Alamanos, EPJ Web of Conferences, to be published in the present volume (2021).
4. I. Mardor et al., Eur. Phys. J. A **54**, 91 (2018).
5. H. Ngoc Tran et al., EPJ Web of Conferences **231**, 01007 (2020).
6. http://www.ucans.org/
7. http://www.jcans.net/nuans.html
8. iThemba LABS Long range plan (2018) https://tlabs.ac.za/wp-content/uploads/pdf/finalLRP_may2018c.pdf
9. https://home.infn.it/en/projects-2/progetti-foe/spes
10. https://ceem.indiana.edu/lens/
11. https://exp-astro.de/index.php?id=expFRANZ
12. O. Sgouros et al., Eur. Phys. J. A **57**, 125 (2021).
13. A. Pakou et al., Eur. Phys. J. A **57**, 25 (2021).
14. A. Pakou et al., Phys. Rev. C **94**, 014604 (2016).
15. A. Pakou et al., Phys. Rev. C **95**, 044615 (2017).
16. A. Pakou et al., Phys. Rev. C **96**, 034615 (2017).
17. N. Keeley et al., Phys. Rev. C **99**, 014615 (2019).
18. A. Pakou et al., Phys. Rev. C **101**, 024602 (2020).
19. V. Soukeras et al., Phys. Rev. C**102**, 064622 (2020).
20. A. Pakou et al., Nucl. Phys. A **1008**, 122155 (2021).
21. V. Soukeras et al., Phys. Rev. C **91**, 057601 (2015).
22. Ch. Betsou et al., Eur. Phys. J. A **51**, 86 (2015).
23. V. Soukeras et al., Phys. Rev. C **95**, 054614 (2017).
24. Evaluated Nuclear Data File (ENDF) https://www.nndc.bnl.gov/exfor/endf00.jsp
25. H. H. Andersen, J. F. Ziegler, *Hydrogen Stopping Powers and Ranges in all elements*, vol. 3, Pergamon Press (1977).

EPJ Web of Conferences **252**, 06003 (2021)
HINPw6
https://doi.org/10.1051/epjconf/202125206003

Estimation of neutron and γ-rays flux at the MAGNEX facility via FLUKA simulations

O. Sgouros[1,*], *F.* Cappuzzello[1,2], *M.* Cavallaro[1], and *L.* Pandola[1] for the NUMEN collaboration

[1]INFN - Laboratori Nazionali del Sud, Catania, Italy
[2]Dipartimento di Fisica e Astronomia "Ettore Majorana", Università di Catania, Catania, Italy

Abstract. Simulations for estimating the neutron and γ-rays radiation background in the forthcoming NUMEN experiments at the upgraded MAGNEX facility were performed with the simulation package FLUKA. Three main radiation sources were considered in our simulations namely, the beam-target interaction, the moderation of the beam particles inside the beam dump and a hypothetical 10W loss in the beam intensity during the beam transport. In the present contribution, the results of this preliminary analysis are presented and discussed.

1 Introduction

The upgrade of the K800 Superconducting Cyclotron at the Istituto Nazionale di Fisica Nucleare – Laboratori Nazionali del Sud (INFN-LNS) is still in progress. This improvement was driven by the NUMEN (NUclear Matrix Elements for Neutrinoless double beta decay) project [1], which aims to study Double Charge Exchange (DCE) reactions, characterized by low cross-sections, as a mean to shed some light on the determination of the Nuclear Matrix Elements (NMEs) of the $0\nu\beta\beta$ decay [2–4]. Once the Cyclotron upgrade is completed, in the case of NUMEN experiments, the use of ^{18}O and ^{20}Ne ion beams with an intensity of (1-10)kW is foreseen. The interaction of such high intensity beams with the target may yield a substantial increase in the radiation level inside the MAGNEX [5] experimental hall. Furthermore, one should keep in mind that the beam particles must be stopped into a shielded area, like a beam dump. The moderation of the beam particles will trigger a large production of neutrons and gamma rays. This is an important aspect to take into account during the R&D of the electronic devices as well as of the detectors in the new Focal Plane Detector (FPD) of the MAGNEX spectrometer [6, 7]. Taking into consideration all the above, Monte Carlo simulations for the neutrons and γ-rays induced radiation background at the MAGNEX facility were performed using the FLUKA code [8].

2 The simulation

The simulation was performed following the prescription of [9] using a schematic geometry which appears in Figure 1. In more details, a ^{20}Ne^{10+} beam at the energy of 60 *A*MeV with a flux of 1.05x10^{13} pps was impinged into a ^{76}Ge target 214 μg/cm^2 thick, followed by a 2 μm ^{12}C layer [10, 11]. The simulated neutron and γ-ray energy spectra, measured at the FPD (see Figure 1), are presented in Figures 2 and 3, respectively, with the dashed green-line. Further on, a loss of 10 W during the transport of the beam towards the beam dump was also considered in the simulations. In this case, a substantial increase in the radiation level in the vicinity of the FPD is inferred. In the last part of our simulations, the ^{20}Ne^{10+} beam was moderated onto a thick Ag target located inside the beam dump. The beam dump is confined into a (5x5x3)m^3 portland concrete box, providing thus a shield against neutron and gamma radiation.

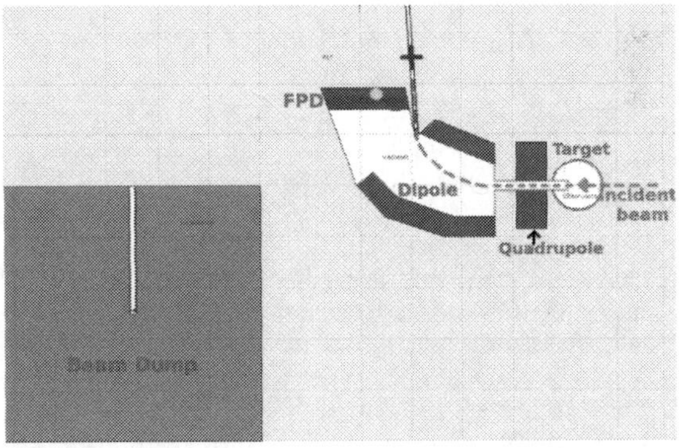

Figure 1. A schematic representation of the expected geometry, after the facility upgrade, at the MAGNEX experimental apparatus considered in the FLUKA simulations. The detector used in the simulation is indicated by a filled green circle, while the position where the 10W loss occurs is denoted with the blue cross (see text for details).

*e-mail: onoufrios.sgouros@lns.infn.it

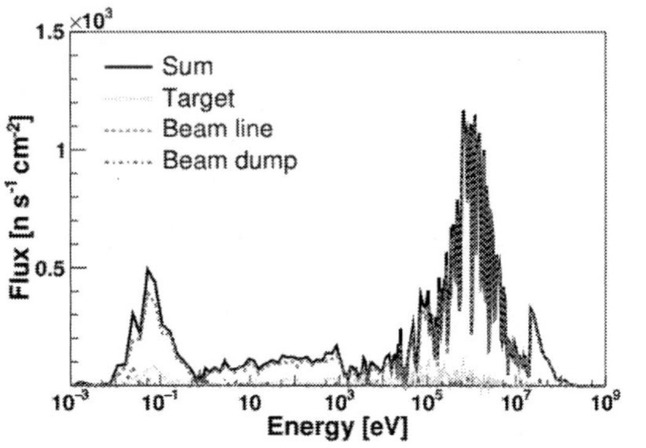

Figure 2. The simulated neutron energy spectra at the MAGNEX FPD region. The spectrum corresponding to neutrons emerging from the beam-target interaction, the beam dump and the hypothetical 10W loss are denoted with the dotted green, dotted-dashed blue and dashed red lines, respectively.The sum of the three contributions is presented with the solid black line. Figure from Ref. [7].

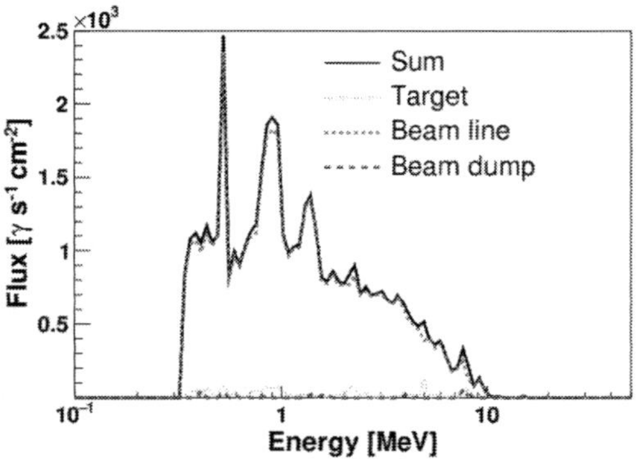

Figure 3. Same as in Figure 2, but for the simulated γ-rays energy spectrum.

3 Results and discussion

Taking into consideration the radiation sources presented above, the simulated energy distribution of the neutron flux at the FPD region is presented in Figure 2. The total integrated neutron flux at the FPD region is 7.2×10^4 n s^{-1} cm^{-2}, dominated by fast neutrons above 100 keV. These results are also reported in a recent publication of the NUMEN collaboration [7]. The simulated γ-ray energy distribution is presented in Figure 3 corresponding to an integrated yield at the FPD zone of about 4.8×10^4 γ s^{-1} cm^{-2}.

Both Figures 2 and 3 show that the main sources of neutrons and γ-rays in the FPD region come from the possible scattering of the beam into the beam pipe close to the FPD. Therefore, a detailed study of the beam transport is in progress in order to quantify the beam losses in the FPD region and reduce them to tolerable values. Moreover, a significant reduction of radiation level is expected by the introduction of proper shields to the detectors and electronics, which are also currently under study.

4 Summary

Systematic simulations for the neutron and γ-rays induced radiation background were performed with the simulation code FLUKA. Three different radiation sources were considered in our simulations namely, the beam-target interaction, the moderation of the beam particles inside the beam dump and a hypothetical 10W loss in the beam intensity during the beam transport. It was found that the bulk of the produced neutron and γ-rays is originated from the possible beam scattering into the pipe close to the FPD. The introduction of shielding materials around the FPD and their effect on the simulated energy spectra is currently under study. In the near future, we also intend to implement a more realistic geometry of the MAGNEX FPD in our simulations.

References

[1] F. Cappuzzello, C. Agodi, M. Cavallaro et al., Eur. Phys. J. A **54**, 72 (2018)
[2] H. Sagawa and T. Uesaka, Phys. Rev. C **94**, 064325 (2016)
[3] E. Santopinto, H. García-Tecocoatzi, R. I. Magaña Vsevolodovna et al., Phys. Rev. C **98**, 061601(R) (2018)
[4] N. Shimizu, J. Menéndez and K. Yako, Phys. Rev. Lett. **120**, 142502 (2018)
[5] F. Cappuzzello, C. Agodi, D. Carbone et al., Eur. Phys. J. A **52**, 167 (2016)
[6] P. Finocchiaro, L. Acosta, C. Agodi et al., Universe **6**, 129 (2020)
[7] F. Cappuzzello, L. Acosta, C. Agodi et al., Front. Astron. Space. Sci. **8**, 668587 (2021)
[8] T. T. Böhlen, F. Cerutti, M. P. W. Chin et al., Nucl. Data Sheets **120**, 211 (2014)
[9] D. Lo Presti, N. H. Medina, M. A. Guazzelli et al., Rev. Sci. Instr. **91**, 083301 (2020)
[10] F. Pinna, F. Iazzi, R. Introzzi et al., J. Phys. G: Conf. Ser. **1056**, 012046 (2018)
[11] F. Pinna, D. Calvo, M. Campostrini et al., Phys. Scr. **95**, 094002 (2020)

Two-stage description of charge distributions in ^{56}Fe+p spallation reactions at 0.3-1.5 GeV/A

N.G. Nicolis [1,*], *G.A. Souliotis*[2] *and A. Bonasera*[3,4]
[1]*Department of Physics, The University of Ioannina, Ioannina 45110, Greece*
[2]*Laboratory of Physical Chemistry, Department of Chemistry, University of Athens, Athens, Greece*
[3]*Cyclotron Institute, Texas A & M University, College Station, Texas, USA*
[4]*Laboratori Nazionali del Sud, INFN, Catania, Italy*

Abstract. Spallation reactions are studied in the framework of a two-stage model. The intra-nuclear cascade (INC) stage of the reaction is simulated with the code ISABEL and a Constrained Molecular Dynamics (CoMD) code. The de-excitation of the highly excited pre-fragments is described with the multi-sequential binary decay code MECO, based on a generalized Weisskopf-Ewing formalism. Emission of nucleons, gamma rays and IMFs in their ground, excited bound and unbound states is considered. Calculated cross sections are compared with the experimental charge distributions of $^{56}Fe + p$ spallation reaction products studied at GSI with the fragment separator FRS in the energy-range $0.3 - 1.5\ GeV/A$. At all bombarding energies, a good description of the experimental data is obtained with ISABEL-MECO using a global set of parameters. A preliminary calculation with CoMD-MECO at 1GeV/A is discussed.

1 Introduction

In recent years, proton-induced spallation reactions have received a great deal of attention in basic and applied Nuclear Physics research. In basic research, they provide a framework for the development and testing of high-energy nuclear reaction models [1]. In Applied Nuclear Research they provide useful information for the development of accelerator-driven systems (ADS), the transmutation of nuclear waste, spallation neutron sources and the production of radioactive beams. In the case of targets with mass A~50-60, spallation reactions are used for the production of isotopes of medical interest, studies concerning the interaction of cosmic-rays with interstellar bodies (radiation damage) and radiation protection in space [2].

A proton-induced spallation reaction (SR) proceeds in two stages [3]. In the first stage, the incident particle induces a sequence of collisions transferring energy and momentum to the target nucleons. As a result, we have the formation of an intranuclear cascade (INC) of high-energy (greater than 20 MeV) protons, neutrons and pions within the nucleus. This is a fast process lasting approximately $10^{-22} s$. During this process some of the energetic hadrons may escape from the target. Others deposit their kinetic energy in the nucleus leaving it in an excited state.

In the second stage, the produced hot pre-fragments equilibrate and deexcite. Sequential evaporation is assumed to occur with a typical time scale of $10^{-1}\ s$. Emission of nucleons, protons, alpha-particles and gamma-rays is dominant. Emission of heavier nucleon clusters in their ground or excited states is also possible. If the target is heavy enough, high-energy fission may compete with sequential evaporation. The deexcitation products of target-like and/or fission fragments may be radioactive.

Recently, detailed experimental and theoretical studies have been devoted on the spallation of ^{56}Fe with protons. In a high precision experiment, spallation residue cross sections were measured in $^{56}Fe + p$ bombardments at 0.3 – 1.5 GeV/A. The experiment

* Corresponding author, email: nnicolis@uoi.gr

employed reverse-kinematics with a liquid hydrogen target and the fragment separator FRS at GSI (Darmstadt) [4]. High resolution measurements of fully identified fragments and coincidence measurements of residues and light particles at 1.0 GeV/A have also been reported [5,6]. Efforts are made for the understanding of the odd-even structure observed in the charge distributions [7,8] and the possibility of a multifragmentation decay mechanism at the highest bombarding energies [6,9].

In the present work, we employ two different models for the INC stage coupled with the same evaporation code for the description of the experimental charge distributions of $^{56}Fe + p$ spallation reaction data at 0.3-1.5 GeV/A [4,5]. In the next Section we describe our reaction models. There follows a comparison of our calculations with the experimental data, a brief discussion, and a summary.

2 Reaction models

We describe the INC stage with the Monte-Carlo code ISABEL [10,11]. ISABEL describes the target nucleus as a continuous medium with a diffuse surface. Nucleon-nucleon collisions may be elastic or inelastic. Linear trajectories are assumed between collisions, which occur with a criterion based on the mean-free-path. Free N-N cross sections are assumed. A full Pauli blocking mechanism forbids nucleons from falling below the Fermi surface.

We have also employed the Constrained Molecular Dynamics (CoMD) model described in Refs. [12-15]. CoMD is a semi-classical model originally designed for reactions near and below the Fermi energy. Nucleons are considered as Gaussian wave packets interacting with a phenomenological N-N interaction (Skyrme). Pauli principle is taken into account through an appropriate restriction in phase-space. Angular momentum conservation is obeyed. Recognition of fragment formation is made at each step of the time evolution.

The deexcitation of the pre-fragments is calculated in the context of equilibrium compound nucleus decay. Possible pre-equilibrium emissions between the two stages are not considered. For the equilibrium decay process, we employ the Multi-sequential Evaporation COde MECO [16]. MECO is a Monte-Carlo code sufficient to describe low-spin, high-excitation energy reaction systems. It describes the equilibrium decay of excited nuclei as a sequence of binary processes involving emission of fragments in their ground, excited bound and unbound states. Any number of user-defined channels may be considered. Emission of nucleons and heavy fragments up to symmetric mass divisions is described in the framework of a generalized Weisskopf-Ewing evaporation formalism. The applicability of this formalism is limited to systems with an effective fissility below the Businaro-Gallone point [17], where the saddle point becomes unstable toward asymmetric divisions.

In the statistical decay calculations, each fragment emitted in an excited-bound state is considered as a separate decay channel. We performed calculations with a total of 180 decay channels consisting of 42 ground states of fragments with Z=0 (neutrons) up to Z=12, 100 excited states for fragments with $Z \geq 5$. The heaviest 28 fragments were allowed to decay in the continuum. We employ phenomenological level densities described with the composite Gilbert and Cameron formula [18]. Level-density parameters were considered as energy-independent according to Gilbert and Cameron. Adjustments were made in the energy shifts of the Fermi-gas formula, thus improving the reproduction of the charge-distribution yields. Inverse cross sections were calculated with the optical model for Z<4. For heavier fragments with $Z \geq 5$, barrier penetration calculations were performed with the Christensen and Winther nuclear potential [19].

At each bombarding energy of the present study, 20000 events were generated from the INC models following a triangular distribution in the impact parameter range $b = 0 - 6.0fm$. These events were fed into MECO and their sequential decay was calculated. There followed a second stage of equilibrium decay calculations for the unbound binary decay fragments. Events from both equilibrium decay calculations were sorted to produce the isotopic distributions of spallation residues to be compared with the data.

Figure 1 shows the excitation energy distributions of pre-fragments calculated with ISABEL for $p + {}^{56}Fe$ at 0.3 and 1.0GeV. These distributions involve contributions from many different isotopes, each formed with an excitation energy distribution.

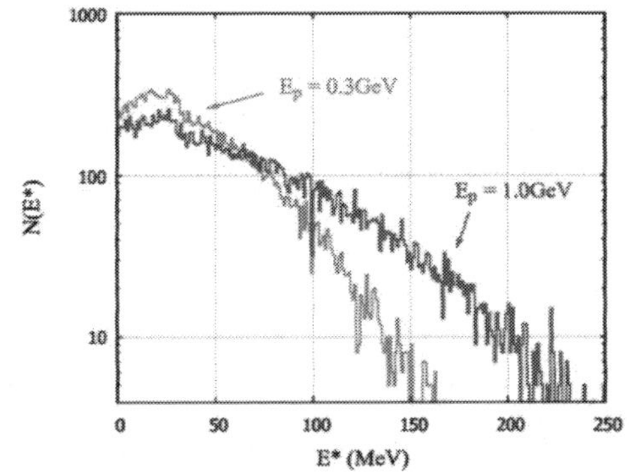

Figure 1. (Color online) Excitation energy distributions of pre-fragments produced in $p + {}^{56}Fe$ reactions at 0.3 and 1.0 GeV/A according to calculations with ISABEL.

Before applying our models to the high-energy data, we compared our procedure to the predictions of the reaction code TALYS [20], known to provide an accurate description of low-energy proton-induced reactions at proton beam energies up to 200MeV. A comparison of the mass and charge distributions predicted by TALYS was made with calculations based on ISABEL/MECO and CoMD/MECO for protons incident on ^{56}Fe with an energy of 50, 100 and 150MeV. Consistency between the three procedures was found for the cumulative A- and Z-distributions.

3 Comparison with experimental data

Symbols in Figure 2 show the experimental charge distributions at 0.3, 0.5, 0.75, 1.0 and 1.5 GeV/A from Ref. [4]. At 1.0 GeV/A, the IMF cross sections of Ref. [5] are also shown. The solid curves show the results of the ISABEL-MECO calculations.

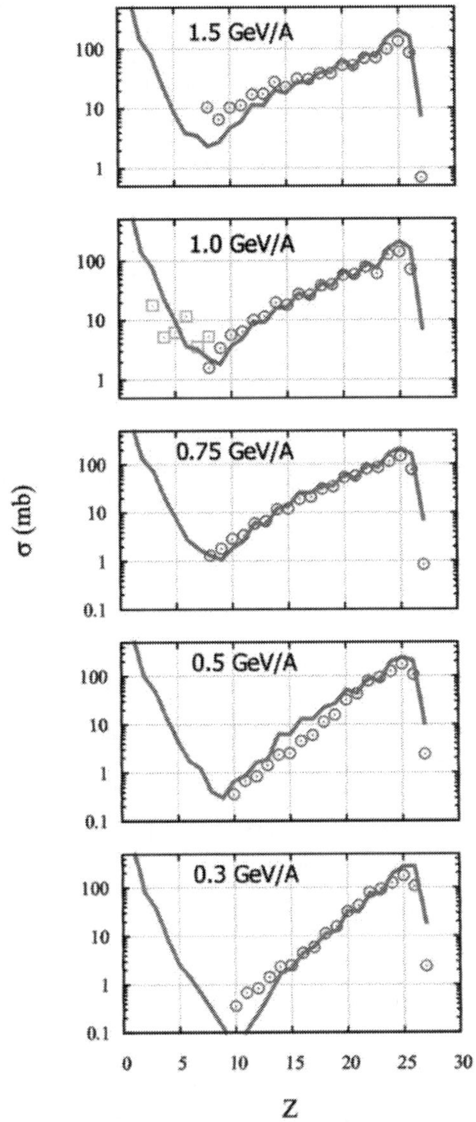

Figure 2. (Color online) Charge distributions of spallation residues in $^{56}Fe + p$ reactions at bombarding energies of 0.3 - 1.5 GeV/A. Experimental data are shown with symbols. Calculations shown with solid curves are described in the text.

We obtain a good overall description of the Z-distributions except for the region with $Z \leq 13$ at the 0.3 and 1.5 GeV/A. At all energies, the odd-even structure observed in the experimental data is reproduced. For the A-distributions we found the same kind of agreement between experimental and calculated cross sections. These calculations reproduce the centroids of the experimental isotopic distributions. The widths of the isotopic distributions are also reproduced for Z>20 and overestimated the lower Z's by less than one mass unit.

The symbols in Figure 3 show the experimental Z-distribution of spallation residues and IMF's at 1.0 GeV/A. The distribution marked with "1" represents the excited pre-fragments predicted by ISABEL. This distribution is structureless. If we allow these events to deexcite by neutron, proton, alpha and gamma emission, we end up with the distribution marked with "2". As Z decreases, the calculation underestimates the experimental cross sections by a lot. If we consider emission of IMFs in their ground states, we obtain the curve marked with "3". This calculation with MECO considers 42 fragments in their ground states with Z up to 12 and accounts for most cross sections. Including emission from excited-bound states and allowing for binary decays in the continuum brings the calculation with close agreement with the experiment (curve marked with "4"). From this Figure, we realize that the odd-even structure of cross sections appears because of the equilibrium decay. Evaporation of just n, p and alpha is sufficient to produce such an effect.

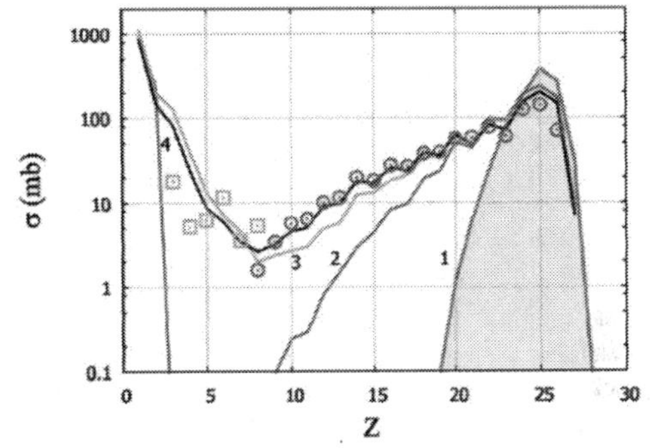

Figure 3. (Color online) Symbols show the experimental spallation residue Z-distribution [4] and emitted IMFs [5] in $^{56}Fe + p$ reactions at 1.0 GeV/A. Solid curves show the calculated distribution with ISABEL (curve 1) and ISABEL-MECO (curves 2, 3 and 4) using various assumptions for the number of equilibrium decay channels described in the text.

We have determined the contributions to the A- and Z-distributions from fragments emitted in their ground or excited states. The thick solid line on Figure 4 shows the total Z- distribution at 1.0 GeV/A calculated with ISABEL-MECO using 180 channels. Experimental data are shown with symbols. The shaded marked "1" shows the contribution from fragments emitted in their ground states. Region marked "2" shows the population from fragments emitted in excited-bound states. The two-bumps marked "3" and "4" correspond to binary decays.

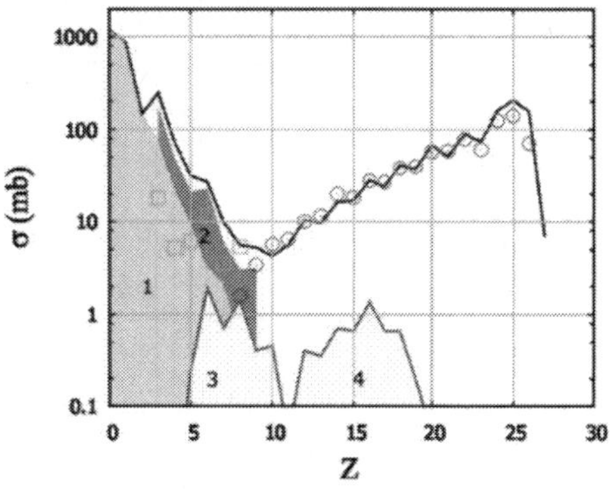

Figure 4. (Color online) Contributions to the calculated total Z-distribution at 1.0 GeV/A. Emission of particles and fragments in their ground states (region 1), excited bound (region 2) and binary decays (regions 3 and 4) are shown. Experimental data [4,5] are shown with symbols.

A preliminary calculation with CoMD for the Z-distribution at 1.0 GeV/A was performed. This is a stringent test for this model since the bombarding energy is well above its range of validity. It was realized that the excitation energy distribution of the pre-fragments, does not extend to energy as high as the ISABEL calculation does (cf. with Figure 1). Applying a correction factor of 1.3 to CoMD produces an excitation energy distribution close to ISABEL. The resulting calculation with CoMD-MECO is shown in Figure 5. We realize that there is a tendency to underestimate the distribution as Z decreases. Furthermore, there is excess of ^{56}Fe events predicted in the primary stage with zero excitation energy. The high-energy behaviour of the CoMD code is currently under investigation.

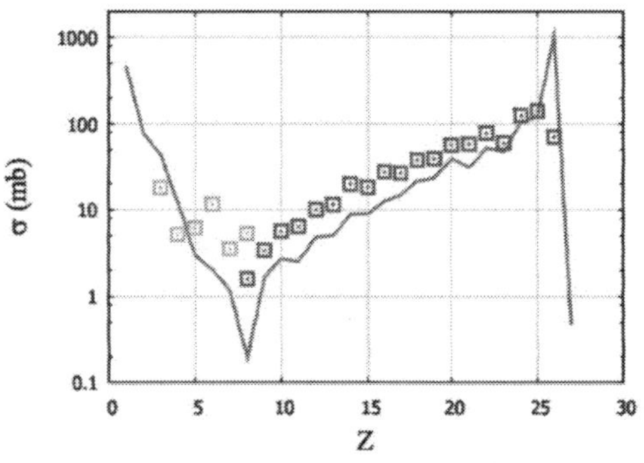

Figure 5. (Color online) Symbols show the experimental Z-distribution of spallation residues [4] and emitted IMFs [5] in $^{56}Fe + p$ reactions at 1.0 GeV/A. The solid curve shows the calculated distribution with CoMD-MECO described in the text.

4 Discussion and Summary

We have obtained a good description of spallation residue cross sections with the combination of the intranuclear cascade model ISABEL coupled with the sequential binary decay code MECO. Assuming that treatment of the primary stage with ISABEL is realistic and neglecting the effect of possible preequilibrium emissions between the two stages of the reaction (not accounted for by ISABEL, if any), we may conclude that the treatment of the equilibrium decay stage is successful. The employed statistical model parameters are global and energy independent. In Figure 2 we see some underestimations in the mid-Z range of the charge distributions at 0.3 and 1.5 GeV/A. They may be related to specific nuclear parameters not accounted for by our average statistical model parameter set. In addition, the underestimation at 1.5 GeV/A may be related to the onset of multifragmentation. Inspection of the ISABEL events shows a small fraction of pre-fragments produced at 1.0 and 1.5 GeV/A with temperatures 3-4 MeV. Calculations with the statistical multifragmentation model SMM [21,22] are under way, keeping in mind that such an interpretation may not be straightforward [6, 9].

Summarizing, spallation products of $^{56}Fe + p$ reactions at 0.3-1.5 GeV/A were described as a two-stage process with the ISABEL INC followed by the MECO sequential binary decay code. A good agreement with the experimental Z- distributions was obtained with an effective set of excitation-energy independent parameters for the equilibrium de-excitation. Contributions from multifragmentation may be considered at the highest bombarding energies (1.0-1.5GeV) in calculations based on more detailed statistical decay parameters and a proper consideration of multifragmentation decays. Improvements in the calculation of the INC stage with the CoMD model are under way.

References

1. J.-C. David, Eur. Phys. J. A51, 68 (2015)
2. J. Benlliure, Spallation Reactions in Applied and Fundamental Research, Lect. Notes Phys. 700, 191–238 (2006)
3. R. Serber, Phys. Rev. 72, 1114 (1947)
4. C. Villagrasa-Canton et al., Phys. Rev. C 75, 044603 (2007)
5. P. Napolitani et al., Phys. Rev. C 70, 054607 (2004)
6. E. Le Gentil et al., Phys. Rev. Lett. 100, 022701 (2008)
7. P. Napolitani et al., Int. J. Mod. Phys. E13, 333 (2004)
8. Jun Su, Long Zhu and Chenchen Guo, Phys. Rev. C 97, 054604 (2018)
9. D. Mancuci et al., Phys. Rev. C 84, 064615 (2001)
10. Y. Yariv and Z. Fraenkel, Phys. Rev. C 20, 2227 (1979); Phys. Rev. C 24, 488 (1981)
11. Y. Yariv and Z. Fraenkel, Phys. Rev. C 24, 488 (1981)

12. M. Papa, A. Bonasera, et al., Phys. Rev. C 64, 024612 (2001)
13. M. Papa, G. Giuliani and A. Bonasera, Jour. Comp. Phys. 208, 403 (2005)
14. N. Vonta et al., Phys. Rev. C 92, 024616 (2015)
15. A. Assimakopoulou et al., J.Phys. (London) G46, 075104 (2019)
16. N.G. Nicolis, Int. J. Mod. Phys. E 17, 1541 (2008)
17. U. L. Businaro and S. Gallone, Nuovo Cimento 1, 629 (1955)
18. A. Gilbert and A.G.W. Cameron, Can. J. Phys. 43, 1446 (1965)
19. P.R. Christensen and A. Winther, Phys. Lett. B 65, 19 (1976)
20. A.J. Koning and D. Rochman, "Modern Nuclear Data Evaluation with The TALYS Code System", Nuclear Data Sheets 113, 2841 (2012)
21. J.P. Bondorf, A.S. Botvina, A.S. Iljinov, I.N. Mishustin and K. Sneppen, Phys. Rep. 257, 133 (1995)
22. G.S. Souliotis, A.S. Botvina et al., Phys. Rev. C 75, 044603 (2007)

Microscopic dynamical description of multinucleon transfer in ^{40}Ar induced peripheral collisions at 15 MeV/nucleon

K. Palli[1], *G. A.* Souliotis[1*], *T.* Depastas[1], *I.* Dimitropoulos[1], *O.* Fasoula[1], *S.* Koulouris[1], *M.* Veselsky[2], *S. J.* Yennello[3], and *A.* Bonasera[3,4]

[1]Laboratory of Physical Chemistry, Department of Chemistry, National and Kapodistrian University of Athens, Athens, Greece
[2]Institute of Physics, Slovak Academy of Sciences, Bratislava, Slovakia
[3]Cyclotron Institute, Texas A&M University, College Station, Texas, USA
[4]Laboratori Nazionali del Sud, INFN, Catania, Italy

Abstract. This paper deals with heavy-ion peripheral reactions in the Fermi energy region for the production of neutron-rich isotopes. Experimental data of projectile fragments from the reactions of an ^{40}Ar beam at 15 MeV/nucleon with ^{64}Ni and ^{58}Ni targets, collected with the MARS spectrometer at the Cyclotron Institute of Texas A&M University, are considered. Momentum distributions, which provide valuable information on the reaction mechanisms, are extracted and compared with two types of calculations: These are, the Deep Inelastic Transfer (DIT) model and the microscopic Constrained Molecular Dynamics model (CoMD). For the latter, the parameters of the original code were systematically varied in order to achieve an overall satisfactory description of the experimental data. Our results will be discussed.

1 Introduction

The production of neutron-rich isotopes and the path toward the neutron drip line are main subjects for modern nuclear physics research [1–3]. Nuclei far from the beta stability line can provide valuable information on the rapid neutron capture process (r-process), responsible for half of the abundance of heavier than iron nuclei in the universe [4, 5]. Therefore it is essential to explore the reaction mechanisms that produce neutron-rich isotopes. The standard routes to generate neutron-rich nuclides are spallation, fission and projectile fragmentation. Another option leading to products with high neutron excess is for the projectile to capture neutrons from the target [2]. This can take place in multinucleon transfer reactions at energies from the Coulomb barrier to the Fermi energy. In our recent work, we studied reactions below the Fermi energy, since this region combines the advantages of both low- and high-energy reactions. Specifically, the interaction of the projectile with the target leads to fragments with enhanced N/Z ratio, having also high velocities, allowing efficient in-flight collection and separation [6–8].

The present work deals with the distributions of projectile-like fragments from the reaction of an ^{40}Ar beam with ^{64}Ni and ^{58}Ni targets at 15 MeV/nucleon. The experimental data are compared with calculations employing the DIT and the CoMD models followed by GEMINI. Mass and momentum distributions are presented. In section 2, we present an overview of the experimental approach and the data collection. In section 3, we give a brief description of the theoretical models. In section 4, we describe the results of our calculations, followed by our conclusions in section 5.

*e-mail: soulioti@chem.uoa.gr

2 Experimental Set-up and Data

The experimental data on neutron rich nuclide production presented in this work were obtained at the Cyclotron Institute of Texas A&M University and are presented in detail in [9, 10]. In this work we give a brief description of the experimental set-up. An ^{40}Ar^{9+} (15 MeV/nucleon) beam, accelerated by the K500 Cyclotron, interacted with ^{64}Ni and ^{58}Ni targets with thickness of 2 mg/cm^2. The MARS recoil separator was used to collect and identify the projectile fragments. The ^{40}Ar beam was sent to the primary target location of MARS with a 4° angle with respect to the optical axis of the separator and the projectile fragments were collected in the polar angular range of 2.2^o-5.5^o covering a solid angle of $\Delta\Omega$=4 msr. The fragments, after the interaction with the target, traversed a parallel-plate avalanche counter (PPAC), which provided information for the position and the magnetic rigidity, as well as the START-time, and they were focused at the end of the separator passing through a second PPAC (for image-size information and STOP-time) and were collected in a $\Delta E-E$ Si detector telescope. To identify the fragments and obtain their atomic number Z, mass number A, velocity and ionic charge, standard techniques of magnetic rigidity, energy-loss, residual energy and time-of-flight were used on an event-by-event basis [8]. Data were obtained in a series of successive magnetic rigidity settings of the spectrometer in the range of 1.1-1.5 Tm. We note that this magnetic rigidity range did not fully cover the neutron deficient side of the product distribution which extends down to 0.8 Tm according to our calculations [10]. The proton-rich isotopes with incomplete magnetic rigidity coverage are to the left of the thin solid lines in Fig. 2 and 4.

In fig. 1, we present the experimental data for the momentum distributions for the reaction ^{40}Ar with ^{64} Ni and ^{58}Ni targets at 15 MeV/nucleon that were extracted as part of this work employing the original data [10]. We note that the horizontal axis of the momentum distributions gives momentum per nucleon, which essentially represents velocity. The resolution of the measured P/A was 0.3%. The vertical axis gives the measured differential cross sections with respect to momentum per nucleon, P/A, obtained in the solid angle window of $\Delta\Omega$ = 4 msr at a reaction angle of 4° as measured with the MARS separator. Each panel represents a different reaction channel. The distributions consist of two peaks, the narrow quasi-elastic peak that corresponds to direct reactions and a wide deep inelastic peak coming from more dissipative collisions.
From the comparison of the momentum distributions for the reactions with the two targets ^{64}Ni and ^{58}Ni, we observe, as expected, lower cross sections for ^{58}Ni, which has a lower N/Z ratio.
We performed binary kinematics calculations and obtained the total excitation energy of the quasi-projectile + quasi-target system. These calculations assume no nucleon evaporation. These excitation energies are reported on some of the peaks of the experimental momentum distributions and help us appreciate the degree of dissipation that is involved. We note that the experimental mass distributions have been obtained and presented in our previous work [10].

3 Description of the Theoretical Models

The theoretical models used to describe the dynamical stage of the collisions were the Deep Inelastic Transfer model (DIT) and the Constrained Molecular Dynamics model (CoMD). Both models follow a Monte-Carlo approach.
The DIT model [11] is a phenomenological model used to describe peripheral collisions. This models considers a di-nuclear configuration of the system allowing stochastic exchange of nucleons, through a "window" that opens in the potential.
We have also used the microscopic model CoMD [12, 13]. This model follows the approach of quantum molecular dynamics [13]. The nucleons are described as Gaussian wavepackets interacting with a Skyrme-type effective interaction with nuclear matter compressibility K=254. The fermionic nature of the system is imposed by a phase-space constraint that enforces the Pauli principle at each time step of the system's evolution. The requirement for the phase space occupation $\bar{f}_i$ is:

$$\bar{f}_i \leq 1 \tag{1}$$

with

$$\bar{f}_i \equiv \sum_j \delta_{\tau_i,\tau_j}\delta_{s_i,s_j}\int_{h^3} f_j(\mathbf{r},\mathbf{p})d^3rd^3p \tag{2}$$

where s_i and τ_i are the z component of the spin and isospin of nucleon i, respectively . The integration is performed in a hypercube of volume h^3 in phase space centered around the point $(\langle r_i\rangle, \langle p_i\rangle)$, in the r and p space, respectively. Within the Gaussian wavepacket representation, it is found necessary to empirically scale the occupation fraction $\bar{f}_i$ as:

$$\bar{f}_i \rightarrow \frac{128}{paulm}\bar{f}_i \tag{3}$$

employing a mass-dependent Pauli constraint parameter, paulm, with the value 94 for the mass range of interest to this work. For each particle i and for each time step, the phase space occupation $\bar{f}_i$ is checked. If the value is greater than 1, an ensemble K_i of nearest particles within distances $3\sigma_r$ and $3\sigma_p$, is determined. Then, for the particles in the ensemble, the momenta are changed in a way that the total momentum is conserved. The ensemble is accepted only if it reduces the value of $\bar{f}_i$. In the present work, the Pauli constraint was enforced further in our calculations, by lowering the value of the parameter paulm from 94 to 87, as will be presented in section 4.

The system is allowed to evolve for time t=600 fm/c (2 zs). The ground state configurations of the nuclei were obtained with a simulated annealing approach. After the annealing, the possible configurations evolve with the model equations of motion and they are accepted if stability is achieved for a time interval of 1000 fm/c. Each configuration, representing initial phase space coordinates of the nucleons, is characterized by a set of model parameters as K, ρ_0, the Pauli constraint parameter ($paulm$) and a set of parameters averaged for each configuration ("achieved parameters"): the binding energy per nucleon $\overline{BE/A}$, the root mean squared radius (R_{rms}), the average $\overline{paulm}$ and the average density $\bar{\rho}$. In the calculations presented in this work, configurations with optimized parameters were employed. In table 1, we present the values of the parameters for ^{40}Ar and ^{64}Ni. In all calculations presented in this work, the value of the surface parameter was set at C_{sup} = −1 [12–14].

	$\frac{BE}{A}$ (MeV)	σ_r	$paulm$	$\rho_0(fm^{-3})$
^{40}Ar	8.68	1.30	94	0.165
^{64}Ni	8.75	1.30	94	0.165

Table 1. Parameters of optimized configurations for ^{40}Ar and ^{64}Ni with K = 254 MeV.

We note that the experimental values of the binding energy per nucleon of the projectile and target are 8.60 MeV and 8.78 MeV, respectively, very close to the achieved values as reported in table 1.

The de-excitation of the hot projectile-like fragments produced in the dynamical stage was realized by the

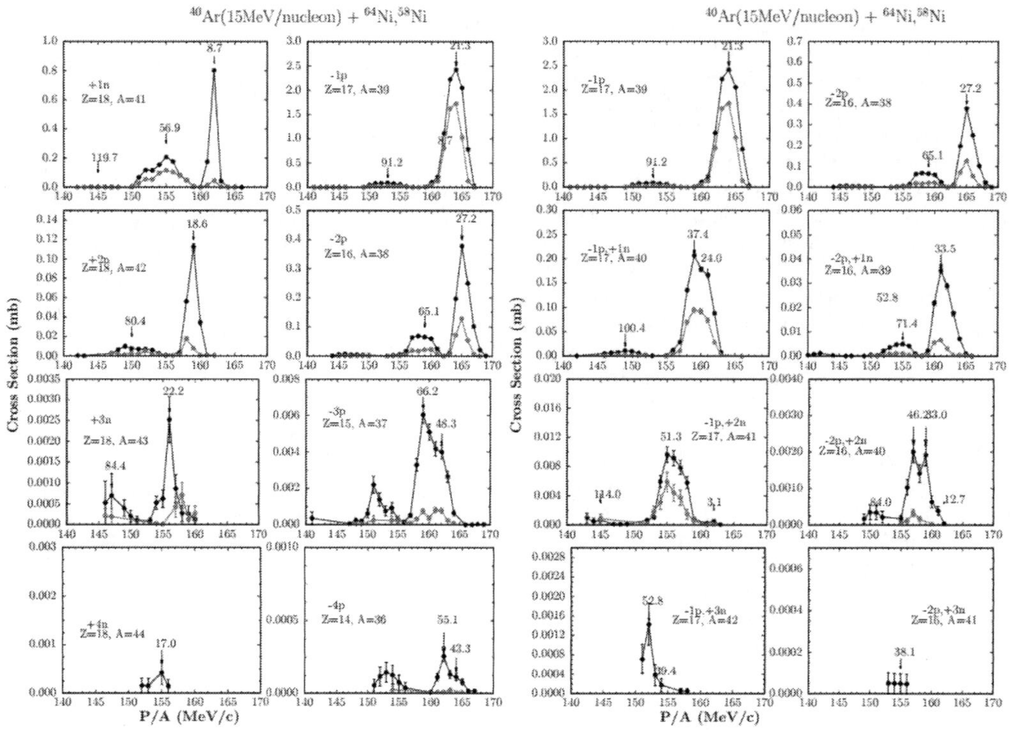

Figure 1. (Color online) Experimental momentum distributions of projectile fragments from the reaction ^{40}Ar (15 MeV/nucleon) with ^{64}Ni, ^{58}Ni. Black points show the data for the reaction with ^{64}Ni target and red points the reaction with ^{58}Ni target.

binary-decay code GEMINI. This statistical de-excitation code [16] uses Monte Carlo techniques and the Hauser-Feshbach formalism to calculate the probabilities for fragment emission with $Z < 2$. Heavier fragment emission probabilities are calculated via a transition state formalism. Within this model, the final partition of products is generated by a succession of fragment emission in binary decays.

4 Calculations and Comparisons

In this section, we present comparisons of DIT and CoMD calculations with the experimental data. Furthermore, we analyze the effect of some parameters of the CoMD model on the calculations. We chose to focus on a systematic attempt to describe the experimental data with the CoMD model due to its many-body approach, which contains no assumptions on the dynamics, and therefore, provides a microscopically-based description of the process. We present calculations only for the reaction of ^{40}Ar with the neutron-rich target ^{64}Ni, since our goal is to study the production of neutron-rich isotopes. We note that the results with the ^{58}Ni target have similar characteristics as those presented here for the ^{64}Ni target.

4.1 Comparison of the models

For the DIT calculations, after extensive tests and following our previous experience, we scaled the excitation energy of the primary quasi-projectiles to 75% of the original value produced by DIT. For the CoMD calculation we chose to run with the optimized configurations and enhanced Pauli constrained (paulm=87). The effect of the Pauli constraint will be discussed below.

In fig. 2, we present the mass distributions for the reaction ^{40}Ar (15 MeV/nucleon) with ^{64}Ni. The experimental data [10] are shown with the full black points. The region of isotopes to the left of the thin solid lines corresponds to incomplete data coverage as explained in section 2. The vertical dashed green line shows the beginning of neutron pick-up. Each panel represents a specific element. The DIT calculation is shown with the dashed red line and the CoMD calculation with the full green line. By comparing the two calculations, we observe that the DIT model describes reasonably well the neutron rich part of the yield, especially the tails. The CoMD code achieves a rather detailed description of the distributions, however slightly overestimating the n-rich tails.

Regarding the momentum distributions, before comparing the calculations to the experimental data, we had to filter the theoretical distributions for the angular and the momentum acceptance of the experimental setup. We

 https://doi.org/10.1051/epjconf/202125207002

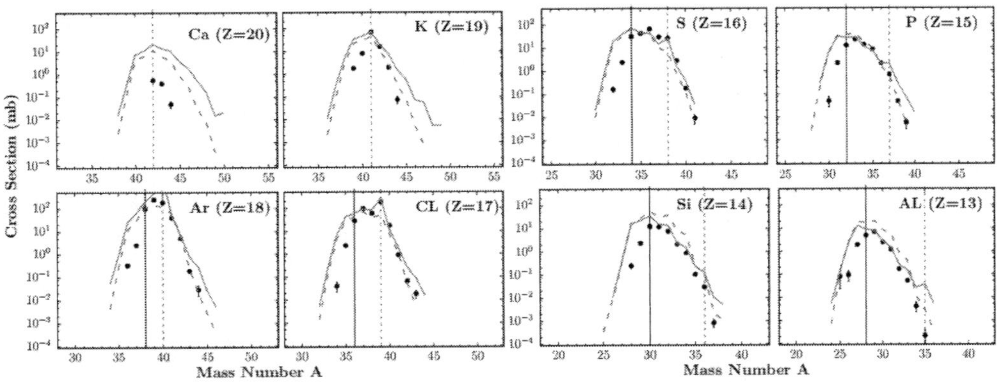

Figure 2. (Color online) Mass distributions of projectile fragments from the reaction of ^{40}Ar (15 MeV/nucleon) with ^{64}Ni. The experimental data are shown by solid black points. The DIT calculation (0.75E*) is shown with the dashed red line and the CoMD calculation (optimum configuration, enhanced Pauli constraint) with the full green line. The region of isotopes to the left of the thin solid lines corresponds to incomplete data coverage. The vertical dashed green line shows the beginning of neutron pick-up.

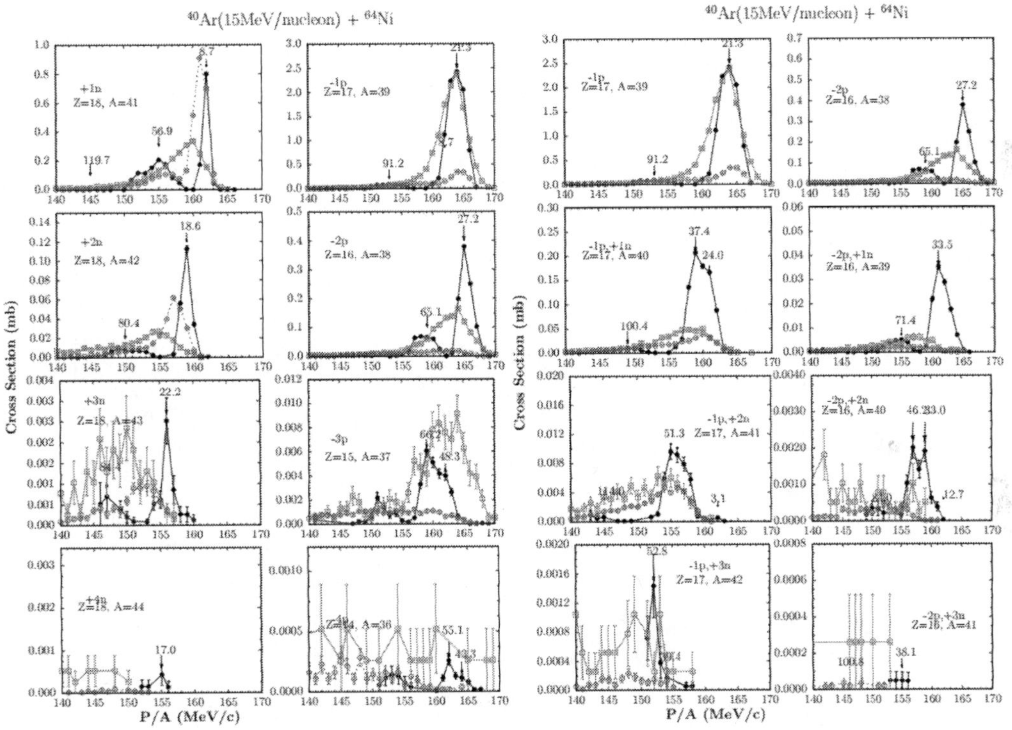

Figure 3. (Color online) Momentum distributions of projectile fragments from the reaction of ^{40}Ar (15 MeV/nucleon) with ^{64}Ni. The experimental data are shown by solid black points. The DIT calculation (0.75E*) is shown with the dashed red line and the CoMD calculation (optimum configuration, enhanced Pauli constraint) with the full green line.

found that this filtering process required the application of a scaling factor of $\sim 1/20$ to the unfiltered calculations, in order to reproduced the experimental acceptance. (The above factor corresponds to the combination of 1/3 and 1/7, respectively, for the polar and azimuthal angular acceptance of the spectrometer.) For the momentum distributions, shown in Fig. 3, the DIT calculations result in low cross sections especially for the proton removal products. The CoMD calculations lead to higher cross sections and wider distributions with respect to the experimental data. It appears that while both models provide a rather reasonable description of the mass distri-

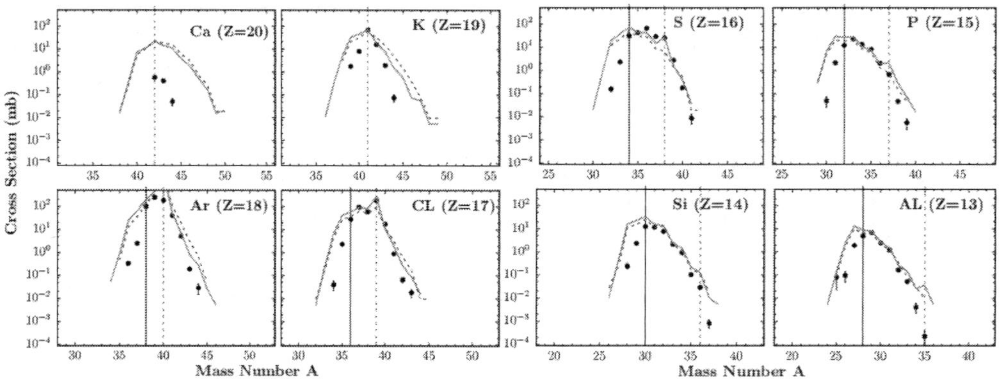

Figure 4. (Color online) Mass distributions of projectile fragments from the reaction of ^{40}Ar (15 MeV/nucleon) with ^{64}Ni. The experimental data are shown by solid black points. The standard CoMD calculation is shown with the dashed purple line and the calculation with enhanced Pauli constraint with the full green line. The region of isotopes to the left of the thin solid lines corresponds to incomplete data coverage. The vertical dashed green line shows the beginning of neutron pick-up.

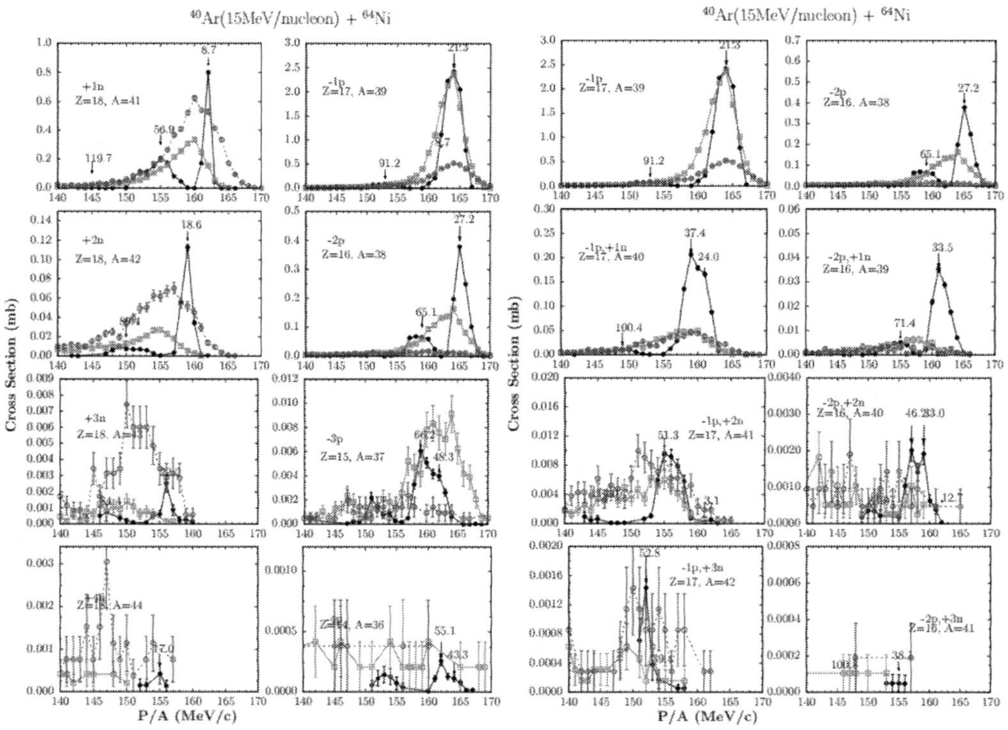

Figure 5. (Color online) Momentum distributions of projectile fragments from the reaction of ^{40}Ar (15 MeV/nucleon) with ^{64}Ni. The experimental data are shown by solid black points. The standard CoMD calculation is shown with the dashed purple line and the calculation with enhanced Pauli constraint with the full green line.

butions, the CoMD model does a better job in describing the experimental momentum distributions in the whole isotope range. Of course, further improvements and tests are necessary.

4.2 Optimization of the CoMD calculations

As already mentioned, in this work we tried to optimize the parametrization of the effective nucleon-nucleon interaction in the CoMD model. The first step was the generation and use of optimized configurations that were discussed in section 3. We applied a more strict enforcement of the Pauli principle by enhancing the Pauli constraint, setting paulm=87, a value lower than the standard value paulm=94 for this mass range (see also table 1). In fig. 4 and 5, we present the effect of this enhancement in the mass and momentum distributions compared with the standard Pauli constraint. With the black points we present the experimental data, with the dashed purple line the standard calculation and with the full green line the calculation with the enhanced Pauli constraint.

The use of enhanced Pauli constraint in the calculations resulted in a rather significant overall improvement. Specifically, for the description of the yields, fig. 4, the calculation with the lower value describes some points of the distribution missed by the calculation with the standard value. For the momentum distributions, fig. 5, we observe better agreement with the data, with higher cross sections for the proton removal products and lower cross sections for the neutron pick-up products. A significant effect on the momentum distributions, is the difference of the width of the peaks, especially in proton removal channels. The calculations with the enhanced Pauli constraint produce narrower peaks, closer to the experimental quasi-elastic peak, compared to the standard calculations. This effect is stronger in the one-proton removal channel. We think that the present calculation with the enhanced Pauli constraint is the best that we have achieved in the context of a systematic variation of CoMD parameters to describe the yields and the momentum distributions of the projectile fragments from the studied reaction.

5 Summary and Conclusions

In continuation of our previous work on the production of neutron-rich nuclides in peripheral collisions below the fermi energy, we studied projectile-like fragments from the reaction an ^{40}Ar beam with ^{64}Ni, ^{58}Ni targets at 15 MeV/nucleon. In this work we obtained and presented the experimental momentum distributions of the ejectiles. Subsequently, we compared both the mass and momentum distributions of projectile fragments from the reaction ^{40}Ar+^{64}Ni with calculations employing the theoretical models DIT and CoMD followed by the deexcitation code GEMINI.

We tried to optimize the parameters of the CoMD model to improve its ability to describe the experimental mass and momentum distributions. We found that the characteristics of the initial configurations used in the CoMD (most importantly, the saturation density and binding energy) are crucial for the outcome of the CoMD calculations. However, within this work, we found that a significant effect comes from an additional enforcement of the Pauli principle. Further exploration and possible improvements of the CoMD model are in line in order to sufficiently describe peripheral reactions with medium-mass heavy ions that lead to neutron-rich products.

6 Acknowledgements

Financial support for this work was provided, in part, by the Special Account for Research Grants of the National and Kapodistrian University of Athens. M.V. was supported by the Czech Science Foundation (GACR) grant No. 21-24281S. SJY and AB were supported by the U.S. Department of Energy under Grant No. DE-FG03-93ER40773 and NNSA Grant No. DENA0003841 (CENTAUR).

References

[1] S. R. Stroberg, J. D. Holt, A. Schwenk, J. Simonis, Phys. Rev. Lett. **126**, 022501 (2021).
[2] G. G. Adamian, N. V. Antonenko, A. Diaz-Torees, S. Heinz, Eur. Phys. J. A **56:47** (2020).
[3] J. Erler, N. Birge, M. Kortelainen, et al, Nature **486**, 509 (2011).
[4] K. Langanke and M. Wiescher, Rep. Prog. Phys. **64**, 1657 (2001).
[5] M. Arnould, S. Goriely, Prog. Part. Nucl. Phys. **112**, 103766 (2020)
[6] G. A. Souliotis, M. Veselsky, G. Chubarian et al, Phys. Lett. B **543**, 163 (2002).
[7] B. Borderie, M.F. Rivet, L. Tassan-Got, Ann. Phys. Fr., 15:4 pp. 287-390 (1990).
[8] G. A. Souliotis, M. Veselsky, S. Galanopoulos, et al., Phys. Rev. C **84**, 064607 (2011).
[9] G. A. Souliotis et al., Progress in Research 2009, Cyclotron Institute, Texas A&M University.
[10] A. Papageorgiou, G. A. Souliotis, K. Tshoo et al., J. of Phys. G **45**, 095105 (2018).
[11] L. Tassan-Got and C. Stephan, Nucl. Phys. A **524**, 121 (1991).
[12] M. Papa, T. Maruyama, A. Bonasera, Phys. Rev. C **64**, 024612 (2001).
[13] M. Papa, G. Giuliani, and A. Bonasera, J. Comput. Phys. **208**, 403 (2005).
[14] N. Vonta, G. A. Souliotis, M. Veselsky, A. Bonasera, Phys Rev. C **92**, 024616 (2015).
[15] A. Bonasera, M. Di Torro, F. Gulminelli, Phys. Rev. C **42**, 3 (1990).
[16] R. J. Charity, M. A. McMahan, G. J. Wozniak et al., Nucl. Phys. A **483**, 371 (1988).
[17] C. A. Bertulani, C. De Conti, Phys. Rev. C **81**, 064603 (2010).
[18] B. Chen, F. Sammarruca, C. A. Bertulani, Phys. Rev. C **87**, 054616 (2013).

EPJ Web of Conferences **252**, 07003 (2021)
HINPw6
https://doi.org/10.1051/epjconf/202125207003

A Constrained Molecular Dynamics (CoMD) study of nuclear near-ground-state properties

T. Depastas[1], *G.A.* Souliotis[1,*], *K.* Palli[1], *A.* Bonasera[2,3], and *H.* Zheng[4]

[1]Department of Chemistry, National and Kapodistrian University of Athens, Athens, Greece
[2]Cyclotron Institute, Texas A&M University, College Station, Texas, USA
[3]Laboratori Nazionali del Sud, INFN, Catania, Italy
[4]School of Physics and Information Technology, Shaanxi Normal University, Xi'an, China

Abstract. The Constrained Molecular Dynamics (CoMD) model is used to describe the properties of nuclear systems near the ground state. A procedure for global optimization of the initial configurations of the nuclei is developed. In addition, the neutron skins of various nuclear systems are calculated. Finally, the GDR and GMR spectra of medium-mass nuclear systems are studied. The effect of the model parameters to the spectra is explored. We conclude that an increased compressibility of $K = 308\ MeV$ results in increased GDR energy and decreased skin, while the total energy and the GMR energy remain almost unaltered.

1 Introduction

The nuclear N-body problem is intricate due to the underlying complicated nuclear interaction. The properties of the resulting low lying states have been studied with a wide variety of methods. These methods include various mean-field approaches, such as Skyrme-Hartree-Fock (SHF) [1] and semiclassical quantum and antisymmetrized molecular dynamics (QMD & AMD) models [2–4]. In recent years, the Constrained Molecular Dynamics (CoMD) model has been used to describe numerous phenomena of low-energy dynamics. Some examples incorporate studies of near Fermi energy reactions and fission of heavy nuclei [5]. The CoMD model forms the basis of the present work and some of its details are presented in the next section.
In this work we study some important near ground state properties of several nuclei. These properties include the neutron skin, radial density, GDR and GMR spectra. The organization of the paper is as follows. First the theoretical framework and computational details of the model are given. Then, we present a study of the initial configurations and the optimization process. After that, we approach the problem of the Giant Resonances. In particular, the Giant Dipole and Giant Monopole Resonances, along with the corresponding soft modes for several nuclear systems, are presented.

2 Theoretical Framework

The CoMD model is based on an effective two-body interaction with Skyrme potential characteristics, as described in [3, 5], with two-body potential terms as follows

$$V_{vol} = \frac{T_0}{\rho_0}\delta\left(\boldsymbol{r}_i - \boldsymbol{r}_j\right), \tag{1}$$

$$V_3 = \frac{2T_3\rho^{\sigma-1}}{(\sigma+1)\rho_0^{\sigma}}\delta\left(\boldsymbol{r}_i - \boldsymbol{r}_j\right), \tag{2}$$

$$V_{sym} = \frac{a_{sym}}{\rho_0}\delta\left(\boldsymbol{r}_i - \boldsymbol{r}_j\right)\left(2\delta_{\tau_i,\tau_j} - 1\right), \tag{3}$$

$$V_{sur} = \frac{C_s}{\rho_0}\nabla^2_{\langle \vec{r}_i\rangle}\delta\left(\boldsymbol{r}_i - \boldsymbol{r}_j\right), \tag{4}$$

$$V_{coul} = \frac{e^2}{\left\|\boldsymbol{r}_i - \boldsymbol{r}_j\right\|} \tag{5}$$

In the above relations, V_{vol} corresponds to the volume term, V_3 to the three-body term, V_{sym} to the symmetry term, V_{sur} to the surface term, while V_{coul} describes the Coulomb repulsion of the protons. The parameters ρ_0, a_{sym} and C_s are the saturation density, the symmetry and surface parameter, respectively. These are fixed to $\rho_0 = 0.165\ fm^{-3}$, $a_{sym} = 32\ MeV$ and $C_s/\rho_0 = -1$, unless otherwise stated. The constants T_0, T_3 and σ are related to the compressibility of nuclear matter (NM) and are determined by the enforcement of the saturation condition of NM with the use of the effective interaction with terms as in eq. (1–5), as described in [3]. The trial one-body wavefunctions that are used in the model, are Gaussian wave-packets and, in the Wigner representation, have the form

$$f_i(\boldsymbol{r},\boldsymbol{p}) = \frac{1}{\left(2\pi\sigma_r\sigma_p\right)^3} e^{-\frac{(r-\langle r_i\rangle)^2}{2\sigma_r^2}} e^{-\frac{(p-\langle p_i\rangle)^2}{2\sigma_p^2}} \tag{6}$$

The σ_r and σ_p are the widths, while $\langle \boldsymbol{r}_i\rangle$ and $\langle \boldsymbol{p}_i\rangle$ are time dependent centroids of the wave-packet of nucleon i in phase space. In the calculations presented below,

*e-mail: soulioti@chem.uoa.gr

$\sigma_r = 1.30\, fm$ and $\sigma_p = \hbar/2/\sigma_r$, unless stated otherwise. The application of the Time Dependent Variational Principle (TDVP) to the Lagrangian of the model with the wavefunctions of eq. (6) leads to the Hamilton's equations of motion in the Wigner representation.
The many-body wavefunction of the model is considered to be a direct product of the one-body wavefunctions of eq. (6). The Pauli Principle is enforced in the model through a constraint of the phase space volume of each nucleon and through Pauli-blocking of nucleon-nucleon (NN) scattering. The total occupation fraction of a phase space hypercube with volume h^3 is constrained through

$$\bar{f}_i = \sum_{j \subset L} \delta_{\tau_i,\tau_j} \delta_{s_i,s_j} \int_{h^3} f_j(\boldsymbol{r},\boldsymbol{p})\, d^3r d^3p \le \frac{paulm}{128} \qquad (7)$$

where L is an ensemble of nucleons j neighboring i. The parameter *paulm* signifies the strength of the Pauli correlations. Higher *paulm* values result in less strict enforcement of the Pauli Principle.

3 Computational Details

In our computational procedure, every calculation with CoMD consists of three steps, the initialization, the evolution and the data processing. The collection of the phase space centroids for each nucleon make up the initial nuclear configuration. These configurations result from a Monte Carlo Simulated Annealing algorithm, that minimizes the total energy of the nucleus. This process takes place in the initialization step [6, 7].

After an initial configuration is chosen, the nucleus evolves for a determined time interval. To account for the approximate character of the equations, before the evolution, the phase space centroids are randomly rotated. This rotation respects the conservation of energy and angular momentum and produces a set of different possible nucleon trajectories in phase space. Each of these is defined as an "event". The aforementioned calculations comprise the evolution step.

The final step depends on the nuclear phenomenon that is studied. First, the nucleon trajectories for each event are properly averaged. Then, we extract information regarding various nuclear properties, such as the mass density or the neutron skin. In order to study the Giant Resonances (GR), one must transform a nuclear property from the time to the frequency domain. This property forms the response function and the transformation is achieved through the use of a Fast Fourier Transform (FFT) code.

4 Initial Configurations and Ground State Properties

4.1 Global Optimization Algorithm

The initialization algorithm produces a set of various possible initial configurations, as described in the previous section. Thus, the question regarding the choice of the initial configuration arises naturally. Each configuration is characterized by the total energy per nucleon, the mass radius and average density and *paulm* parameter. In some of our previous works, the configurations with minimum energy per nucleon have been chosen for the evolution calculation. Another option would be the use of configurations with energies near the experimentally measured value of the binding energy. Both approaches can yield good predictions for the dynamical aspects of reactions and other nuclear phenomena, but usually have higher radii and lower densities than expected.

When one of the characteristics of the initial configurations is improved, the other characteristics become worse. To settle this problem, we developed an algorithm for global optimization of the configurations. In the algorithm, these "optimum" states are chosen based on the values of total energy per nucleon, rms radius, average density and occupation fraction. For the relative error evaluation, the experimental binding energies [8] are used along with the *paulm* values for the occupation fractions. Additionally, the radius choices are evaluated according to the formula $R = 1.2A^{1/3}$. The nuclei tend to shrink in the evolution code, with lower radii that are closer to the experimental value, rather than the empirical one. Finally the average densities are evaluated according to their proximity to Skyrme-Hartree-Fock (SHF) calculations. The evaluation process is performed by a dedicated code that extracts the configurations according to aforementioned criteria.

4.2 Ground State Properties and the Effect of Model Parameters

In the present work, we used the optimum configurations to study the properties of the nuclear ground states, such as the total energy per nucleon. We note that the model reproduces the experimental values for the energies with fair accuracy, while it predicts higher values for the average radii.

The radial density is another property that was studied. The optimum states yield improved densities, that have close proximity with those calculated by SHF mean-field approaches. In figure 1 the mass densities (top row), the proton densities (central row) and the neutron densities (bottom row) of ^{40}Ar (left) and ^{64}Ni (right) are plotted against the radial coordinate. The CoMD calculated density is represented by a continuous line, while the dashed line corresponds to the SHF density. In the present preliminary calculations, the approximate treatment of the Pauli principle and the gaussian profile of the wavepacket result in a dip in the nuclear density at low radii that we expect to remove in subsequent calculations.

To understand the dependence of the equation of state (EOS) on the parameters of the effective interaction, we studied total energies, radii, average densities and occupation fractions of the optimum configurations. In figure 2 we show the variation of the minimum energy per nucleon of the ^{68}Ni configuration space with respect to the *paulm* parameter (a), compressibility (b), saturation density (c) and effective mass ratio (d). These parameters and their effects are discussed in the following paragraphs.

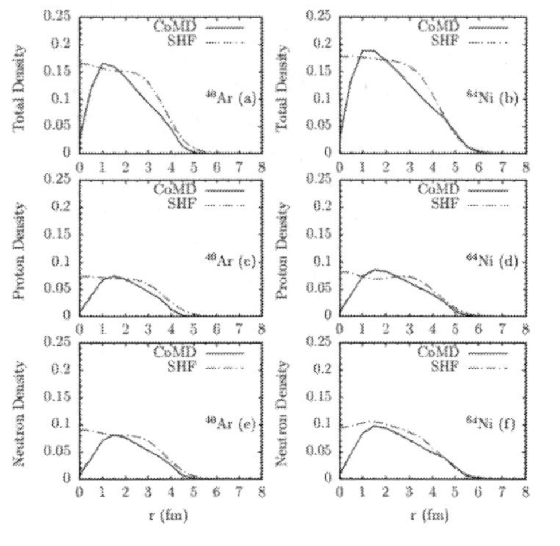

Figure 1. Radial densities of ^{40}Ar (left column) and ^{64}Ni (right column) plotted against the radial coordinate. The top row gives the plots of the total density, the middle row of the proton density and the bottom row of the neutron density. The solid black line corresponds to the CoMD calulation and the purple dashed line to the SHF calculation.

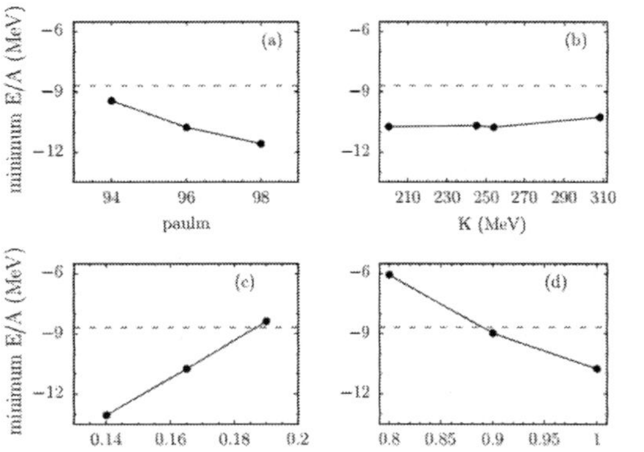

Figure 2. Minimum Energy per nucleon of ^{68}Ni plotted against paulm (a), compressibility (b), saturation density (c) and effective mass ratio (d). The points connected with the solid black lines correspond to the CoMD calulation and the red dashed line to experimental data from [8].

The *paulm* parameter determines the strength of the Pauli Principle, as it is already discussed. The smaller values increase the radius and decrease the average density of the nucleus. Additionally, the energy of the system is increased, as seen in figure 2(a). This is understood to be a consequence of the degeneracy pressure of a Fermi Gas. Thus, mostly the kinetic part of the Hamiltonian contributes to the increase of the energy. We wish to comment that the enforcement of the Pauli principle in CoMD via the constraint corresponds to a purely quantal effect and may correspond to the emergence of shell effects in the energy spectrum of the system. We have made such a preliminary exploration, but we do not discern energy gaps corresponding to shells in our calculations. In this vein, the possibility of including a spin-orbit term in the potential will also be explored.

The compressibility K is an important parameter of the EOS, that signifies the response of the nuclear system to compression. For the effective interaction of the CoMD model (eq. 1–5), it can be expressed as:

$$K = 9T_3\frac{\sigma(\sigma-1)}{(\sigma+1)} - 2\bar{\epsilon}_F \tag{8}$$

where $\bar{\epsilon}_F$ is the average Fermi energy of NM. As expected, the increase of the compressibility results in a decrease of the radius and an increase of the density. Interestingly, the energy of the system seems to be independent of the compressibility choice, as seen in figure 2(b). This results from a simultaneous increase of the kinetic component and a decrease of the potential component of the Hamiltonian.

The saturation density (ρ_0) of the system signifies the maximum density of a nucleus and results from the repulsive part of the effective interaction. As shown in 2(c), this parameter is proportional to the system's energy. Another important characteristic of the nuclear interaction is the inclusion of momentum dependent terms. These terms represent the finite range of the interaction. In the current version of the model, we include the momentum dependence in the low energy limit, as an effective mass (m^*). The smaller effective mass results in an increase of the kinetic energy of the system, and thus to an increase of the total energy. This tendency can be clearly seen in figure 2(d). In the discussion below, the effective mass is $m^* = m$, unless otherwise stated.

4.3 Neutron Skin Calculations

The neutron rich nuclei are stabilized by distributing the excess of neutrons near the nuclear surface that is distinct from the N=Z symmetric core. This extends the neutron density further from the proton density by a few tenth of fm, creating the characteristic neutron skin (r_{np}), expressed as the difference of the rms radii of protons and neutrons. The neutron skin of various nuclides has been investigated with both theoretical and experimental approaches [4]. The CoMD model generally predicts neutron skins with higher values, compared to experimental data or empirical expectations. In general, increased compressibility and *paulm* result in decreased values for the skin. We report herein, that the ^{208}Pb calculation with $K = 254$ MeV, $paulm = 102$ and $m^*/m = 1$ yields 0.43 fm neutron skin, somehow higher than the experimental value of $0.33^{+0.16}_{-0.18}$ fm.

The neutron skin, of course, depends on the relative number of protons and neutrons, that may be expressed by the isospin asymmetry factor, $m_\chi = \frac{N-Z}{A}$. As described in [4], the neutron skin depends linearly on the isospin asymmetry. In figure 3, we present the correlation of the neutron skin with the m_χ for various isotopic and isobaric chains of medium-mass nuclei. The negative neutron skin values correspond to proton-rich nuclei, as expected. The

https://doi.org/10.1051/epjconf/202125207003

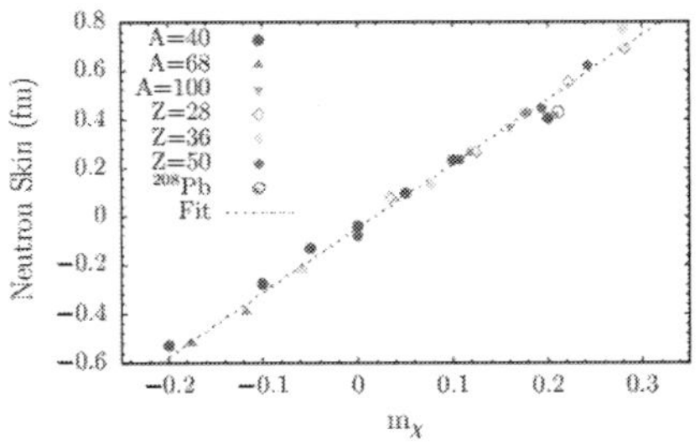

Figure 3. Neutron skin plotted against the isospin asymmetry parameter. The points represent CoMD calculations of various isotopic and isobaric chains, according to the key. The dashed red line represents our linear fit, eq. (9).

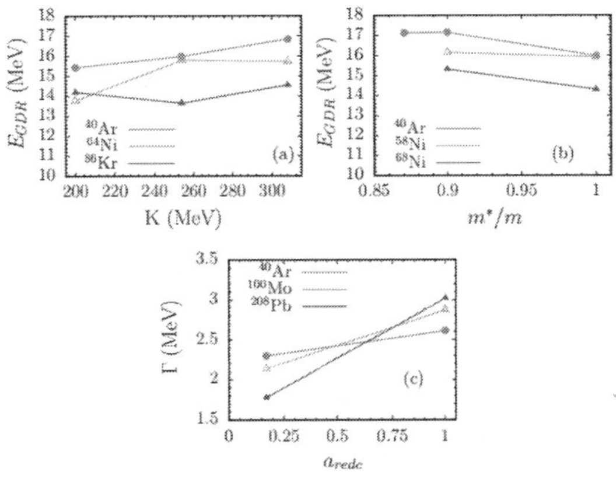

Figure 4. Effects of important model parameters to the GDR spectrum of various nuclides, according to the key. The top row presents the variation of the GDR energy with respect to compressibility (a) and effective mass ratio (b). The bottom panel (c) shows the variation of the width with respect to the reduction parameter of the NN scattering cross section.

calculations shown in fig. 4 are fitted with the formula

$$r_{np} = (2.65 \pm 0.051)\, m_\chi + (-0.047 \pm 0.081) \quad (9)$$

5 Isovector Giant Dipole Resonances

The Giant Resonances (GR) are some of the most interesting phenomena of nuclear dynamics. They consist of several collective modes of motion, that can be considered as multipole excitations of the nuclear radius. By studying these resonances, one can extract useful information with regard to the EOS and effective interaction.
In this work, we study the Isovector Giant Dipole (IVGDR or GDR). This is in effect the off-phase oscillation of the proton and neutron fluids, around their equilibrium positions. This resonance is one of the most thoroughly investigated GRs (e.g. [7, 11**?**]). It has been described by a wide variety of theoretical and computational methods ranging from liquid-drop and BUU approaches [9] to RPA and TDHF [8].

5.1 Theoretical Framework and the Effect of Model Parameters

In this study, the GDR problem is approached as follows. We consider an initial perturbation of the nucleon centroids in either cartesian or momentum space. The coordinate space perturbation affects the nucleon's z-coordinate, while conserving the center of mass of the system. The momentum space perturbation consist of a quadrupole perturbation of the Fermi sphere that conserves the total volume in momentum space. Both these perturbations induce the GDR effect. The spectrum is extracted by the Fourier transform of the distance between the centers of mass of protons and neutrons. In the coordinate space, the z-axis perturbation is defined as $D = \frac{A}{Z} d_n = \frac{A}{N} d_p$, where d_n, d_p are the neutron and proton perturbations respectively.

By considering only the variation of the symmetry term in the low energy limit, we constructed a simple model in order to understand the dependencies of the full CoMD model. The corresponding collective equations of motion represent the damped GDR oscillation and their solution gives the following approximate relation for the centroid energy

$$E_{GDR} \approx \sqrt{\frac{A}{NZ} \frac{a_{sym}\rho_{np}}{m\rho_0 \sigma_r^2}} \quad (10)$$

where ρ_{np} is the neutron-proton interaction density.

In figure 4, we present the effect of the compressibility (a) and effective mass (b) to GDR energy, for a few nuclei. The increased compressibility corresponds to higher 3-body interaction (eq. 8) and yields nuclei with a more bound potential well. The main contribution to the GDR energy comes from the variation of the nuclear potential, thus higher compressibility values yield higher energy centroids. This can be seen in figure 4 (a). Additionally, the momentum dependence has a major role in the determination of the GDR [4]. In the approximation of eq. (10), the mass is present in the denominator. Thus lower effective mass yields higher GDR energies, a tendency that is confirmed in figure 4 (b).

The width of the GDR spectrum is mostly determined by the NN scattering cross section. In the CoMD model, the in-medium NN scattering cross section is approximated as the free cross section times a reduction parameter. In our implementation, this factor can have a constant value or it can be parameterized as [10]:

$$a_{redc}(T_{cm}, \overline{\rho}) = e^{-a_R \frac{\overline{\rho}/\rho_0}{1+(T_{cm}/T_0)^2}} \quad (11)$$

where $T_0 = 150 MeV$ and $a_R = 1.8$. For the present study, we may set $\overline{\rho} \approx \rho_0$ and $T_{cm} \ll T_0$, to obtain $\langle a_{redc} \rangle \approx 0.17$. The reduction parameter is given by eq. (11) in the following calculations, unless stated otherwise. In figure 4 (c) the width is plotted against the reduction parameter for different nuclei. As expected, increased cross section values result in increased GDR widths.

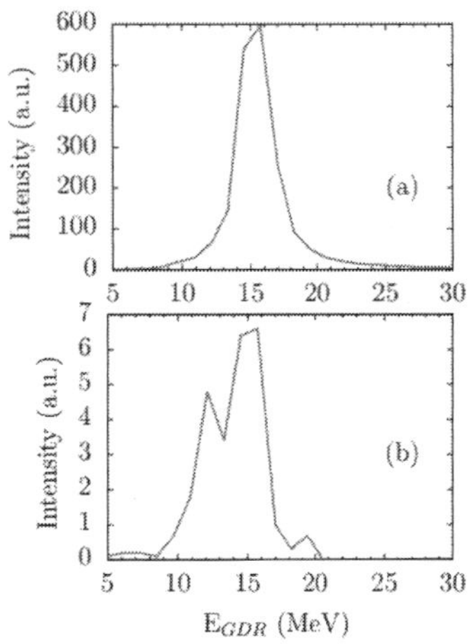

Figure 5. GDR spectra of ^{68}Ni. The top panel (a) represents the GDR response in cartsian space and the bottom (b) represents the momentum space response.

Apart from GDR, other isovector modes may be present in the model's isovector spectrum. The most important is the Pygmy resonance, with corresponds to an off-phase oscillation of the neutron skin of a neutron rich nucleus against its symmetric core. In the CoMD model, a momentum space perturbation can excite these soft degrees of freedom. In figure 5 we present the r-space (a) and p-space (b) GDR spectrum of ^{68}Ni for K = 308 MeV. We note that the main peak at 15 MeV corresponds to the GDR and is lower by 2 MeV from the experimental value, of 17.84 MeV [13]. On the other hand, we may identify the secondary peak at 12 MeV as the Pygmy resonance. This is close to the experimental value of 11 MeV [13].

5.2 Isotopic and Isobaric Results

It is known that an effective interaction with an approximate treatment of the momentum dependence, appears to underestimate the GDR energies [4]. In figure 6, we present the variation of the GDR energy with mass number A. The dashed red curve corresponds to an experimental parameterization [4]. The points connected with lines show calculations with two EOS. The green points correspond to the standard "soft" EOS with K = 254 MeV, m^*/m = 1 and a_{sym} = 32 MeV, while the cyan points correspond to a "hard" EOS with K = 308 MeV, m^*/m = 0.9 and a_{sym} = 38 MeV. The soft EOS calculations differ from the experimental values by $3-4$ MeV, while this energy gap becomes somewhat narrower for the hard EOS. We wish to comment here that the above tendency to favor larger K values does not agree with results supporting a softer EOS with K~250 (e.g. [14]). We think that the inclusion of a proper momentum dependence in the effective interaction of the CoMD model may reconcile this issue and is one of our future goals.

In the literature, empirical expressions for the GDR energy usually do not contain an explicit dependence on

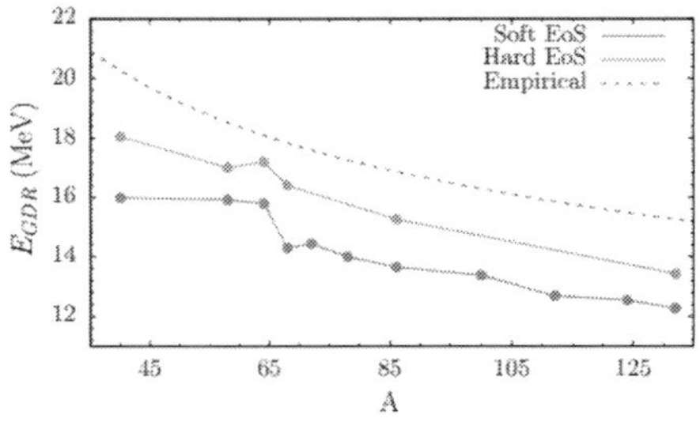

Figure 6. GDR energies of various nuclides with respect of the mass number. The green curve represents the soft EOS calculations, the cyan the hard EOS calculations and the red curve is an experimental parameterization from [4].

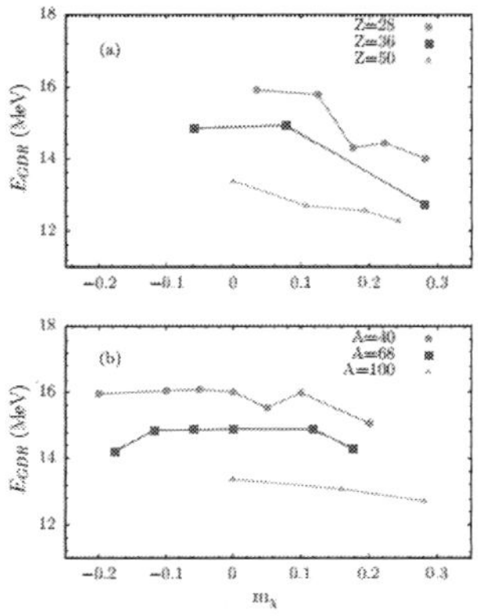

Figure 7. GDR energies plotted against the isospin asymmetry parameter. The top panel (a) shows the variation of various isotopic and the bottom (b) of various isobaric chains, according to the key.

the isospin asymmetry (e.g. [4, 11, 12]), as the observed isospin effects are rather small. Here we confirm this behavior, but we note some decreasing trend for the neutron rich $m_\chi > 0$ nuclei. This can be seen in figure 7, where we show the variation of the GDR energy with respect to A, for different isotopic chains (a) and with respect to Z, for different isobaric chains (b).

6 Isoscalar Giant Monopole Resonances

The Isoscalar Monopole Resonance (ISGMR or GMR) is another important GR. It consists of the in-phase breathing mode of both neutrons and protons. This GR can constrain the compressibility of EOS and give important information regarding the effective interaction. Additionally, there exists a soft monopole mode for neutron rich nuclei, analogous to the Pygmy resonance. This GR corresponds to an isoscalar breathing mode of the neutron skin against the symmetric core.

In the CoMD model, the isoscalar modes are studied via a monopole perturbation in momentum space. The spectrum results from an FFT of the nuclear rms radius.

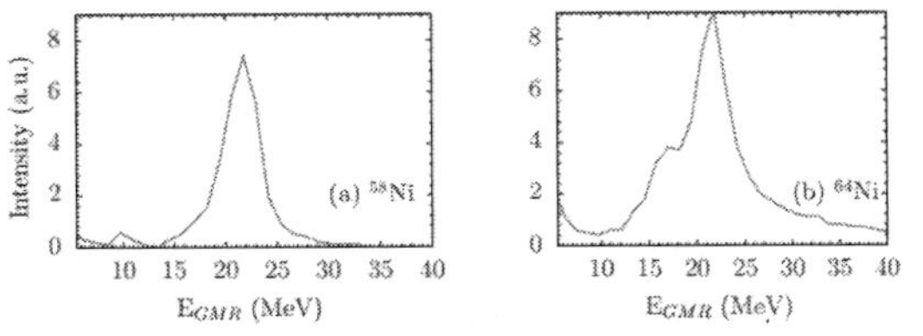

Figure 8. GMR spectrum of ^{58}Ni (a) and ^{64}Ni (b). Both are extracted by an momentum space GMR perturbation.

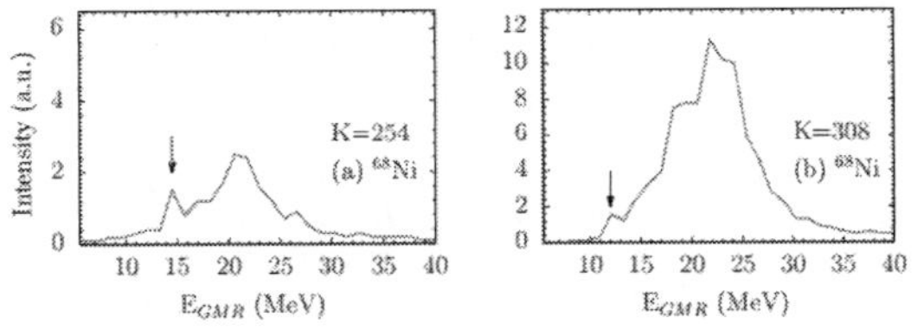

Figure 9. GMR spectrum of ^{68}Ni for two different compressibilities. The left panel (a) represents K=254 MeV and the right panel (b) represents K=308 MeV calculation. The soft monopole mode is indicated by an arrow.

In this work, we studied the isoscalar response of ^{A}Ni isotopes, with $A = 58, 64, 68$. In figure 8 we present the GMR spectrum of ^{58}Ni (a) and ^{64}Ni (b). Furthermore, in figure 9 we present the GMR spectrum of ^{68}Ni with $K = 254$ MeV (a) and $K = 308$ MeV (b). We note that, the main GMR peak of the nickel isotopes is found at 21.5 MeV, which is close to the experimental value of 21.1 ± 1.9 MeV [15]. As for the ^{68}Ni isotope, the soft monopole appears at 14.5 MeV, which is also near the experimental value of 12.9 ± 1.0 MeV [15]. An interesting observation regarding the effect of the compressibility can be made from figure 9. The increase of K from 254 MeV to 308 MeV leads to larger total cross section (strength) of the peak and an increase of the GMR centroid by ~ 1 MeV. Finally, the soft mode peak retains its strength and width, while the width of the main GMR peak is increased.

7 Discussion and Conclusions

In this work, we used the CoMD model to describe properties of nuclear systems near the ground state. First, we developed a procedure to choose an initial phase-space configuration with optimum characteristics. Furthermore, we studied the effective interaction and the effect of its parameters. We found that an increased compressibility results in an increased GDR energy and decreased skin, while the total energy and the GMR energy remain almost unaltered. The neutron skins of various nuclei were calculated. Our CoMD calculations appear to overestimate the skin compared to empirical values or experimental results.

An important part of our work was devoted to the study of GDR. In the CoMD model, the GDR energy is underestimated by a few MeV. These results can be partially improved by choosing a "harder" EOS with $K = 308$ MeV. The GDR can be improved further by the inclusion of a momentum dependent interaction, that we plan to do in the near future. Finally, it appears that our CoMD results agree better with the experimental values, as the mass of the nuclear system increases. This suggests that the CoMD model in its current implementation provides overall better results for heavier nuclei.

8 Acknowledgement

Financial support for this work was provided by the Special Account for Research Grants of the National and Kapodistrian University of Athens. AB and HZ were supported by the U.S. Department of Energy under Grant No. DE-FG03-93ER40773 and NNSA Grant No. DENA0003841 (CENTAUR).

References

[1] D. Vautherin, D.M. Brink, Phys. Rev. C **5**, 3 (1971).
[2] Y. Kanada-En'yo, M. Kimura, A. Ono, Prog. Theor. Exp. Phys. **1**, 202 (2012).
[3] M. Papa, T. Maruyama, A. Bonasera, Phys. Rev. C **64**, 024612 (2001).
[4] G. Giuliani, H. Zheng, A. Bonasera, Prog. in Part. and Nucl. Phys. **76**, 116 (2014).
[5] N. Vonta, G.A. Souliotis, M. Veselsky, A. Bonasera, Phys. Rev. C **92**, 024616 (2015).
[6] M Papa, J. Phys.: Conf. Ser. **420**, 012082 (2013).
[7] M. Papa, G. Giuliani, A. Bonasera., J. Comput. Phys. **208**, 403 (2005).
[8] J. L. Basdevant, J. Rich, M. Spiro, *Fundamentals in nuclear physics: From nuclear structure to cosmology* (Springer, New York, 2005).
[9] G. Bonasera, M. R. Anders, S. Shlomo, Phys. Rev. C **98**, 054316 (2018).
[10] R. Wang, Z. Zhang, et al Phys. Let. B **807**, 135532 (2020).
[11] S. Shlomo, G. Bertsch, Nucl. Phys A **243**, 507 (1975).
[12] J.S. Wang, W.Q. Shen et al, Eur. Phys. J. A **7**, 355 (2000).
[13] D.M. Rossi, et al, Phys. Rev. Lett. **111**, 242503 (2013)
[14] M. Veselsky et al, Phys. Rev. C **94**, 064608 (2016).
[15] X. Sun, Phys. Rev. C **103**, 044603 (2021).

Total kinetic energy release in the fast neutron induced fission of actinide nuclei

Ashley Pica[1], *Alexander* Chemey[1], and *Walter* Loveland[1]

[1]Dept, of Chemistry, Oregon State University, Corvallis, OR 97331, USA

Abstract. The total kinetic energy (TKE) release in the fast neutron-induced fission of various actinide nuclei was measured for neutron energies for E_n=2.6-100 MeV at the Weapons Neutron Research facility of the Los Alamos National Laboratory. The data are compared to the GEF model of the fission process. The variances of the TKE distributions appear to decrease with increasing Z and A of the fissioning systems.

1 Introduction

The largest part of the prompt energy release in fission (169.1 MeV out of 185.6 MeV for ^{235}U(n_{th},f)) is in the form of the total kinetic energy (TKE) of the fission fragments. Fission is a large scale collective motion of a few hundred nucleons and as such, is difficult to understand. The magnitude of the TKE release depends on the Coulomb forces between the nascent fragments at scission and the possible transfer of collective energy into either nucleonic excitation or fragment motion at scission. In this paper we describe the measurement of the TKE release in the fast neutron induced fission of ^{232}Th, ^{235}U and ^{237}Np. Some of these data have been published previously [1–3]. We comment on common features of these data and suggest directions for further research.

2 Experimental

The experimental arrangements for these measurements have been described in detail in [1–3]. The measurements were made at the Los Alamos National Laboratory using the Weapons Neutron Research Facility (WNR) at the Los Alamos Neutron Science Center (LANSCE). The measurements typically took 5-7 days. The targets were bombarded with "white spectrum" neutron beams generated from an unmoderated tungsten spallation source using the 800 MeV proton beam from the LANSCE linear accelerator. The experiment was located on the 15R beam line. The fast neutron beam intensities were $10^5 - 10^6$ neutrons/s for E_n =2-100 MeV. The proton beam was pulsed allowing one to measure the time of flight of neutrons striking the actinide fission targets. The neutron energies were determined with an uncertainty of 4.7-5%. Because of the low beam intensities and relatively short measurement times, large solid angles had to be employed for the detection of any fission fragments produced.

The actinide targets were mounted at the center of an evacuated Al scattering chamber. Fission fragments from the fast neutron-actinide reactions were detected by large arrays of Si PIN diode detectors. The actinide targets were ~ 195 μg/cm^2 (^{232}Th) , ~ 175 μg/cm^2 (^{235}U), and 34 μg/cm^2 (^{237} Np). The combination of similar measurement times and the thinner ^{237}Np target lead to larger uncertainties in the ^{237}Np measurements. The actinides were deposited on ~ 100 μg/cm^2 C foil by vapor deposition [4].

We measured, on an event by event basis, the pulse heights of coincident fission fragments. To transform these pulse heights into energies, we corrected for the pulse height defect of our Si detectors and the energy loss of the fission fragments in the target and backing materials, iteratively. We used the 2E method to treat the data. In the 2E method, we corrected iteratively for (a) pre-equilibrium neutron emission (b) pre-fission neutron emission (c) the dependence of the post neutron emission on fragment mass and (d) the mean atomic number associated with each post neutron emission fragment. The pulse height defect is taken into account using the Schmitt procedure [5] and the energy losses in the target deposit and the target backing material are calculated using the Northcliffe-Schilling tables [6]. The GEF model [7] is used to estimate ν_{pre-eq}, ν_{pre}, ν_{post}, and Z(A).

To benchmark the experimental method, the TKE release in the ^{235}U(n_{th},f) reaction was measured at the Oregon State University 1 MW TRIGA Reactor. The measurement was made using the same apparatus and target as in the LANSCE experiment. The thermal neutron-induced fission data at Oregon State University was analysed using the same methods and corrections as with the fast-neutron induced fission data, except ν_{tot}= 2.5 and ν_{pre}= 0. The measured post-neutron emission TKE was 169.8 ± 0.4 MeV. The deduced pre-fission TKE was 170.7 ± 0.4 MeV. This result is consistent with previous measurements of 171.9 ± 1.4 MeV [5], 172.0 ± 2.0 MeV [8] and 170.1 ± 0.5 MeV [9].

The measurements of the TKE release in this work are thus absolute measurements with no normalizations to theory or other measurements.

3 Data

3.1 ^{232}Th(n,f)

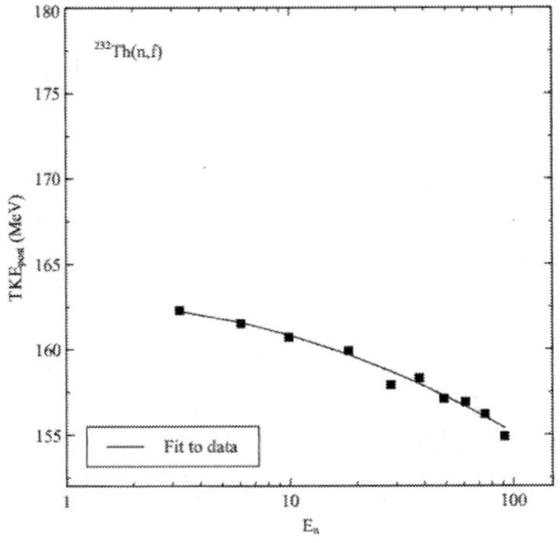

Figure 1. The average post-fission TKE measured for the ^{232}Th(n,f) reaction as a function of neutron energy.

In Figure 1, we show the mean post-fission TKE release for the fast neutron induced fission of ^{232}Th for the energy range from 3-91 MeV. For this energy range , the mean total kinetic energy decreases non-linearly from 162.3 ± 0.3 MeV to 154.9 ± 0.3 MeV. A polynomial fit was made to these data. The resulting polynomial is TKE_{post} (MeV) =162.8-0.1884 $\log_{10}E_n$-1.866 $(\log_{10}E_n)^2$.

3.2 ^{235}U(n,f)

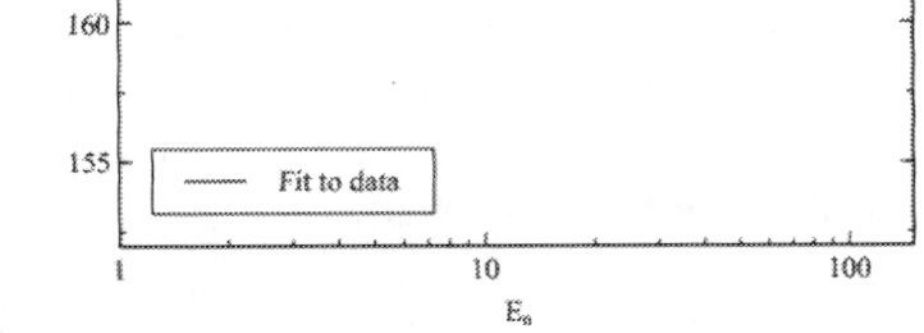

Figure 2. The average post-fission TKE measured for the ^{235}U(n,f) reaction as a function of neutron energy.

In Figure 2, we show the mean post-fission TKE release for the fast neutron induced fission of ^{235}U for the energy range from 3.0 to 89.7 MeV. As observed for ^{232}Th(n,f), the mean TKE values decrease non-linearly with increasing neutron energy. A polynomial fit was made to these data. The resulting polynomial is TKE_{post} (MeV) =170.92 - 3.73$\log_{10}E_n$ – 0.65$(\log_{10}E_n)^2$.

3.3 ^{237}Np(n,f)

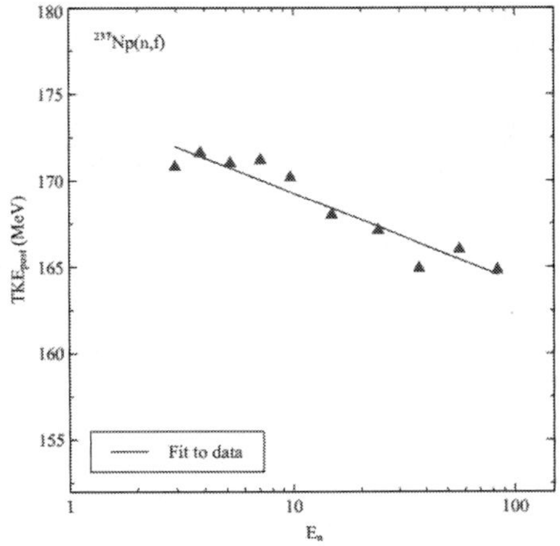

Figure 3. The average post-fission TKE measured for the ^{237}Np(n,f) reaction as a function of neutron energy.

In Figure 3, we show the mean post-fission TKE release for the fast neutron induced fission of ^{237}Np for the energy range from 2.4 to 83.8 MeV. As observed for ^{232}Th(n,f), the mean TKE values decrease non-linearly with increasing neutron energy. A polynomial fit was made to these data. The resulting polynomial is TKE_{post} (MeV) =174.38 - 5.11$\log_{10}E_n$ for E_n > 1 MeV.

It is clear from Figures 1-3 that the post-neutron emission TKE plots for the three systems we have studied are similar apart from a shift in the TKE values with increasing Z and A of the fissioning systems. The data from Figures 1-3 and similar plots for other fissioning systems can be used to reliably predict the TKE values in fast neutron induced fission.

3.4 TKE distribution variances

Another interesting feature of the TKE distributions is the variances of the TKE distributions. In Figures 4-6 we show the variances of the TKE distributions for the fast neutron induced fission of ^{232}Th (Fig. 4), ^{235}U(Fig. 5) and ^{237}Np (Fig. 6). What is striking about these distributions is that the variances become smaller as the Z,A of the fission system increases. While there is some "spatter" in the data, this trend seems clear and is not expected in a simple model of fission

3.5 Models for the TKE release in fission.

To understand the data we have discussed up to this point, it is useful to compare our observations with predictions

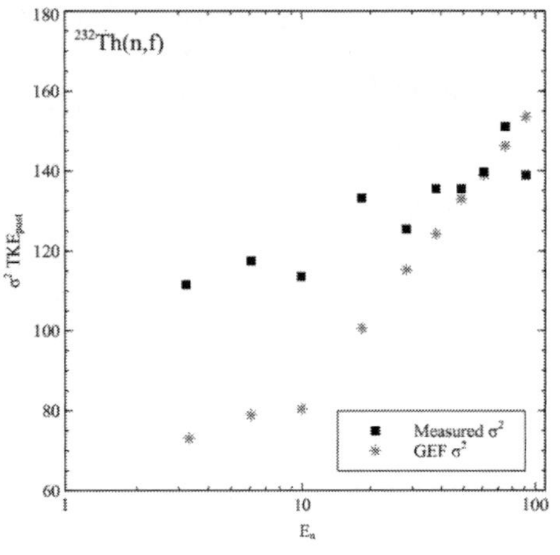

Figure 4. Measured and GEF calculated variances of the TKE distributions for the fast neutron induced fission of ^{232}Th.

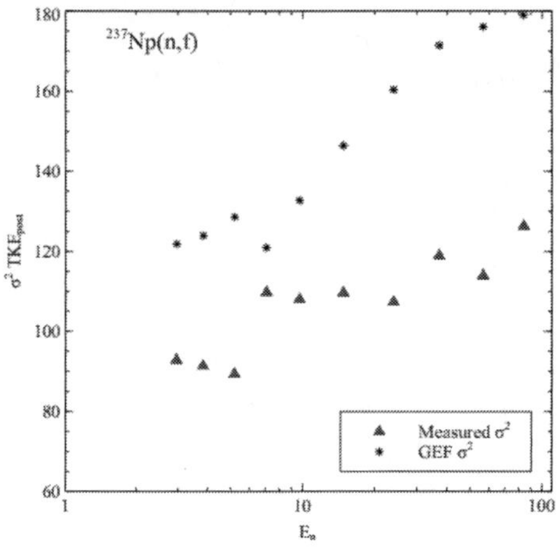

Figure 6. Measured and GEF calculated variances of the TKE distributions for the fast neutron induced fission of ^{237}Np.

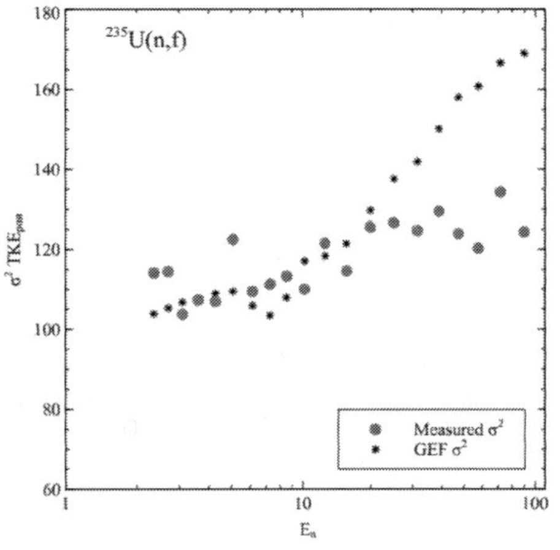

Figure 5. Measured and GEF calculated variances of the TKE distributions for the fast neutron induced fission of ^{235}U.

(not "post-dictions") of these quantities. One such prediction of the observables in fast neutron induced fission is the general description of fission observables **(GEF)**, which is a semi-empirical model for the calculation of fission observables, with 50 adjustable parameters. The version of the code used in this study is GEF 2019/1.3, downloaded May 2019 [7]. The main ingredients of the GEF model are:

(a) The mass division and the charge distribution are calculated assuming a statistical population of states in the fission valleys at freeze-out. The freeze-out time considers the influence of fission dynamics and is not the same for different collective variables.

(b) The compound nucleus and fission fragment properties are assumed to be separable.

(c) Five fission channels are considered. The strengths of the shells in the fission valleys are identical for all fissioning systems. The mean positions of the heavy fragments in the asymmetric fission channels are essentially constant in atomic number.

(d) The stiffness of the macroscopic potential with respect to mass asymmetry is deduced from the widths of the measured mass distributions.

(e) An excitation energy sorting mechanism determines the prompt neutron yields and the odd-even effect in fission fragment yields of even Z and odd Z systems.

(f) Neutron evaporation is calculated with a Monte Carlo statistical code using level densities from empirical systematics and binding energies with theoretical shell effects with gamma competition included.

A notable aspect of the GEF code is that it postulates that as E* increases, the excess energy goes to the heavy fragment only.

In Figures 7-9, we show the measured TKE distributions from this work and the predictions of the GEF model. For the ^{235}U(n,f) reaction, there is good agreement between the GEF predictions and the measured data. For the ^{232}Th(n,f) reaction, the agreement between the GEF predictions and the data is not good. For the ^{237}Np(n,f) reaction, the agreement between the predicted and measured values of the TKE is abysmal. It is puzzling to us as to why a "50 parameter model" such as GEF does not do a better job of accounting for the measured data.

4 Summary and Conclusions

We have presented measurements of the TKE release in the fast neutron induced fission of the actinide nuclei, ^{232}Th, ^{235}U and ^{237}Np. Our measured values of the TKE release are in good agreement with the semi-empirical GEF model for the case of ^{235}U(n,f) but not for ^{237}Np or ^{232}Th. In addition, we find the variances of the TKE distributions become smaller as the Z,A of the fissioning system increase, an unexpected finding.

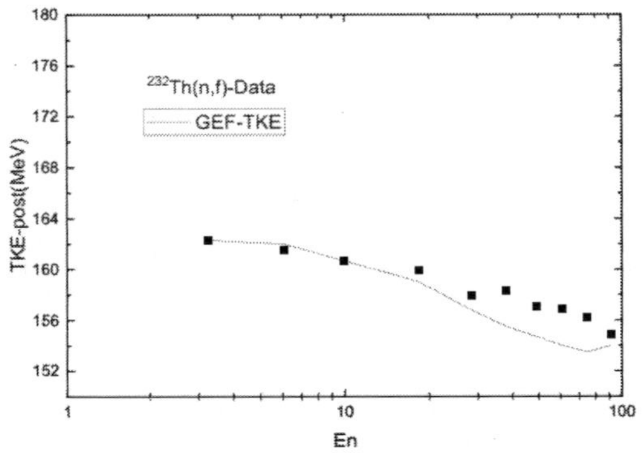

Figure 7. A comparison between the measured values of the TKE and those predicted by the GEF model for the fast neutron induced fission of ^{232}Th.

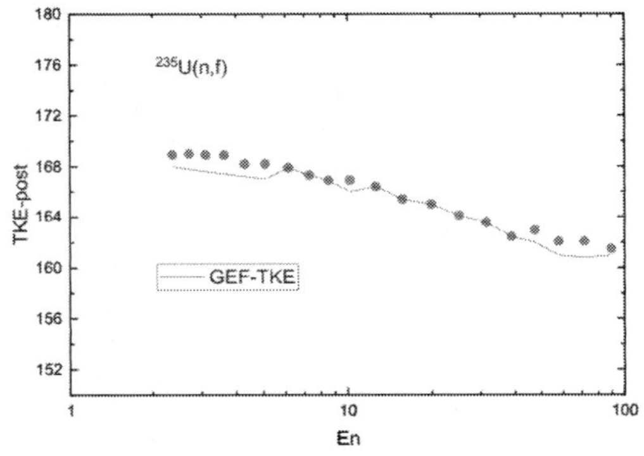

Figure 8. A comparison between the measured values of the TKE and those predicted by the GEF model for the fast neutron induced fission of ^{235}U.

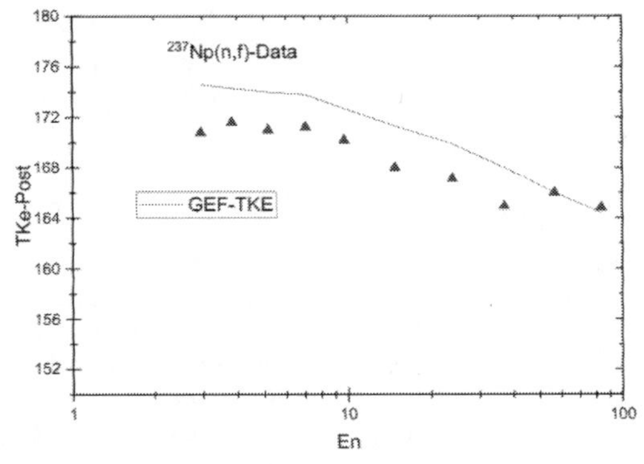

Figure 9. A comparison between the measured values of the TKE and those predicted by the GEF model for the fast neutron induced fission of ^{237}Np.

Acknowledgments

We gratefully acknowledge the financial support of the U.S. Dept. of Energy, NNSA, under grant DE-NA0022926 and the U.S. Dept. of Energy, Office of Science, Nuclear Physics Division, under grant DE-FG06-97ER41026.

References

[1] J. King, R. Yanez, W. Loveland, J.S. Barrett, B. Oscar, N. Fotiadis, F. Tovesson, and H.Y. Lee, Eur. Phys. J. A **53**, 238 (2017).

[2] R. Yanez, J. King, J.S. Barrett, W. Loveland, N. Fotiadis and H.Y. Lee, Nucl. Phys. A **970**, 65 (2018).

[3] A. Pica, A.T. Chemey, L. Yao, W. Loveland, H.Y. Lee and S. A. Kuvin, Phys. Rev. C **102**, 064612 (2020).

[4] M.J. Silveira, A. Pica and W. Loveland, Nucl. Instru. Methods Phys. Res. Sect. A **982**, 164750 (2020)

[5] H.W. Schmitt, J.H. Neiler and F.J. Walter, Phys. Rev **141**, 1146 (1966).

[6] L.C. Northcliffe, and R. Schilling, At. Data Nucl. Data Tables 7, 233 (1970).

[7] K.H. Schmidt, B. Jurado, C. Amouroux, and C. Schmitt, Nuclear Data Sheets **131**, 107 (2016).

[8] M.J. Bennett and W.E. Stein, Phys. Rev. **156**, 1277 (1967)

[9] G.F. Bertsch, W. Loveland, W. Nazarewicz. and P. Talou, J. Phys. G. Nucl. Part. Phys. **42**, 077001 (2015).

Study of the reaction ^{70}Zn (15 MeV/nucleon) + ^{64}Ni with the MAGNEX spectrometer for the production of neutron-rich isotopes

S. Koulouris[1], *G. A.* Souliotis[1*], *F.* Cappuzzello[2,3], *D.* Carbone[3], *A.* Pakou[4], *C.* Agodi[3], *G.* Brischetto[2,3], *S.* Calabrese[3], *M.* Cavallaro[3], *I.* Ciraldo[2,3], *O.* Fasoula[1], *J.* Klimo[5], *K.* Palli[1], *O.* Sgouros[3], *V.* Soukeras[3], *A.* Spatafora[2,3], *D.* Torresi[3], and *M.* Veselsky[6]

[1]Laboratory of Physical Chemistry, Department of Chemistry, National and Kapodistrian University of Athens, Athens, Greece
[2]Dipartimento di Fisica e Astronomia "Ettore Majorana", Universita di Catania, Italy
[3]Laboratori Nazionali del Sud, INFN, Catania, Italy
[4]Department of Physics and HINP, The University of Ioannina, Ioannina, Greece
[5]Institute of Physics, Slovak Academy of Sciences, Bratislava, Slovakia
[6]Institute of Experimental and Applied Physics, Czech Technical University, Prague, Czech Republic

Abstract. We describe our efforts to study the production of neutron-rich isotopes from peripheral reactions of medium-mass heavy ions with the MAGNEX spectrometer at the INFN-LNS in Catania, Italy. Experimental data were obtained for the ^{70}Zn+^{64}Ni reaction at 15 MeV/nucleon. For the analysis of the data, we developed a new procedure for the reconstruction of both the atomic number Z and the ionic charge q of the ions, which is then followed by the identification of the mass. Preliminary results and the analysis plan will be discussed.

1 Introduction

To date, approximately one half of the theoretically estimated 7000 bound nuclei have been produced and investigated. Exotic nuclei located far away from the line of beta stability are not present in nature and need to be prepared in laboratory using proper nuclear reactions and separation techniques [1, 2]. One of the main concurrent challenges of the nuclear community in rare isotope beam facilities around the world is the production of neutron-rich nuclides to the limit of the neutron dripline. The investigation of neutron-rich nuclei could lead to a better understanding of important astrophysical nucleosynthesis processes, such as the rapid neutron capture process (r-process), which is responsible for half of the abundance of the nuclides heavier than iron. To have access to nuclides with high neutron-excess, besides the traditional routes of spallation, fission and projectile fragmentation, it is necessary to pick up neutrons from the target. Such nucleon transfer mechanisms mainly take place in peripheral nucleon-exchange reactions at beam energies from the Coulomb barrier [3] to the Fermi energy (20–40 MeV/nucleon) [4, 5].

Our interest was initially focused on the study of heavy-ion reactions at 25 MeV/nucleon, which led to a substantial production of neutron-rich nuclides [4]. Encouraged by these results, we moved on to studies at 15 MeV/nucleon [6, 7] that were mainly based on the use of the medium-acceptance MARS spectrometer [8]. Our studies at this energy, in particular with heavier targets such as Sn and Pb, underlined the importance of a large acceptance spectrometer for the access of nuclides with high neutron excess, which are concentrated at angles near the grazing angle. For this reason, we started a project to identify projectile-like fragments with the MAGNEX large acceptance spectrometer at the INFN-LNS from the reaction ^{70}Zn (15 MeV/nucleon) + ^{64}Ni.

The structure of the paper is as follows. In section 2, theoretical calculations are briefly presented. In section 3, the experimental setup and measurements are described. In section 4, the analysis of the data is presented with emphasis on the identification of heavy projectile-like fragments. Finally, a discussion and conclusions are given in section 5.

2 Calculations

Theoretical calculations carried out with the models CoMD [9] and DIT [10], followed by the statistical deexcitation code GEMINI [11], indicated substantial production of neutron-rich projectile fragments. In Fig. 1, we show the calculated mass distributions of projectile fragments with Z = 27–32 from the reaction of ^{70}Zn (15 MeV/nucleon) with ^{64}Ni obtained by DIT/GEMINI (solid red line), and by CoMD/GEMINI (dashed blue line). The results are encouraging concerning the production of neutron-rich isotopes especially near the projectile (e.g. Ni, Cu, Zn).

Furthermore, in Fig. 2 we present the elemental angular distributions of the same projectile fragments with

*e-mail: soulioti@chem.uoa.gr

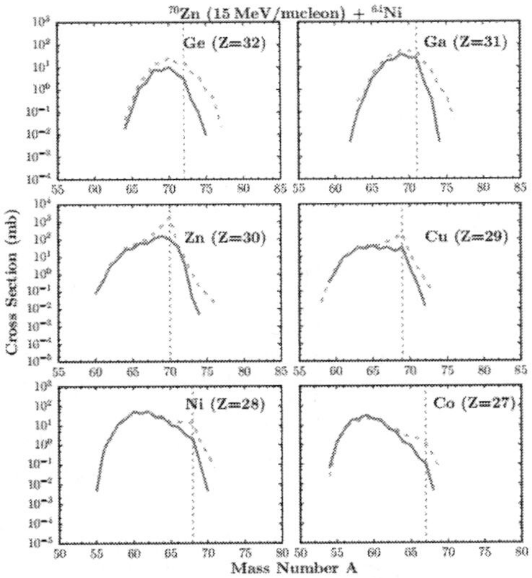

Figure 1. (Color online) Calculated mass distributions (lines) of projectile fragments with Z = 27–32 from the reaction of ^{70}Zn (15 MeV/nucleon) with ^{64}Ni. The calculations shown are: DIT/GEMINI (solid red line), CoMD/GEMINI (dashed blue line). The green vertical line indicates the starting point of neutron pickup.

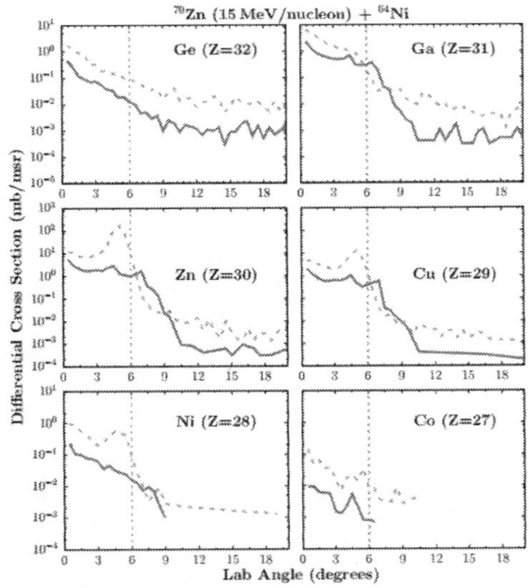

Figure 2. (Color online) Calculated elemental angular distributions of projectile fragments with Z = 27–32 from the reaction of ^{70}Zn (15 MeV/nucleon) with ^{64}Ni. The calculations shown are: DIT/GEMINI (solid red line), CoMD/GEMINI (dashed blue line). The green vertical line indicates the grazing angle for this reaction (θ_{gr} = 6.0°).

Z=27–32 from the above ^{70}Zn+^{64}Ni reaction. We note that for this system, the grazing angle is θ_{gr} = 6.0°. (A vertical dashed green line indicates this angle in Fig. 2.) The general feature of the calculated angular distributions with both the DIT and the CoMD models is that they peak near the grazing angle for elements close to the projectile (Cu, Zn, Ga). However, elements further away from the projectile are characterized by a rather monotonically decreasing angular distribution. As we see in Fig. 1 and Fig. 2, the CoMD model appears to give higher yields compared to DIT, especially for the neutron-rich products. We are currently trying to understand these differences. Of course, the plan is to compare these calculations with the experimental data from our MAGNEX experiment after the analysis will be completed in the near future.

3 Experimental Setup and Measurements

The experimental measurements were performed at the MAGNEX facility of the Istituto Nazionale di Fisica Nucleare, Laboratori Nazionali del Sud (INFN-LNS) in Catania, Italy. MAGNEX is a high-acceptance device which makes use of both the advantages of the traditional magnetic spectrometry and those of a large angular and momentum acceptance detector [12, 13]. A beam of $^{70}Zn^{+15}$ at 15 MeV/nucleon, delivered by the S800 superconducting cyclotron, bombarded a 1.18 mg/cm^2 ^{64}Ni target foil. The ejectiles coming out from the target passed through a 6 μm stripper foil and then were detected by the spectrometer's Focal Plane Detector (FPD). In this pilot experiment, only one-half of the active area of FPD was used and the vertical angular acceptance was restricted in the range $-0.8°$ to $0.8°$, resulting in the use of only seven Si detectors of FPD. Finally, the time-of-flight (TOF) of the ions was measured via a start signal from the silicon detectors of FPD and a stop signal from the radiofrequency of the cyclotron.

4 Particle Identification Procedure and Data Analysis

The particle identification is based on a new technique that we developed in ref. [14] in combination with the approach of [12]. Herein, we give an overview of the main points. According to [12], the atomic number of the ejectiles is determined by the ΔE_{cor}–E_{resid} correlation involving the residual energy measured by the silicon detectors (E_{resid}) and the total energy loss (ΔE_{cor}) in the gas section of FPD corrected for path length differences depending on the angle of incidence. Although this correlation is adequate for the determination of Z of light ions (see, e.g., [15]), in the present case of medium-mass heavy ions, we could not obtain a clear Z separation. We thus developed an approach to reconstruct the atomic number Z based on the measured and properly calibrated quantities ΔE_{cor} , E_{resid} and TOF. Using these calibrated quantities, and guided by the relation $Z \propto \upsilon \sqrt{\Delta E}$ from Bethe-Bloch equation, we developed an approach for the Z reconstruction, following [16], that is based on the velocity, ΔE_{cor} and E_{tot}, where E_{tot} is the total kinetic energy of the ions reaching FPD and determined from the expression:

$$E_{tot} = \Delta E_w + \Delta E_{tot} + E_{resid} \quad (1)$$

where ΔE_{tot} is the sum of the measured energy loss in the gas section of the FPD, E_{resid} the residual energy, as already mentioned, and ΔE_w a calculated correction for the energy loss in the entrance window of FPD. In our procedure, Z is reconstructed according to the following expression:

$$Z = c_0(\upsilon) + c_1(\upsilon)\sqrt{\Delta E_{cor}E_{tot}} + c_2(\upsilon)(\sqrt{\Delta E_{cor}E_{tot}})^2 \quad (2)$$

To determine the functions $c_0(\upsilon)$, $c_1(\upsilon)$ and $c_2(\upsilon)$ in the velocity range of interest, we used the energy loss data of Hubert et al. [17] to determine the coefficients of eq. (2) for the Z range 6–36 and in the energy range of 8–18 MeV/nucleon via a least-squares fitting procedure at each energy in steps of 0.5 MeV/nucleon. The values of each coefficient at the various energies were subsequently fitted with polynomial functions of velocity.

After the Z reconstruction, we proceeded to reconstruct the ionic charge state of the ejectiles from the equation:

$$q = \frac{2E_{tot}}{B\rho}\frac{TOF}{L} \quad (3)$$

employing the total energy E_{tot}, the TOF measurement, as well as the trajectory length L of each particle, obtained from the trajectory reconstruction [9]. The magnetic rigidity $B\rho$ was obtained from the equation:

$$B\rho = B\rho_0(1+\delta) \quad (4)$$

where δ is the fractional deviation from the magnetic rigidity $B\rho_0$ of the central trajectory obtained from the trajectory reconstruction procedure.

Our approach for particle identification involves a correlation of the reconstructed atomic number Z with the reconstructed ionic charge state q of the ejectiles in a two dimensional plot [14]. Adequate separation of the Z and q groups was achieved. The resolutions (FWHM) of Z and q are ~0.8 and ~0.7 units, respectively.

For the mass determination, we first set proper gates on the atomic number Z and the charge state q plot. Furthermore, we implement the standard approach of particle identification for large acceptance spectrometers [13]. This approach is based on the relationship between the total kinetic energy of the ions and the magnetic rigidity, according to the equation:

$$B\rho = \frac{\sqrt{m}}{q}\sqrt{2E_{tot}} \quad (5)$$

In a $B\rho$ vs E_{tot} correlation, the ejectiles are distributed on bands according to their $\frac{\sqrt{m}}{q}$. We note that in the Z vs q plot, we have identified and separated products with specific Z and q. Since q is given, the dependence is only on $\sqrt{m}$. Consequently, these bands represent various masses for a specific Z and q. In Fig. 3, we present a $B\rho$ versus E_{tot} plots of Cu^{28+} (Z=29, q=28) events. These events were selected by proper graphical contours of Z=29 and q=28 in the Z–q correlation [14]. The gap in the $B\rho$ range is due to a software gate applied to exclude the elastically scattered $^{70}Zn^{29+}$ ejectiles. Adequate separation between the different mass bands of Cu^{28+} products is achieved. In this plot, the selection of the various masses can be performed by setting the respective graphical cuts, as shown in Fig. 3. Thus, we conclude that our goal to produce and identify medium-mass neutron-rich nuclides is achieved with a large acceptance spectrometer as MAGNEX.

5 Discussion and Conclusions

In this work, we presented our effort to produce and identify medium-mass ejectiles from a heavy-ion reaction with the use of the large acceptance magnetic spectrometer MAGNEX at INFN-LNS. We investigated the reaction of a ^{70}Zn beam at 15 MeV/nucleon with a ^{64}Ni target. In the data analysis, we found that the ΔE_{cor} vs E_{resid} correlation was not adequate for the Z identification of the medium-mass ejectiles. For this reason, we developed a new identification approach employing a measurement of TOF and

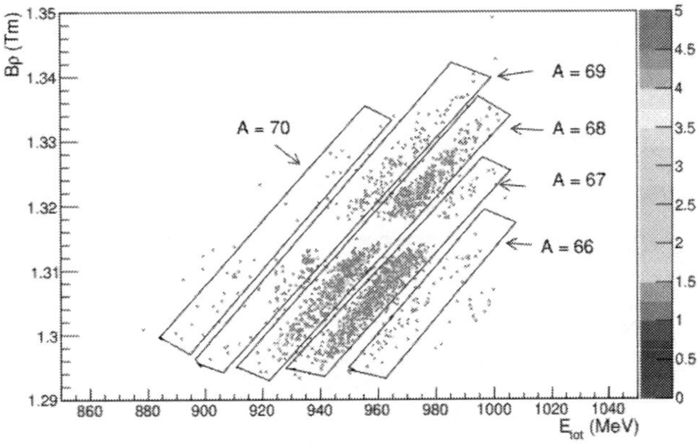

Figure 3. (Color online) Magnetic rigidity vs E_{tot} correlation of ejectiles from the ^{70}Zn (15 MeV/nucleon) + ^{64}Ni reaction corresponding to the gate of Z=29, q=28. The gap in $B\rho$ is due to a gate set to exclude elastically scattered $^{70}Zn^{29+}$ ejectiles. Graphical contours are drawn to separate the masses A = 66–70 of Cu^{28+}.

the reconstruction of both the atomic number Z and the ionic charge state q of the ejectiles. Our Z reconstruction approach is based on the energy loss, the residual energy and the TOF measurements after appropriate absolute calibration. Moreover, a reconstruction of q was performed using the magnetic rigidity and the trajectory length of the ions, obtained from the ion optical trajectory reconstruction. Finally, the mass identification was performed with specific gates on Z and q via a magnetic rigidity vs. total energy correlation. The present analysis of medium-mass projectile-like fragments indicates the necessity for our new procedure that has to be applied in relation to the use of a large acceptance spectrometer like MAGNEX.

We expect that with our newly developed PID procedure, a good identification of the neutron-rich ejectiles will be achieved. After identification, we plan to obtain their angular and momentum distributions and then their production cross sections. The experimental data will be compared with our model calculations, as those shown in Figs. 1 and 2. We hope that the extracted experimental distributions, along with comparisons with the theoretical models will shed light to the complex reaction mechanisms leading to the production of neutron-rich nuclides at this energy regime.

6 Acknowledgements

We are grateful to the support staff of INFN–LNS for providing the primary beam. This project has received funding from the European Research Council (ERC) under the European Union's Horizon 2020 research and innovation programme (grant agreement No 714625). G.S. acknowledges support from the Special Account for Research Grants of the National and Kapodistrian University of Athens. M.V. was supported by the Czech Science Foundation (GACR) grant No. 21-24281S.

References

[1] Y. Blumenfeld, T. Nilsson, P.V. Duppen, Phys. Scr. T **152**, 014203 (2013).
[2] G.G. Adamian, N. V. Antonenko, A. Diaz-Torees, S. Heinz, Eur. Phys. J. A **56**, 47 (2020).
[3] V.V. Volkov, Phys. Rep. **44**, 93 (1978).
[4] G.A. Souliotis et al., Phys. Lett. B **543**, 163 (2002).
[5] G.A. Souliotis et al., Phys. Rev. C **84**, 064607 (2011).
[6] P.N. Fountas, G.A. Souliotis, M. Veselsky and A. Bonasera, Phys. Rev. C **90**, 064613 (2014).
[7] A. Papageorgiou, G.A. Souliotis et al., Journal of Physics G **45**, 095105 (2018).
[8] G.A. Souliotis et al., Nucl. Instrum. Methods B **204**, 166 (2003).
[9] M. Papa et al., Phys. Rev. C **64**, 024612 (2001).
[10] L. Tassan-Got and C. Stephan, Nucl. Phys. A **524**, 121 (1991).
[11] R. Charity et al., Nucl. Phys. A **483**, 371 (1988).
[12] F. Cappuzzello et al., Eur. Phys. J. A **52**, 167 (2016).
[13] F. Cappuzzello et al., Nucl. Instr. and Meth. A **621**, 419 (2010).
[14] G.A. Souliotis, S. Koulouris, F. Cappuzzello et al. Nuclear Instrum. and Methods A (submitted).
[15] M. Cavallaro, et al., Nucl. Inst. and Meth. B **463**, 334 (2020).
[16] G.A. Souliotis, K. Hanold, W. Loveland et al., Phys. Rev. C **57**, 3129 (1998).
[17] F. Hubert, R. Bimbot and H. Gauvin, At. Data Nucl. Data Tables **46**, 1 (1990).

EPJ Web of Conferences **252**, 08001 (2021) https://doi.org/10.1051/epjconf/202125208001
HINPw6

Texas A&M US Nuclear DATA Program

Ninel Nica[1,*]

[1]Cyclotron Institute, Texas A&M University, College Station, Texas 77843-3366, USA

Abstract. Nuclear data evaluation is an independent century-long expert activity accompanying the development of the nuclear physics science. Its goal is to produce periodic surveys of the world literature in order to recommend and maintain the set of the best nuclear data parameters of common use in all basic and applied sciences. After WWII the effort extended and while it became more international it continued to be supported mainly by the US for the benefit of the whole world. The Evaluated Nuclear Structure Data File (ENSDF) is the most comprehensive nuclear structure database worldwide maintained by the United States National Nuclear Data Center (NNDC) at Brookhaven National Laboratory (BNL) and echoed by the IAEA Vienna Nuclear Data Services. Part of the US Nuclear Data Program since 2005 the Cyclotron Institute is one of the important contributors to ENSDF. Since 2018 we became an international evaluation center working in a consortium of peers hosted traditionally by prestigious national institutes as well as universities. In this paper the main stages of the evaluation work are presented in order to facilitate a basic understanding of the process as a guide for our potential users. Our goals are to maintain a good productivity vs. quality performance assuring the currency of the data and participating in the effort of modernizing the structure of ENSDF databases in order to make them compatible with the data-centric paradigms of the future.

1 Introduction

Nuclear data evaluation fills a century-long chapter of nuclear science. A search in the Nuclear Science Reference (NSR) database maintained at the National Nuclear Data Center (NNDC) (https://www.nndc.bnl.gov/nsr/) on the author M. Curie produces a paper titled The Radioactive Constants as of 1930 [1]. The introduction to this paper states that "the need has arisen for the publication of special Tables of the Radioactive Constants" and continues: "This responsibility has been assumed by the International Radium Standards Commission chosen in Brussels in 1910 (...)." Here we have the origin of what today is known as Nuclear Data Evaluation.

After WWII Nuclear Data Evaluation got sustained mainly by the USA being initiated by Katharine (Kay) Way who in the late 1940s worked on the Manhattan project [2]. In 1948 she founded the Nuclear Data Project at the U.S. National Bureau of Standards, which in 1950 published the first Nuclear Data Report. In 1953 the Nuclear Data Project relocated to the U.S. National Academy of Sciences - National Research Council in Washington D.C. publishing its first nuclear data AEC report in the form of loose leaf pages called *Nuclear Data Sheets*. In 1964 the Nuclear Data Project moved to the Oak Ridge National Laboratory (ORNL) in Oak Ridge Tennessee following the need to embed the data program in an active physics research environment, thus making nuclear data evaluation a basic science domain. In February 1966 the first issue of the Nuclear Data Sheets in a journal format rather than as loose leaf sheets of data appeared in Section B of the Nuclear Data journal published by Academic Press, while Section A had been released an year earlier as Atomic Data Tables. In August 1973 the two journals merged into the Atomic and Nuclear Data Tables with Kay Way as the editor for both.

The main evaluation work was organized in the frame of the Nuclear Data Project (NDP) at ORNL where the data were still entered by hand directly on sheets of paper with drawings done by hand, which were typed and photographed together with the professional draftsman redrawings in order to generate the the final publishable version. An important landmark was the implementation of the so-called 80-column computer format for ENSDF designed in 1977 by Bruce Ewbank and Marcel Schmorak of the NDP staff, described in February 1978 in the ORNL-5054/R1 report. The 80-column format was done judiciously and with some changes added by NNDC served for many years being still in use today, which is now in a vast process of modernization under the lead of NNDC. The same report describes the original versions of data computer codes (Logft, Alpha HF, GTOL, HSICC (Hager-Selzer), Medlist, and plot computer programs), which have ever since been modified and extended at NNDC, with many additional analysis and utility programs added.

The evaluation activity extended and became international with the establishment in 1974 of the Nuclear Structure and Decay Data Network, NSDD [3], under the auspices of the IAEA, Nuclear Data Section, putting together the evaluation efforts in several countries and having the US as the main contributor. Although NNDC at

*e-mail: nica@comp.tamu.edu

Figure 1. USNDP evaluation centers (inside the circle) together with the two international institutions of which IAEA assumes the international coordination role. The main goals of the USNDP are also listed at the top.

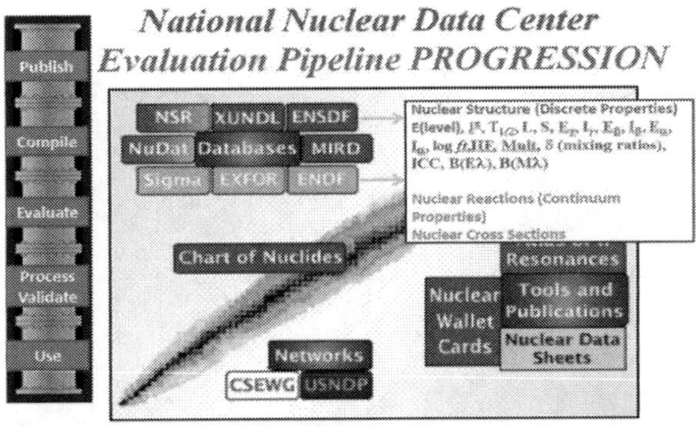

Figure 2. Main portal page of the NNDC internet site (https://www.nndc.bnl.gov/) together with the "Evaluation Pipeline" drawing (on the left side) which illustrates the way the nuclear structure data evaluation is conducted as described in the text.

BNL became the coordinator of both the national (USNDP) and the international (NSDD) eforts for the US, the leading contribution in editing and processing of the evaluation effort was still assumed by Oak Ridge. In 1981, the NNDC took over production of Nuclear Data Sheets, and completely computerized the entire operation. NDP and NNDC jointly edited the journal, with Murray Martin as Editor-in-Chief and Jagdish Tuli as Editor, with Tuli taking over as sole editor after June 1988, when Martin retired. The present editor, E. A. (Libby) McCutchan, took over upon Tuli's retirement in April, 2016, with NNDC still being the most prominent evaluation center in the US and abroad [4].

2 US Nuclear Data Program

Nuclear science research since its beginings has been producing a great variety of experimental data published in a multitude of publications worldwide. However each such piece of information is difficult to use by the data users because its context needs to be understood and quite often it can be different or even discrepant relative to the same type of data reported by different authors. The situation is similar to a jungle where there is an abundance of useful fruits and seeds which are difficult to use unless a group of experts is assuming the task to retrieve, categorize and make the data available for the basic nuclear science and applications communities.

As we saw in the Introduction, this mission was gradually undertaken by the small group of scientists constituting the nuclear data evaluation community, who developed the skills and standard procedures suitable for it.

According to the USNDP Mission Statement, our goal is to provide current, accurate, authoritative data for workers in pure and applied areas of nuclear science and engineering. This is accomplished primarily through the compilation, evaluation, dissemination, and archiving of extensive nuclear datasets. USNDP also addresses gaps in the data, through targeted experimental studies and the use of theoretical models.

The nuclear data evalution community is organized by evaluation centers located in the historic BNL and ORNL national labs plus the newer ones as shown in Fig. 1, in total six national labs and four universities of which Texas A&M University (ATM) joined in 2017.

Figure 2 presents the main portal page of the NNDC internet site (https://www.nndc.bnl.gov/) together with the "Evaluation Pipeline". The group of buttons situated at the top delineates the two main components of the low energy nuclear data evaluation, one for Nuclear Structure and other for Nuclear Reactions, each of which is maintained by two different groups of specialized evalutors. As a rule the Nuclear Structure group takes care of the discrete nuclear properties, a list of which is given in the figure, while the Nucleare Reactions group evaluates the continnum properties - the cross sections. The evaluated databases for each direction are called ENSDF (Evaluated Nuclear Structure Data File) for Nuclear Structure and ENDF (Evaluated Nuclear Data File) for Nuclear Reactions, respectively. In what follows we will limit our presentation to the discrete nucleare structure properties only.

Despite the fact that nuclear data are of great need and extensively used, most of the data users have no knowledge of the evalution process, which can result in the underperformance or even wrongful usage of the databases, reason for which a short description of the evaluation pipe progression is going to be done in the next subsections.

Thus there are two important things useful for a user to know. First, that there is a complete transparency of the evalution process all over the evalution pipe, as well as in the evalution procedures themselves. Secondly, that all the evaluation tools are equally available to the evalutor as well as to the data user, which can use them for scientific purposes.

In the following subsections we are going to overview each segment of the evaluation pipe.

2.1 Publish

The first segment of the evaluation pipe is that of retrieving the relevant publications. Before starting the evaluation an extensive bibliographic search is conducted by the evaluator for each nucleus in order to retrieve virtually all the articles published for a particular nucleus worldwide, including both primary and secondary publications.

The next step consists in a first selection of those publications that contain relevant experimental data and any piece of relevant information, followed by sorting them based on the reaction or decay process by which the nucleus of interest was populated and studied. For this purpose the NSR bibliographic database is the best tool which was developed and is maintained on daily basis by the NNDC and collaborators [5]. The great advantage of the NSR database is that it contains a detailed list of keywords which allow a reliable retrieval of all relevant publications.

2.2 Compile

The second segment of the evaluation pipe is that of compilation, which is extracting the most relevant data from one publication and placing it in the 80 columns ENSDF format. The compilation files are stored by the so-called XUNDL database (XUNDL is the acronym of Experimental Unevaluated Nuclear Data List) which has its own search engine activated by the XUNDL button of the NNDC portal. Each XUNDL file is named after the type of nuclear reaction or decay process and is a collection of data tables and comments related to the publication together with some general descriptors of the experimental procedures. A pdf copy or a copy in the ENSDF format of the XUNDL file can be downloaded at the end of the search.

The main purpose of the XUNDL compilation is to collect and check for the correctness of the data before the evaluation process, allowing the evalutor to concentrate on the generally more difficult operations and decisions related to the evaluation process itself. In the last years the XUNDL compilation started to be used as a tool to vet the data structure of articles submitted for publication to various scientific journals as a preliminary step of the peer review process, leading to a substantial increase of the quality of the published papers.

Although the compilation is a specialized operation, the ENSDF format is relatively easy to deal with for the non-expert users (the ENSDF manual can be downloaded from https://www-nds.iaea.org/public/documents/ensdf/) allowing the researchers to use the powerful evalution tools (see subsection 2.4) alone or under the guidance of an expert evalutor.

2.3 Evaluate

The evaluation segment is the most important step of the evaluation pipe where properly speaking the evaluation process takes place. As for the previous segments full access to the evaluation database is given to the users through the search engine activated by the ENSDF button in Fig. 2.

The basic ideea for evalution is prefigured by the precedent steps: once all suitable publications and XUNDL files are prepared, the evaluator builds one particular dataset for each individual reaction or decay study. For example if for a particular α-decay process ten articles were collected, of which the two most important were compiled into XUNDL files, the evaluator uses the two XUNDL-coded data files together with the remaining eight publications to build the α-decay dataset by assembling together in a coherent way all the individual pieces of data therein. When multiple data exists for one parameter, as for example several $T_{1/2}$ measured values for one particular level, a unique value is adopted by reviewing how each individual value was measured and deciding its weight in the averaging procedure. If there are several measured α energies in between the same pair of parent-daughter levels another averaging procedure is specifically required, and so on until all data are included, the discrepancies are resolved and the dataset is fully built.

The work of composing a particular dataset is similar to reconstructing an archeologial artifact from individual fragments: when many fragments are found, the reconstruction is almost certain; but when fewer fragments are found the reconstruction becomes very tentative. This way, although the evaluation procedures are generally exact, as a rule one cannot develop fully standard methodologies; all depends on how the evaluator can integrate redundant or incomplete types of information coming from diverse sources in order to make coherent decisions to help buiding the dataset, even in cases when data from different sources are contradictory.

Here comes also the concept of *data sheets* which from the beginning conceptualized data evaluation as described in the Introduction, where a data sheet hosts one particular reaction or decay dataset, built out of the "leaves" of infinitesimal (individual) data collected from publications. Once the particular data sheets are built, the second level of evalution takes place by building up the whole "tree" of data - the *Adopted Levels, Gammas dataset* - out of the individual reaction and decay datasets for each nucleus. For each level and transition process data is collected from the individual datasets and "interweaved" together to build the most comprehensive *Adopted* dataset by recurring the procedures used for the individual datasets. Finally the most difficult operation follows, of adopting the spins and parities of the levels based on measured properties known as "strong" arguments in combination with the more theoretical/qualitative "weak" arguments.

The most important methodology tool at the evaluator's disposal is the *comment* by which different data patterns are discussed, the bibliographic references are introduced, and the way the evalution sequence was conducted is permanently disclosed. Thus an "open standard" con-

IAEA Vienna Nuclear Data Services, https://www-nds.iaea.org/public/ensdf_pgm/	
Analysis Codes	Utility Codes
1) ALPHAD, alpha HF's, theoretical half-lives.	1) ADDGAM, add gammas to adopted level dataset
2) ALPHAD-RadD, as ALPHAD plus r_0 parameter	2) AveTools, calculate averages with uncertainties
3) Bricc, conversion electron, E0 electronic factors	3) CheckKeynumber, NSR key numbers
4) BriccMixing, mixing ratios	4) PANDORA, basic physics checks
5) DELTA, $\gamma\gamma$ angular correlations	5) ConsistencyCheck, as PANDORA
6) GABS, γ-ray absolute intensity and normalization	6) FMTCHK, format and syntax checking
7) GTOL, level energies, net feedings	7) JAVA_NDS, publishing code
8)GLSC, JAVA version of GTOL and GABS	8) TREND, data in a tabular form with optional screen display
9) HSICC, Hager-Seltzer ICC	9) V.AveLib, calculate averages with uncertainties
10) LOGFT, log ft, average E(b), capture fractions	10) xls2ens, ENSDF formatted file from Excel
11) RadD, r_0 parameter odd-odd and odd-A nuclei	11) ens2xls, ENSDF-formatted file to Excel
12) RadList, atomic radiations, energy balance	12) JGAMUT, generates Adopted Levels, Gammas

Figure 3. ENSDF Analysis and Utility Codes.

cept gets operational by combining rigorous standard evalution rules with ad hoc decisions consensually taken to solve non-standard aspects and get an optimum decision in every context. This is the best aspect of the ENSDF evalution, which makes the process transparent for the public up to detailes by fully disclosing data, bibliographic references and decisional thinking for each evaluation sequence.

The *Adopted Levels, Gammas* dataset is the nuclear data evaluation unique contribution to nuclear science: that is building for each nucleus the "Tree of Nuclear Knowlwdge" through Nuclear Structure RESTAURATION.

2.4 Process & Validate

The Process & Validate segment is the computerized stage of the evalution pipe where the formatted datasets are passed through computer codes in order to calculate new parameters from the evaluated data as well as to multiply check, process and prepare the data for publication. There are two families of codes for each category, the Analysis Codes and the Utility Codes, as listed in Fig. 3 where a short description is given for each code. The programs that were initially designed and implemented in the Fortran

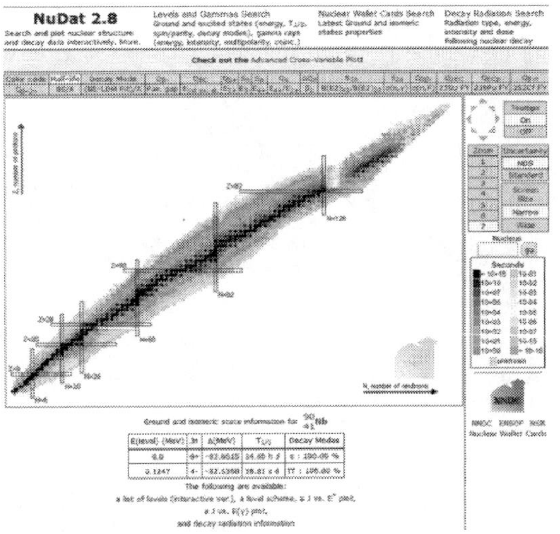

Figure 4. NuDat data dissemination interface.

language are in a vast program of redesign and modernization, the newer versions being implemented in JAVA. They are publicly available for free downloading in the IAEA Vienna Nuclear Data Services site where executable versions are given for the main computing platforms, together with user's manuals. The codes are higly reliable and can be run not only by evaluators but by any users provided the data are put in the ENSDF format.

2.5 Use

Once the evalution is completed the data are ready to use through several dissemination ways. The main dissemination channel is through the NNDC internet site which is mirrored by the IAEA Vienna Data Services. As already mentioned the NNDC portal has under the XUNDL and ENSDF buttons the search engines where the interested user can retrieve and download data in the pdf format or in the database format, the latter being of interest for the users that for example want to sample the formatted data to use as templates for their own purposes.

The Nuclear Data Sheets (NDS), an Elsevier publication, is the traditional dissemination tool of the ENSDF database which is also fully available through the internet (however the access is free only through an institutional subscription). The main difference is that the NDS published articles are done usually only for the so called full mass chain evalutions, where the whole set of isobar nuclei are re-evaluated about each ten years (so the new data published in this interval are put together with the historic data). If after the full evalution was published some datasets are separately revisited and re-evaluated, the updated version is added only to the ENSDF database which thus is maintained as the most current version of the database.

The most visited nuclear structure data dissemination tool is the NuDat interface designed and maintained by

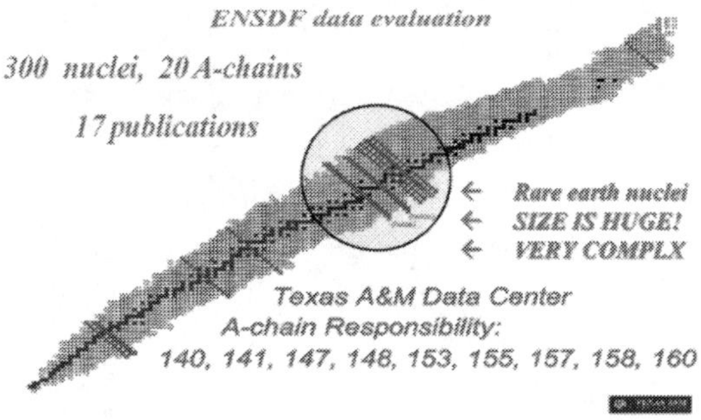

Figure 5. Texas A&M Data Center A-chain responsibility and the distribution on the nuclear chart.

the NNDC. This is a standalone data environment of the ENSDF pool the main characteristic of which is that it delivers only numerical data (with no comments), which makes it very popular for the majority of the more casual users who need "just a number" to use in their applications, but are not interested to follow how that piece of data was obtained. The easiness in delivering the data combined with the general trend of enhanced needs for nuclear data brought the number of data retrivals at over five million per year. However for the expert users the full ENSDF data structure is the most suitable because it makes the evaluation process transparent and such users can evaluate by themselves the quality of the data, and conduct their private evalution sequences once they get the data and the bibliographic references.

3 Texas A&M US Nuclear Data Program

3.1 Data evaluation

At Texas A&M University we assumed the scope of the US Nuclear Data Program to promote and accomplish mass-chain nuclear structure data evaluation as well as to address gaps in experimental data through targeted experiments. We took this mission since 2005 when we started to work under contract with BNL/NNDC; since 2018 we extended our activity as an international NSDD evalution center of the ENSDF/US Nuclear Data Program financed by the Cyclotron Institute DOE grant.

Figure 5 presents the performance of the Texas A&M Evaluation Center and our actual responsibility assumed in the evaluation network. We covered a large region from the relatively low mass numbers A=34, 35 and 37, then A=77 and A=97, the rare earth region marked distinctly on the figure, up to the superheavy mass chain A=252. Each of these regions has its own specificity (for example abundant resonance reactions at low A numbers, among others, and sparse and fragmented data at the extreme superheavy elements). For this reason nuclear data evalution was done traditionally by experimentalists with a large domain of expertise.

Finally our focus became the rare earths region of nuclei with A=140-160, where there is the highest abundance of data, both as number of isobars (the nuclear chart being the widest in this region, typically 17 nuclei), as well as number of reactions and decays datasets and their size. Texas A&M Center got specialized in dealing with ones of the most complex datasets of the nuclear chart. Complexity is expressed by the number of levels and the number of connectiong transitions, over the number of dataset, for these mass chains the typical numbers being 1000-1500 levels, 2200-3000 gamma transitions, and 80-100 datasets. The size of the database is of the largest (typically 20,000-25,000 lines) which generates huge manuscripts that in the unabridged versions are of 600-700 pages per mass chain, while the printed manuscripts get typically 400-500 pages (generally by omitting the level schemes of the reaction datasets, considered redundant).

The mass chains in general and especially those in the rare earth mass region are difficult to bring to publication because of their complexity and size, especially because of the review process that can spread over a couple of years (it is very difficult to review 400-500 pages of very data-dense manuscripts). Moreover once a manuscript gets ready for publication, a couple of years of new publications should be reincluded, which can trigger another round of review, which once more can propagate the delays. Unfortunately the number of experienced data evaluators is close to a critically low limit and the whole activity is being also slowed down because the evaluators are getting overloaded. For these reasons the currency of the ENSDF database tends to be higher then the goal of ten years. Overall we evaluated more than 300 nuclei of 20 mass chains, resulting in a number of Nuclear Data Sheets publication (refs. [6]-[20])

3.2 Precise measurements for data evaluation

An important contribution to the USNDP effort was solving the important problem of the internal conversion coefficients (ICC's). ICC is an important parameter over the entire ENSDF database. They are calculated by the code BrIcc (see Fig. 3) which strongly depends, among others, on the treatment of the atomic vacancy formed by the emission of the conversion electron. The only way to test which of the theoretical approaches is best to describe reality was to precisely measure a set of ICC's and see what theory is in best agreement with the experimental values.

We measured the series of well selected isomeric transitions in ^{93}Nb, ^{103}Rh, ^{125}Te, ^{127}Te, ^{111}Cd, ^{119}Sn, ^{134}Cs, ^{137}Ba, ^{191}Os, ^{193}Ir, and ^{197}Pt and found that the Relativistic Dirac-Fock theory with the so called "frozen orbital" treatment of the atomic vacancy gave the best agreement with the experimental series (refs. [21]-[35]. Based on the results of our ICC measurements, these type of calculations became standard in the ENSDF database.

3.3 New Initiatives & Directions

Derived from nuclear research, evaluated data feeds back to enhance the usefulness of nuclear science and applications. What if reversing the terms one could do nuclear research out of nuclear evalution data? What is the rationale for such an attempt?

Nuclear science collects data in order to add knowledge to nuclear phenomena. Motivated by novelty, basic science gets theoretical interpretations based on some properties of the data that were identified at the moment of the reserch. Once the results are published the data enters the nuclear data repository and pieces of information are then used to aid research and applications, or to study systematic behaviors of some known properties. However it is quite possible that the stored data can be looked for some new "metaproperties" that were not known at the time the data was produced, pretty much like history archives whose documents can sprout new knowledge after centuries of silence, or like a mine that can be reopened for exploitations many years after it was closed.

Thus based on such metaproperty the concept on the level scheme can be revisited with possible good scientific benefits. This research is in progress.

Acknowledgments: This work was supported by the U.S. Department of Energy under Grant No. DE-FG03-93ER40773

References

[1] M. Curie, A. Debierne, A.S. Eve, H. Geiger, O. Hahn, S.C. Lind, S. Meyer, E. Rutherford, E. Schweidler, Revs. Modern Phys. **3**, 427 (1931)

[2] M.J. Martin, Joint ICTP-IAEA Workshop on Nuclear Structure and Decay Data: Theory, Experiment and Evaluation, ICTP Trieste, 15 - 26 October 2018, Trieste, Italy, http://indico.ictp.it/event/8338/session/0/contribution/2

[3] P. Dimitriou, S. Basunia, L. Bernstein, J. Chen, Z. Elekes, X. Huang, A. Hurst, H. Iimura, A.K. Jain, J. Kelley, T. Kibedi, F. Kondev, S. Lalkovski, E. McCutchan, I. Mitropolsky, G. Mukherjee, A. Negret, C. Nesaraja, N. Nica, S. Pascu, A. Rodionov, B. Singh, S. Singh, M. Smith, A. Sonzogni, J. Timar, J. Tuli, M. Verpelli, D. Yang, V. Zerkin, Proc. Intern. Conf. Nuclear Data for Science and Technology (ND2019), Beijing, China, 19-24 May, 2019, Z.Ge, et al. Eds., EPJ Web of Conf. Vol. **239**, 15004 (2020)

[4] E.A. McCutchan, D.A. Brown, A.A. Sonzogni, Nuclear Physics News Vol. **27**, No. 3, 5 (2017)

[5] B. Pritychenko, Nucl. Data Sheets **120** 291, (2014), www.nndc.bnl.gov/nsr

[6] N. Nica, Nucl. Data Sheets **106**, 813 (2005).

[7] N. Nica, Nucl. Data Sheets **108**, 1287 (2007).

[8] N. Nica, Nucl.Data Sheets **122**, 1 (2014).

[9] N. Nica, Nucl. Data Sheets **110**, 749 (2009).

[10] N. Nica, Nucl. Data Sheets **117**, 1 (2014).

[11] N. Nica, Nucl. Data Sheets **111**, 525 (2010).

[12] B. Singh and N. Nica, Nucl. Data Sheets **113**, 1115 (2012).

[13] J. Cameron, J. Chen, B. Singh, and N. Nica, Nucl. Data Sheets **113**, 365 (2012).

[14] N. Nica, J. Cameron, and B. Singh, Nucl. Data Sheets **113**, 1 (2012).

[15] N. Nica and B. Singh, Nucl. Data Sheets **113**, 1563 (2012).

[16] N. Nica, Nucl.Data Sheets **132**, 1 (2016).

[17] N. Nica, Nucl.Data Sheets **142**, 1 (2017).

[18] N. Nica, Nucl.Data Sheets **154**, 1 (2018).

[19] N. Nica, Nucl.Data Sheets **160**, 1 (2019).

[20] N. Nica, Nucl.Data Sheets **170**, 1 (2020).

[21] N.Nica, J.C.Hardy, V.E.Iacob, S.Raman, C.W.Nestor, Jr., M.B.Trzhaskovskaya, Phys.Rev. C **70**, 054305 (2004)

[22] N.Nica, J.C.Hardy, V.E.Iacob, J.R.Montague, M.B.Trzhaskovskaya, Phys.Rev. C **71**, 054320 (2005)

[23] J.C.Hardy, N.Nica, V.E.Iacob, M.B.Trzhaskovskaya, R.G.Helmer, Appl.Radiat.Isot. **64**, 1392 (2006)

[24] N.Nica, J.C.Hardy, V.E.Iacob, W.E.Rockwell, M.B.Trzhaskovskaya, Phys.Rev. C **75**, 024308 (2007)

[25] N.Nica, J.C.Hardy, V.E.Iacob, C.Balonek, M.B.Trzhaskovskaya, Phys.Rev. C **77**, 034306 (2008)

[26] J.C.Hardy, N.Nica, V.E.Iacob, C.Balonek, M.B.Trzhaskovskaya, Appl.Radiat.Isot. **66**, 701 (2008)

[27]) N.Nica, J.C.Hardy, V.E.Iacob, J.Goodwin, C.Balonek, M.Hernberg, J.Nolan, M.B.Trzhaskovskaya, Phys.Rev. C **80**, 064314 (2009)

[28] N.Nica, J.C.Hardy, V.E.Iacob, M.Bencomo, V.Horvat, H.I.Park, M.Maguire, S.Miller, M.B.Trzhaskovskaya, Phys.Rev. C **89**, 014303 (2014)

[29] J.C.Hardy, N.Nica, V.E.Iacob, S.Miller, M.Maguire, M.B.Trzhaskovskaya, Appl.Radiat.Isot. **87**, 87 (2014)

[30] N.Nica, J.C.Hardy, V.E.Iacob, T.A.Werke, C.M.Folden, L.Pineda, M.B.Trzhaskovskaya, Phys.Rev. C **93**, 034305 (2016)

[31] N.Nica, J.C.Hardy, V.E.Iacob, T.A.Werke, C.M.Folden, K.Ofodile, M.B.Trzhaskovskaya, Phys.Rev. C **95**, 064301 (2017)

[32] N.Nica, J.C.Hardy, V.E.Iacob, H.I.Park, K.Brandenburg, M.B.Trzhaskovskaya, Phys.Rev. C **95**, 034325 (2017)

[33] N.Nica, J.C.Hardy, V.E.Iacob, V.Horvat, H.I.Park, T.A.Werke, K.J.Glennon, C.M.Folden, V.I.Sabla, J.B.Bryant, X.K.James, M.B.Trzhaskovskaya, Phys.Rev. C **98**, 054321 (2018)

[34] J.C.Hardy, N.Nica, V.E.Iacob, M.B.Trzhaskovskaya, Appl.Radiat.Isot. **134**, 406 (2018)

[35] V.Horvat, E.E.Tereshatov, J.C.Hardy, N.Nica, C.M.Folden, V.E.Iacob, M.B.Trzhaskovskaya, Phys.Rev. C **102**, 014310 (2020)

EPJ Web of Conferences **252**, 08002 (2021) https://doi.org/10.1051/epjconf/202125208002
HINPw6

A novel approach to medical radioisotope production using inverse kinematics

Márcia Regina Dias Rodrigues[1*], *Victor E.* Iacob[1], *Ninel* Nica[1], *Brian* Roeder[1], *Gabriel* Tabacaru[1], *Kang* Wang[1], *Meixiang* Yu[4], *Paulo* Zanotti-Fregonara[4], *Justin* Mabiala[5], *Jedidiah* Romo[1], *Georgios A.* Souliotis[2] and *Aldo* Bonasera[1,3].

[1]Cyclotron Institute, Texas A&M University, USA
[2]Laboratory of Physical Chemistry, National and Kapodistrian University of Athens, Greece
[3]Laboratori Nazionali del Sud, INFN, Italy
[4]Houston Methodist Research Institute, USA
[5]Prairie View A&M University, Prairie View, USA

Abstract. The inverse kinematics methodology using a gas target has been applied to produce medically important radionuclides at the Cyclotron Institute at Texas A&M University. The production of the theranostic radionuclide ^{67}Cu ($T_{1/2}$ = 62 h) through the reaction of a ^{70}Zn beam at 15 MeV/nucleon with a hydrogen gas target was performed. The activities at end of irradiation and the thick target yield were obtained for ^{67}Cu. A test using the forward-focused neutrons from the primary reaction to irradiate ^{nat}Zn to produce ^{67}Cu is also presented.

1 Introduction

A novel approach to produce medically important radionuclides using inverse kinematics has been developed at the Cyclotron Institute at Texas A&M University. A heavy-ion beam is accelerated at an appropriate energy and focused on a light gas target. A foil catcher is positioned after the target to collect the isotopes of interest, typically emitted in a narrow angle along with the beam direction. Secondary emitted particles such as neutrons from the primary nuclear reaction can be used to irradiate other targets for further radionuclide production. As the quantity of material required to prepare heavy-ion beam is less than that used in the standard solid target approach, material costs are expected to be considerably reduced through this methodology. A detailed discussion about the methodology can be found in [1].

Medical radionuclides are essential in nuclear medicine in the fields of diagnostic imaging and radioimmunotherapy (RIT). RIT in nuclear medicine involves the injection of a radioisotope-labeled monoclonal antibody (mAb) to treat cancer. The radionuclides are carried directly toward the tumor and the emitted radiation breaks the DNA inducing the carcinogenic cell death. Radionuclides emitting low-range highly ionizing radiation are essential for radioimmunotherapy. The longest-lived radioisotope of copper, ^{67}Cu, is ideal for both radioimmunotherapy and imaging, i.e. radioimmunotheranostics [2], for several reasons. First, from a chemical perspective, copper is an essential trace element that can be linked to antibodies, proteins and other biologically important molecules. The half-life of ^{67}Cu (62 h) is appropriate to deliver a high dose rate to the tumor. While the β^- decay (E_{max} = 562 keV) is responsible for the therapeutic part, the emission of the γ radiation of 185 keV (48.7%) and 93 keV (16.1%) is for the diagnostics. The ^{67}Cu is suitable for imaging the radiotracer distribution by single-photon emission computerized tomography (SPECT) and it can be combined with the same type of radiopharmaceuticals as ^{64}Cu ($T_{1/2}$=12.7 h) or ^{61}Cu ($T_{1/2}$=3.3 h), leading to efficient theranostic pairs. The ^{67}Cu is also responsible to the theranostic approach to pretargeted radioimmunotherapy (PRIT) [3]. The main factor limiting preclinical and clinical use is its limited availability [4].

The production of ^{67}Cu has started in the 1960s using nuclear reactors and continues until the present. Recently the interest has shifted towards production at particle accelerators. The main production route is via the reaction ^{68}Zn(p,2p)^{67}Cu [5-8] that makes use of intense medium-energy (E_p = 70–100 MeV) proton beams. Another production route based on lower-energy charged particle reactions is ^{70}Zn(p,α)^{67}Cu [9-11]. The present work uses as primary reaction the inverse kinematic reaction p(^{70}Zn,^{67}Cu)α and the secondary reaction as ^{nat}Zn(n,x)^{67}Cu. The structure of the paper is as follows: in section 2 the experimental setup and the measurements are described. In section 3, the details of the analysis and results are presented including the thick target yields and the use of the neutrons for secondary isotope production. Finally, conclusions and future plans are presented in section 4.

* Corresponding author: mrodrigues@tamu.edu

 https://doi.org/10.1051/epjconf/202125208002

2 Experimental set up and measurements

2.1 Irradiation set up

The $^{70}Zn^{15+}$ beam was produced by the ECR source and accelerated by the K500 superconducting cyclotron at the Cyclotron Institute at Texas A&M University to an energy of 15 MeV/nucleon, then transported to the target chamber of the MARS recoil separator [12]. The beam impinged on a cryogenic gas cell filled with H_2 gas at a pressure of 2.7 atm in contact with a liquid nitrogen reservoir [13]. The cryogenic gas cell had a length of 9.0 cm with 4 μm Havar entrance and exit windows of 1.9 cm diameter. A magnetic stirring system was operated to provide circulation of the gas. The experimental setup is schematically shown in Fig. 1. An aluminum catcher foil, placed after the hydrogen gas cell, was used to collect the ^{67}Cu nuclei, along with other coproduced nuclides. The Al catcher foil was mounted on a rectangular aluminum target frame with a 1.27 cm diameter hole.

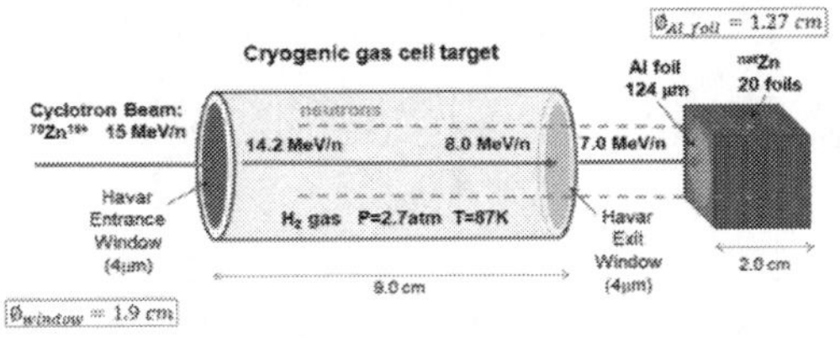

Fig. 1. (Color online) Schematic diagram of the irradiation setup. The energies of the beam are listed as it passes through the entrance window, the gas and the exit window (thick arrows). The dashed arrows represent the neutrons produced via the interactions of the ^{70}Zn beam with the hydrogen gas and the Al catcher [1].

2.2 Irradiation

The irradiation time was 6.5 h with a beam current of 0.19(5) pnA (particle nA). The current was periodically monitored at a Faraday cup positioned after the Cyclotron exit and was nearly constant (within 13%). The measurement of the current after the gas cell was performed by a second Faraday cup mounted on the same target ladder as the Al catcher frame. The measurement of the beam current at this location was 8.0 nA of ^{70}Zn (7.0 MeV/nucleon) at an average charge state of 26^+. The average equilibrium charge-state was calculated with LISE++ Physical Calculator [14] based on model of G. Schiwietz *et al.* [15] for ions with energies lower than 15MeV/nucleon. The beam current was determined considering the beam charge state before and after the gas cell and the transmission ratio between the two Faradays cups. As electron suppression of the Faraday cups was not performed in the present experiment, the sputtering of electrons from the side of the Faraday cup hit by the heavy-ion beam resulted in an increase of the measured positive current. A suppression test was performed for the Faraday cup positioned after the gas target using a ^{21}Ne beam at 28 MeV/nucleon and a decrease of 40% of the beam intensity was observed. The beam intensity for that Faraday cup was corrected by this factor and the uncertainty due to this correction was estimated at about 20%. Considering all uncertainties, beam intensity, 13%; secondary emission electrons suppression, 20%; charge state, 15%; and transmission ratio, 4%; the total uncertainty for the beam current is 28%.

2.3 Off-line γ-spectra measurement

After 36.5 h from the end of the irradiation, the Al catcher foil was placed in front of a high-purity germanium (HPGe) detector for off-line γ-ray analysis. The foil was placed at a distance of d=17.2(10) mm from the end cap of the detector. The dead time of the counting system was about 2–3%, and the energy resolution of the detector system was 2.5–4.0 keV (FWHM) for the peaks of interest. The energy calibration was performed with a ^{152}Eu source and adjusted by using known γ-rays obtained in the spectra. The absolute photopeak efficiencies for the actual source-detector geometry were obtained with the Monte Carlo codes GEANT4 [16] and EGSnrc [17]. The

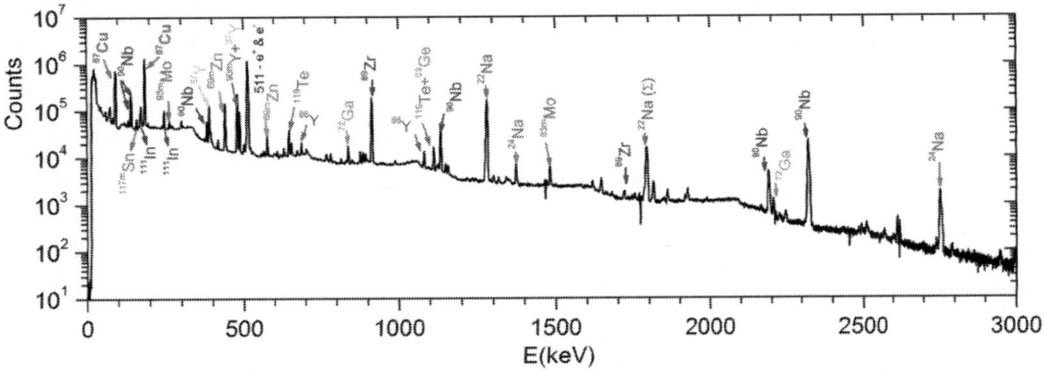

Fig. 2. Al catcher foil background-subtracted γ-ray spectrum after the interaction of a ^{70}Zn (15 MeV/nucleon) beam with the hydrogen gas cell after 36.5 h from the end of the irradiation. The peaks are labelled with the corresponding radionuclides. The peak marked $^{22}Na(\Sigma)$ is the sum peak induced by the simultaneous detection of the β delayed γ–ray (1275 keV) and 511 keV γ-ray from e^+ & e^- annihilation.

uncertainty assumed for the efficiency simulation results is 10%. The spectrum of the room background was measured for 67.3 h. The background-subtracted γ-ray spectrum obtained during an accumulation period of 68h, starting 36.5h after the end of the irradiation, is presented in Fig. 2.

3 Analysis and results

3.1 Activities

The activities at the End of Bombardment (A_{EOB}) were calculated with the equation:

$$A_{EOB} = \frac{C}{\varepsilon(E) t_L I_\gamma} e^{\lambda t_d} \frac{\lambda t_R}{(1-e^{-\lambda t_R})} \quad (1)$$

where C is the number of counts, ε(E) the detector efficiency, I_γ the intensity of the γ-line of interest, λ the decay constant of the radionuclide, t_L and t_R are respectively the live time and real time of the γ-spectroscopy measurement, and t_d is the decay time elapsed from the end of irradiation to the start of the γ-spectroscopy measurement. The activities at the end of bombardment (A_{EOB}) as well as the activities considering a beam intensity of 1pnA and an irradiation time of 1h, the so-called 1pnA-1h activity (H_{EOB}), are presented in Table 1. Other corrections: (i) 5.3 % for the number of ^{67}Cu produced at gas target and not collected by the Al foil due to the solid angle, obtained from the GEANT4 [16] Monte Carlo simulation (see section 3.2 for details); and (ii) the 0.97 % dead time correction for the data acquisition, already accounted for in equation (1) by the ratio t_R/t_L, were applied to the activity values. The A_{EOB} uncertainties include the peak area fit, the background determination and the HPGe detector efficiency uncertainties. For the H_{EOB} values the beam current uncertainty is incorporated. For each radionuclide produced, Table 1 shows the decay mode, the half-life, and the respective gamma-ray energies (E_γ) and intensities (I_γ) used to obtain the activities. The principal production routes are also presented. At the end of the 6.5 h of irradiation, the activity of the ^{67}Cu isotope was 2.16(12) kBq and the 1pnA-1h activity was 1.8(5) kBq/(pnAh). This result includes the Al foil frame contribution. The photopeak associated to E_γ doublet at 91.3 keV and 93.3keV is not resolved in the present measurement. The results were calculated as the weighted average of the activities obtained from the peaks at E_γ = (91.3 + 93.3) keV and 184.6 keV using the respective intensities (relative to 100 beta decays of the parent) of I_γ = (7.0% + 16.1%) and 48.7% as weighting factors. Note that these photopeaks are pure, i.e., the contribution of γ-ray transitions is just from ^{67}Cu decay. Other γ-rays are too weak to be used for analysis and some are contaminated by impurities. The photopeak at 209 keV has contributions from ^{79}Kr ($T_{1/2}$ = 35.04 h, E_γ =208.5 keV, I_γ = 0.797%). The next photopeak at 300 keV is multiply contaminated with ^{79}Kr ($T_{1/2}$ = 35.04 h, E_γ = 299.5 keV, I_γ = 1.54%), ^{101}Pd ($T_{1/2}$ = 8.47 h, E_γ = 296.3 keV, I_γ = 19%) and ^{73}Ga ($T_{1/2}$ = 4.86 h, E_γ = 297.3 keV, I_γ = 79.8%). Finally, the peak at 393 keV also contains ^{88}Zr ($T_{1/2}$ = 83.4 d, E_γ = 392.9 keV, I_γ = 97%). ^{79}Kr and ^{88}Zn are produced from the beam interaction with Al foil, while ^{101}Pd and ^{73}Ga are from beam interaction with Cr from Havar windows.

Another important radionuclide observed that is produced by the reaction ^{70}Zn+p is the ^{69m}Zn ($T_{1/2}$=13.8 h), the activity of which at the end of the irradiation is 2.55(26) kBq, and the 1pnA-1h activity is 2.2(6) kBq/(pnAh). The transition Eγ = 438.6 keV (I_γ =94.77%) was used in the analysis of this isotope. As the ^{69m}Zn half-live is 4.5 smaller than the half-life of ^{67}Cu, its activity can be reduced by an appropriate cooling period.

Other coproduced radionuclides are due to the beam interaction on the Al catcher foil and on the Havar windows. The radionuclides ^{86}Y, ^{87}Y, ^{89}Zr, ^{90}Nb and ^{93m}Mo are fusion-evaporation products of the reaction ^{70}Zn+^{27}Al (with the beam of ^{70}Zn entering the Al catcher at 7.0 MeV/nucleon), of which ^{90}Nb is the main impurity with an A_{EOB} = 2.38(23) kBq. The heavier radionuclides ^{111}In, ^{117m}Sn and ^{119}Te are residues of the reaction of the ^{70}Zn beam with the constituents of the Havar alloy (Co 42.0%, Mo 2.2%, Cr 19.5%, Mg 1.6%, Ni 12.7%, C 0.2%, W 2.7%, and Fe balance) of the gas-cell windows.

Table 1 presents the activities at the end of bombardment for ^{67}Cu and the most intense radioimpurities, together with some important parameters used in the analysis.

Cross section simulations using the code PACE4 [18] for incident ^{70}Zn at 7.0 MeV/nucleon on ^{27}Al, ^{59}Co and ^{52}Cr were performed to support the impurity analysis. The ^{90}Nb production presented a high cross section of σ(^{90}Nb) = 25 mb from the beam interaction with ^{27}Al. Considering the beam range on Al foil is 65 μm, the main contribution is from the reaction ^{27}Al(^{70}Zn,x)^{90}Nb. The results for Havar foil, considering the beam interaction with the most abundante isotopes of Co and Cr are, σ(^{59}Co) = 1.7 mb and σ(^{52}Cr) = 1.2 mb. The other radioimpurities shown in the table are at least a factor 2.5 smaller than ^{67}Cu A_{EOB} with all of them mostly produced by the ^{70}Zn + ^{27}Al reactions.

The radionuclides produced in the proton induced reactions on ^{70}Zn in the incident energy range of 7.8 – 14.3 MeV/nucleon (beam energy at the end of the gas cell and after first Havar window) are ^{67}Cu and ^{69m}Zn.

It is worth mentioning that ^{67}Cu and ^{67}Ga are the beta minus and electron capture parents of ^{67}Zn daughter, using almost the same set of gamma decays, which can make more difficult their separation through gamma ray analysis (moreover they also have similar half-lives, 78.3 h for ^{67}Ga and 61.9 h for ^{67}Cu). However, at this energy range the production of ^{67}Ga is suppressed (^{70}Zn(p,4n)^{67}Ga $E_{threshold}$ = 28.081 MeV). The others usual radioimpurities coproduced with ^{67}Cu, as ^{61,64}Cu, ^{66}Ga, ^{65}Zn are also excluded due to high energy threshold.

3.2 Cross sections

The cross sections up to proton energy of 17 MeV for

Table 1. Activities obtained in the present work as defined in the text (last two columns) for ^{67}Cu and the main contaminants, the radioactive decay data of the respective radionuclides, and the contributing reactions. The listed γ rays were used for the activity calculation.

Radio-nuclide	Decay mode	Half life	Eγ (keV)	Iγ (%)	Principal Production routes	A_{EOB} (kBq)	H_{EOB} $\left(\frac{kBq}{pnA.h}\right)$
^{67}Cu	β^-	61.83 h 12	91.266(5)	7.0(1)	p(^{70}Zn, ^{67}Cu)α	2.16(12)	1.8(5)
			93.311(5)	16.1(2)			
			184.577(10)	48.7(3)			
^{69m}Zn	IT, β^-	13.76 h 2	438.63(2)	94.77(20)	p(^{70}Zn, ^{69m}Zn)d	2.55(26)	2.2(6)
^{90}Nb	ε^+ β^-	14.60 h 5	1129.224(15)	92.7(4)	^{27}Al(^{70}Zn, ^{90}Nb)x	2.38(23)	2.0(6)
			2318.968(10)	82.03(16)	^{59}Co(^{70}Zn, ^{90}Nb)x		
			2186.242(25)	17.96(16)	^{52}Cr(^{70}Zn, ^{90}Nb)x		
^{87m}Y	IT	13.37 h 3	380.79(7)	78	^{27}Al(^{70}Zn, ^{87m}Y)x	0.87(9)	0.74(22)
^{89}Zr	ε^+ β^+	78.41 h 12	908.96(4)	100	^{27}Al(^{70}Zn, ^{89}Zr)x	0.62(6)	0.52(16)
^{22}Na	ε^+ β^+	2.6019 y 4	1.27453(2)	99.944(14)	^{27}Al(^{70}Zn, ^{22}Na)x	0.46(5)	0.41(12)
					^{27}Al(n,x)^{22}Na		
^{86}Y	ε^+ β^+	14.74 h 2	627.72(10)	32.6(10)	^{27}Al(^{70}Zn, ^{86}Y)x	0.38(4)	0.32(10)
			1153.01(4)	30.5(9)	^{59}Co(^{70}Zn, ^{86}Y) x		
			1920.72(13)	20.8(7)	^{52}Cr(^{70}Zn, ^{86}Y) x		
			1854.38(13)	17.2(5)			
			443.14(9)	16.9(5)			
			703.34(10)	15.4(4)			
^{87}Y	ε^+ β^+	79.8 h 3	484.805(5)	89.7(3)	^{27}Al(^{70}Zn, ^{87}Y)x	0.167(16)	0.15(4)
			388.531(3)	82	^{59}Co(^{70}Zn, ^{87}Y) x		
					^{52}Cr(^{70}Zn, ^{87}Y) x		

^{70}Zn(p,x)^{67}Cu and ^{70}Zn(p,x)^{69m}Zn from [19-22] are presented in Fig. 3. For ^{67}Cu production there are the experimental data obtained by Levkovskij *et al.* [19] (triangle) and Kastleiner *et al.* [20] (square) in comparison with the IAEA [21] recommended cross section extracted from a fit of these two data sets (dash-dot line). The experimental data for ^{69m}Zn production from Levkovskij *et al.* [19] are also represented (circle).

The theoretical predictions from the TENDL-2019 library [22], based on the TALYS code [28], are also represented (dot line). These predictions did not reproduce well the experimental data, for ^{67}Cu the values are systematically higher and for ^{69m}Zn the results underestimate the experimental cross sections in the region of 12-14 MeV. The value adopted by Levkovskij *et al.* [19] in the monitor reaction ^{nat}Mo(p,x)^{96}Tc cross section at 30 MeV was 250 mb while more recent cross section evaluation recommended by IAEA [21] is 192.82 mb. The data represented are normalized by a factor 0.77 to take into account the correction in the monitor cross section. The energy region of interest for the present work is within the two dash-dot bars in the graph.

The present measurement is a thick target inverse kinematics measurement as the 15 MeV/nucleon ^{70}Zn beam impinging on 2.7 atm gas target results in ΔE/E = 0.45. To take that into account, a Monte Carlo simulation with the GEANT4 package [16] was performed. The cross sections used for the simulations were the IAEA recommended for ^{67}Cu and the interpolation of the corrected experimental data from Levkovskij *et al.* [19] for ^{69m}Zn. The beam energy loss and the position on gas target where the reactions p(^{70}Zn,^{67}Cu)α and p(^{70}Zn,^{69m}Zn)d occur according to GEANT4 simulation are presented in Fig. 4. The beam energy is represented on the y axis as a function of the target position where the reaction occurs on the x axis. Most of the events are concentrated at the beginning of the target as expected. The simulation results have shown that the number of ^{67}Cu nuclei produced at the gas target differs by 5.3 % from that collected in the Al foil. This result is due to the solid angle and scattering on the gas target.

The experimental energy loss weighted average cross sections in units [cm^3] were calculated from:

$$\langle \sigma R \rangle_{Exp} = \frac{A_{EOB}\, m_{target}\, m_u\, e}{\rho_{target}\, \varphi \left(1 - e^{-\lambda t_{irr}}\right)} \qquad (2)$$

where ρ_{target} is the density of the target [g/cm^3], φ is the beam intensity [pA], t_{irr} is the irradiation time [s], m_{target} is the atomic mass of the target [u], m_u atomic mass constant [g/u] and e is the elementary charge [C].

 https://doi.org/10.1051/epjconf/202125208002

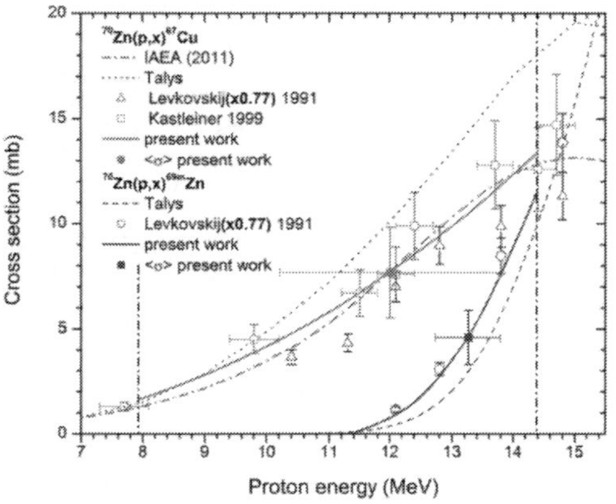

Fig. 3. (Color online) Experimental cross sections for $^{70}Zn(p,x)^{67}Cu$ and $^{70}Zn(p,x)^{69m}Zn$ from Levkovskij *et al.* [19] and Kastleiner *et al.* [20] in comparison with theoretical predictions from the TENDL-2019 library [22], based on the TALYS code and the IEAE [21] recommended cross section for ^{67}Cu. The present work results for the average cross section associated with ^{67}Cu and ^{69m}Zn production and the estimated cross section for the region are also presented.

A_{EOB} is expressed in [Bq] and λ in [s^{-1}].

The cross section can be estimated assuming a parametrization as:

$$\sigma = p_0(E - E_{min})^2 \qquad (3)$$

where E_{min} is the minimum energy that the reaction occurs and p_0 is a parameter to be extracted from experimental data. As the reaction p(^{70}Zn,^{67}Cu)α has zero threshold energy, the minimum energy is associated with the Coulomb barrier potential, E_{min}. = 4.4 MeV. Considering the reaction p(^{70}Zn,^{69m}Zn)d, it has a Q_{value} = -7.5 MeV, thus the cross-section is expected to decrease rapidly with decreasing beam energy. The available experimental data for this reaction indicates a minimum energy of E_{min} = 11.2 MeV (see Fig. 3), that value was adopted here. The energy loss weighted cross-section is given by the following expression:

$$\langle \sigma R \rangle = \int_{E_{min}}^{E_{max}} \frac{\sigma(E)}{\frac{dE}{dx}} dE \qquad (4)$$

where $\frac{dE}{dx}$ is the stopping power. From equation (2) and (3) it is possible to extract the parameter p_0 using the experimental results $\langle \sigma R \rangle_{Exp}$ and the $\frac{dE}{dx}$ obtained from the SRIM code [23] by:

$$p_0 = \frac{\langle \sigma R \rangle_{Exp}}{\int_{E_{min}}^{E_{max}} \frac{(E-E_0)^2}{\frac{dE}{dx}} dE} \qquad (5)$$

The most probable energy where reactions occur was determined using the mean value theorem for the integral of equation (4).

$$\langle \sigma R \rangle_{Exp} = (E_{max} - E_{min}) \frac{p_0(\langle E \rangle - E_{min})^2}{\frac{dE}{dx}(\langle E \rangle)} \qquad (6)$$

Table 2 presents the energy loss weighted average cross sections and the parameter p_0, equation (2) and (5), obtained from the experimental results. The respective average energy $\langle E \rangle$ from equation (6) and the average cross section $\langle \sigma \rangle$ are also shown in Table 2 and Figure 3. The estimated parametrized cross section is represented in Figure 3 with solid lines. The $\langle E \rangle$ uncertainty was obtained from the propagation of $\langle \sigma R \rangle_{Exp}$ and p_0 uncertainties.

Note the results for ^{67}Cu production are in good agreement with experimental cross section from Kastleiner *et al.* [20]. For ^{69m}Zn, the parametrized cross section describes well the experimental values from Levkovskij *et al.* [19] and the average cross section considering the errors bars agrees with the experimental data. Notice that the ansatz, equation (3), could be easily improved if one or more experimental points are added. In fact, in that case the parameters E_{min} and p_0 could be left free . However, the goal is to test a new method to obtain cross-sections using thick targets. The approach, once is suitably adjusted to the experimental case, can be very powerful for instance for the data of ^{99}Mo production [24] where there are three experimental points. A similar methodology is used by P. Tan *et al.* [25] for sub-barrier fusion reactions, but using derivatives of equation (4) instead. The approach here does not require derivatives and can be easily extended to those reactions.

Table 2: The energy loss weighted average cross section and the parameter p_0 obtained from experimental results, and the respective average energy and cross section for ^{67}Cu and ^{69m}Zn.

	$\langle \sigma R \rangle_{Exp}$ ($mb \cdot cm$)	p_0 (mb/MeV^2)	$\langle E \rangle$ (MeV)	$\langle \sigma \rangle$ (mb)
^{67}Cu	53(15)	0.14(4)	12.0(18)	7.7(21)
^{69m}Zn	16(4)	1.3(4)	13.3(5)	4.6(13)

3.3 The thick target yield in inverse kinematics

In Nuclear Data of radioactive isotopes, besides the cross section, the Thick Target Yield (TTY) is an important quantity to be determined. The yield is usually defined as the ratio of the nuclei produced in the reaction over the number of incident particles. In the yield for a radionuclide production, the number of radioactive nuclei is associated to the activity and the number of incident particles to the beam charge. The Thick Target Yield (TTY) is defined for a determined macroscopic energy loss and it is given by the equation [26]:

$$TTY\left[\frac{Bq}{C}\right] = \frac{N_A \lambda \rho_{target}}{Z_{beam} e\, M_{target}} \langle \sigma R \rangle \qquad (7)$$

where N_A is Avogadro's number [mol^{-1}], Z_{beam} is the charge of the fully ionized projectile, e is the elementary charge [C] and M_{target} is the molar mass of the target [g/mol]. The respective units for the following quantities are: λ [s^{-1}], ρ [g/cm^3] and $\langle \sigma R \rangle$[cm^3].

The TTY can be determined experimentally from the A_{EOB} from the following equation:

$$TTY^{exp}[E_{min} \to E_{max}] = \frac{A_{EOB}}{I} \frac{\lambda}{(1-e^{-\lambda t_{irr}})} \qquad (8)$$

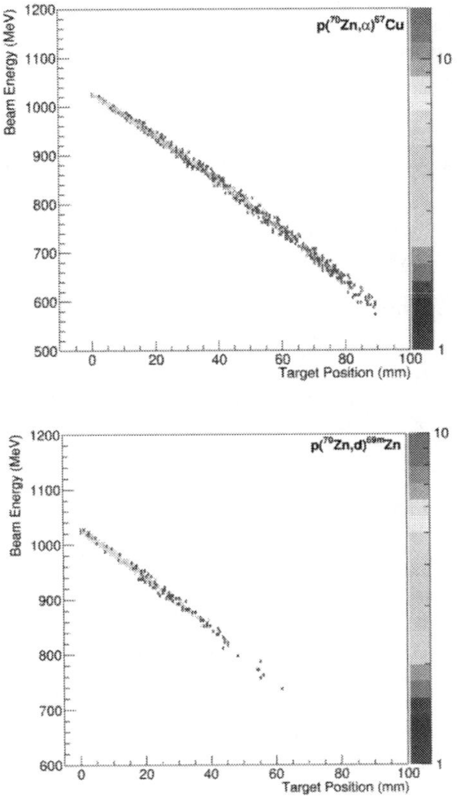

Fig. 4. (Color online) Beam energy as a function of reaction position on the gas target for the reactions p(^{70}Zn, ^{67}Cu) α (upper part) and p(^{70}Zn,^{69m}Zn)α (lower part) according to GEANT simulation [16].

where I is the beam intensity [A]. The TTY_{inv}^{calc} predicted and the TTY_{inv}^{exp} experimentally determined, seen in Table 3, are in good agreement considering the uncertainties. There are several programs and tables in the literature with information about TTY but in most cases, they are forward kinematics reactions [21]. Although in inverse kinematics the TTY definition does not change, the stopping power $\frac{dE}{dx}$ does. The conversion from TTY_{inv} to TTY_{for} can be done, as described in the work of M. Aikawa *et al.* [27], with the ratio *R(E)* between the differential yields:

$$R(E) = \frac{Z_{target}}{Z_{beam}} \frac{M_{beam}}{M_{target}} \frac{\rho_{target}}{\rho_{beam}} \frac{\frac{dE}{dx}_{inv}}{\frac{dE}{dx}_{for}} \quad (9)$$

$$TTY_{for}(E) \cong R(E) TTY_{inv}(E) \quad (10)$$

For the reaction ^{70}Zn + p in the energy range ($E_{c.m.}$ = 7.8-14.3 MeV) the ratio is $R(E) \cong 18.4$. In this way the experimental TTY was converted to forward kinematics. The results for the ^{67}Cu are also in good agreement with the integral yields recommended by IAEA [21], $Y(E_{max}) - Y(E_{min})$ as seen in Table 3.

Table 3: The thick target yield in inverse kinematics determined from cross section and stopping power TTY_{inv}^{calc}, the thick target yield in inverse kinematics determined with the experimental activities and beam current TTY_{inv}^{exp}, the conversion of the experimental TTY to forward kinematics TTY_{for}^{exp} and the integral yield, $Y(E_{max}) - Y(E_{min})$, recommended by IEAE [21].

	TTY_{inv}^{calc} (GBq/C)	TTY_{inv}^{exp} (GBq/C)	TTY_{for}^{exp} (GBq/C)	$Y(E_{max}) - Y(E_{min})$ (GBq/C)
^{67}Cu	0.0161(16)	0.021(5)	0.38(10)	0.314
^{69m}Zn	0.0214(21)	0.027(7)	0.50(14)	

3.4 Use of the neutrons from the primary reaction for secondary isotope production

A test using the forward-focused neutrons from the primary reaction to irradiate ^{nat}Zn in order to produce ^{67}Cu was also performed. The interaction of ^{70}Zn beam at 15 MeV/nucleon with the proton target produces about 1.6 neutrons per reaction, as calculated with the code TALYS [28]. These neutrons are kinematically focused in the forward direction and can be used for further radioisotope production. In the present experiment, a block of twenty 25.4 × 25.4 mm^2 foils of ^{nat}Zn with 1 mm thickness was placed behind the Al catcher (Fig. 1).

The γ-ray spectrum obtained during an accumulation period of 40 h, starting 7.5 d after the end of the irradiation, is presented in Fig. 5 in comparison with the background spectrum normalized for the measurement time. Note that both spectra are identical, except for the low energy region that is zoomed in the bottom panel of Fig. 5. It is possible to observe that the source spectrum is lower than the background spectrum with this difference increasing for lower energies, this being due to the shielding effect of the source on the background radiation. On the other hand, the photopeak associated to 185 keV energy presents an enhancement. The background contribution is associated to ^{226}Ra ($T_{1/2}$ = 1600 y, E_γ = 186.2 keV, I_γ = 3.59 %) almost coinciding with the ^{67}Cu source ($T_{1/2}$= 61.9 h , E_γ = 184.6 keV, I_γ = 48.7 %). The respective areas for source and background were determined and subtracted. The activity of ^{67}Cu produced by the neutrons from the primary reaction at the end of the 6.5 h of irradiation was 5.3(8) Bq. Considering the cross sections and isotopic abundance, the main production of ^{67}Cu comes from ^{67}Zn(n,p)^{67}Cu reaction [29]. Considering the neutron angular distribution produced from the primary reaction according to CoMD predictions [30-32], the distribution presents a peak around θ_{LAB} ~ 10-20°. The experimental set up for ^{nat}Zn target covers the angular range 0° - 3° which corresponds to collecting of about 2% of the flux of the neutrons. Other important corrections are the γ attenuation in the source of about 50% and the isotopic natural abundance of ^{67}Zn, 4.04%. Note that these results are an estimation and more detailed experimental investigation is needed.

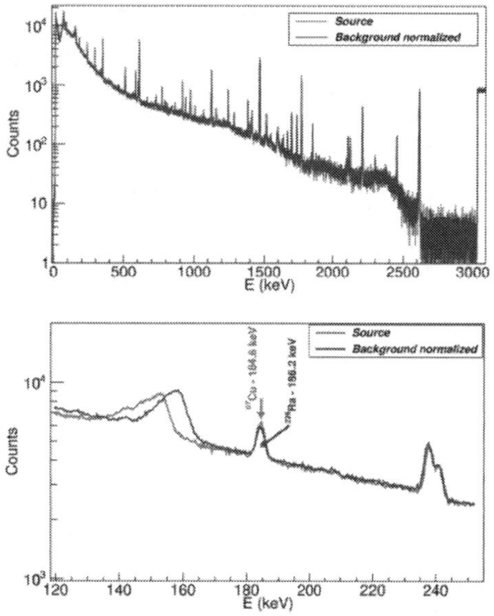

Fig. 5. (Color online) The γ-ray spectrum obtained during an accumulation period of 40 h, starting 7.5 d after the end of the irradiation in comparison with the background spectrum normalized for the measurement time.

4 Conclusion and future plans

The methodology to produce medically important radionuclides using inverse kinematics [1] has been tested with success. The production of the radionuclide ^{67}Cu ($T_{1/2}$=62 h) through the reaction of a beam of 15 MeV/nucleon ^{70}Zn with a cryogenic hydrogen gas target was performed. The activity of the ^{67}Cu isotope after 6.5 h of irradiation was A_{EOB} = 2.16(12) kBq and the 1pnA-1h activity was H_{EOB} = 1.8(5) kBq/(pnAh). The thick target yield in inverse kinematics is TTY_{inv}^{exp} = 0.021(5) (GBq/C) and the correspondent in forward kinematics is TTY_{for}^{exp} = 0.38(10) (GBq/C). The main radioimpurity in the Al catcher coming from the ^{70}Zn + p reaction is ^{69m}Zn ($T_{1/2}$=13.8 h), the activity of which at the end of irradiation is A_{EOB} = 2.55(26) kBq and 1pnA-1h Activity H_{EOB} = 2.2(6) kBq/(pnAh). The ^{90}Nb is the main impurity from the beam interaction on the Al catcher foil with an activity of A_{EOB} = 2.38(23) kBq. The present successful test indicates the possibility of producing important non-standard radionuclides of high radionuclide purity with the approach of inverse kinematics. The secondary neutrons from the primary reaction were used to irradiate a secondary target of Zn for further radioisotope production with promising results. The main requirement necessary to achieve the production of activities appropriate for preclinical studies is the availability of high-intensity (particle μA) heavy-ion primary beams.

Acknowledgements

We are grateful to the support staff of the Cyclotron Institute for providing the primary beam. Financial support for this work was provided, in part by the U.S. Department of Energy under Grant No. DE-FG03-93ER40773, by the NNSA under grant no de-na0003841 (CENTAUR), and by the Special Account for Research Grants of the National and Kapodistrian University of Athens. M.Y. and P.Z.F. acknowledge the support of the Houston Methodist Research Institute (HMRI).

References

1. G. A. Souliotis *et al.*, *A novel approach to medical radioisotope production using inverse kinematics: A successful production test of the theranostic radionuclide ^{67}Cu*, Appl. Radiat. Isotopes **149**, 89 (2019)
2. G. Hao *et al.*, *Copper-67 radioimmunotheranostics for simultaneous immunotherapy and immuno-SPECT*, Sci. Rep. **11**, 3622 (2021)
3. O. Keinänen,Harnessing *et al.*, *^{64}Cu/^{67}Cu for a theranostic approach to pretargeted radioimmunotherapy*, P. Natl. Acad. Sci. **117**, 28316 (2020)
4. N.A. Smith *et al.*, *The production, separation, and use of ^{67}Cu for radioimmunotherapy: a review*, Appl. Radiat. Isot. **70**, 2377–2383 (2012)
5. D. G. Medvedev *et al.*, *Development of a large scale production of ^{67}Cu from ^{68}Zn at the high energy proton accelerator: closing the ^{68}Zn cycle*, Appl. Radiat. Isot. **70**, 423–429 (2008)
6. T. Katabuchi *et al.*, *Production of ^{67}Cu via the ^{68}Zn(p,2p)^{67}Cu reaction and recovery of ^{68}Zn target*. J. Radioanal. Nucl. Chem. **277**, 467–470 (2008)
7. G. Pupillo *et al.*, *New production cross sections for the theranostic radionuclide ^{67}Cu*. Nucl. Instrum. Methods Phys. Res. B **415**, 41–47 (2018)
8. T. Stoll *et al.*, *Excitation functions of proton induced reactions on ^{68}Zn from threshold up to 71 MeV, with specific reference to the production of ^{67}Cu*, Radiochim. Acta **90**, 309–313 (2002)
9. K. Hilgers *et al.*, *Cross section measurements of the nuclear reactions ^{nat}Zn(d,x)^{64}Cu, ^{66}Zn(d,α)^{64}Cu, ^{68}Zn(p,αn)^{64}Cu for production of ^{64}Cu and technical developments for small scale production of ^{67}Cu via the ^{70}Zn(p,α)^{67}Cu process*, Appl. Radiat. Isot. **59**, 343–351 (2003)
10. D.J. Jamriska SR *et al.*, *Activation rates and chemical recovery of ^{67}Cu produced with low energy proton irradiation of enriched ^{67}Zn targets*. J. Radioanal. Nucl. Chem. **195**, 263–270 (1995)
11. S. Kastleiner *et al.*, *Possibility of production of ^{67}Cu at a small-sized cyclotron via the (p, α) reaction on enriched ^{70}Zn*, Radiochim. Acta **84**, 107–110 (1999)
12. R. E. Tribble *et al.*, *MARS: a momentum achromat recoil spectrometer*, Nucl. Instrum. Methods A **285**, 441–446 (1989)
13. J. F. Brinkley *et al.*, Progress in Research. Cyclotron Institute, Texas A&M University (2002-

2003), p. V-9 (2009) accessible at: ttp://cyclotron.tamu.edu/ publications.html.

14. O. B. Tarasov, D. Bazin, *LISE++: Radioactive beam production with in-flight separators,* Nucl. Instrum. Methods Phys. Res. B **266,** 4657–4664 (2008)
15. G. Schiwietz *et al., Improved charge state formulas,* Nucl. Instrum. Methods B **175-177,** 125-131 (2001)
16. Agostinelli, S., *et al., Geant4 a simulation toolkit.* Nucl. Instrum. Methods Phys. Res. **506**, 250–303 (2003)
17. I. Kawrakow, *Accurate condensed history Monte Carlo simulation of electron transport.* I. EGSnrc, the new EGS4 version. Med. Phys. 27, 485 (2000). Kawrakow, I., Rogers, D.W.O., NRCC Report PIRS-701, NRC, Ottawa (2003) https://nrc-cnrc.github.io/EGSnrc/.
18. O.B.Tarasov, D.Bazin Nucl. Instrum. Methods B **266**, 4657-4664 (2008); A.Gavron, Phys.Rev. C **21**, 230-236 (1980) http://lise.nscl.msu.edu/pace4
19. Levkovskij, Act. Cs. By Protons and Alphas, Moscow, *Cross sections of medium mass nuclide activation (A=40-100) by medium energy protons and alpha-particles (E=10-50 MeV)*, (1991)
20. S. Kastleiner *et al., Possibility of Production of ^{67}Cu at a Small-Sized Cyclotron via the (p,α)-Reaction on Enriched ^{70}Zn*, Radiochim. Acta **84**, no. 2, 107-110 (1999)
21. INTERNATIONAL ATOMIC ENERGY AGENCY, *Nuclear Data for the Production of Therapeutic Radionuclides*, Technical Reports Series No. **473**, IAEA, Vienna (2012)
22. D. Koning *et al., TENDL: Complete Nuclear Data Library for Innovative Nuclear Science and Technology*, Nuclear Data Sheets **155**, 1-55 (2019)
23. J. Ziegler, J. P. Biersack, M. Ziegler, *SRIM: the Stopping and Range of Ions in Matter*, http://www.srim.org
24. J. Mabiala *et al., Enhanced production of ^{99}Mo in inverse kinematics heavy ion reactions*, HINPw6 proceedings, EPJ-woc proceedings (article 08003 in this volume).
25. W. P. Tan *et al., New Measurement of ^{12}C +^{12}C Fusion Reaction at Astrophysical Energies*, Phys. Rev. Lett. **124**, 192702 (2020)
26. M. Bonardi, *The contribution to the nuclear data for biomedical radioisotope production from Milan Cyclotron Laboratory*, Data Requirements for Medical Radioisotope Production, INDC(NDS)-**193**, IAEA, Vienna (1988)
27. M. Aikawa , S. Ebata, S. Imai, *Thick-target yields of radioactive targets deduced from inverse kinematics*, Nucl. Instrum. Methods Phys. Res. B **353,** 1–3, (2015)
28. A.J. Koning, D. Rochman, *Modern nuclear data evaluation with the TALYS code system*, Nucl. Data Sheets **113** (12), 28412934 (2012) https://tendl.web.psi.ch/tendl_2019/talys.html
29. Z. Karimi *et al., Experimental production and theoretical assessment of ^{67}Cu via neutron induced reaction*, Annals of Nuclear Energy **133**, 665-668 (2019)
30. M. Papa, T. Maruyama, A. Bonasera, *Constrained molecular dynamics approach to fermionic systems*, Phys. Rev. C **64**, 024612 (2001)
31. M. Papa, G. Giuliani and A. Bonasera, *Constrained molecular dynamics II: A N-body approach to nuclear systems*, J. Comput. Phys. **208,** 403-415 (2005)
32. G. Giuliani, H. Zheng, A. Bonasera, *The many facets of the (non relativistic) Nuclear Equation of State*, Prog. in Part. and Nucl. Phys. **76**, 116 (2014)

Enhanced production of ^{99}Mo in inverse kinematics heavy ion reactions

Justin Mabiala[1,*], *Marcia R.* D. Rodrigues[2], *Georgios A.* Souliotis[3], *Victor E.* Iacob[2], *Ninel* Nica[2], *Brian* Roeder[2], *Gabriel* Tabacaru[2], *Kang* Wang[2], *Jedidiah* Romo[2], *Dustin* Scriven[2], *Nolan* Tenpas[2,5], and *Aldo* Bonasera[2,4]

[1]Department of Chemistry and Physics, Prairie View A&M University, USA
[2]Cyclotron Institute, Texas A&M University, USA
[3]Laboratory of Physical Chemistry, National and Kapodistrian University of Athens, Greece
[4]Laboratori Nazionali del Sud, INFN, Italy
[5]Texas Lutheran University, USA

Abstract. The reaction of a ^{100}Mo beam at 12 MeV/nucleon impinging on a ^{4}He gas-cell target was performed. The ^{99}Mo alongside other coproduced isotopes were collected after the gas target on an aluminum catcher foil and their respective radioactivities were measured by offline γ-ray analysis. In this contribution, preliminary experimental results which are used to discuss the possibility of optimal large-scale production conditions of the produced radioisotopes are presented.

1 Introduction

Today, radioisotopes are commonly used in medicine, both in diagnosis and therapy. A novel method for the production of important medical radioisotopes has been developed at the Cyclotron Institute at Texas A&M University. The approach is based on performing the nuclear reaction in inverse kinematics, namely sending a heavy-ion beam of appropriate energy on a light target and isotopes of interest are collected. The production of the theranostic radionuclide ^{67}Cu ($T_{1/2}$ = 62 h) was performed through the reaction of a ^{70}Zn beam at 15 MeV/nucleon with a hydrogen gas target [1]. The ^{67}Cu radionuclide alongside other coproduced isotopes were collected after the gas target on an aluminum catcher foil and their radioactivities were measured by offline γ-ray analysis. The production of additional radioisotopes is also possible by making use of the forward-focused neutrons from the reaction and allowing them to interact with a secondary target. The main requirement to obtain activities appropriate for preclinical studies is the development of high-intensity heavy-ion primary beams.

The next goal was the study of the production routes for the formation of medically interesting ^{99}Mo (used as ^{99}Mo/^{99m}Tc generator) using cyclotron accelerators, as the production in reactors is not enough to cover the world demand [2]. The ^{99m}Tc as a 140 keV γ-ray emitter (I_γ = 89%) with a half-life of $T_{1/2}$ = 6.01 h, is considered to be an ideal radiotracer and it is estimated to be used in approximately 85% of all nuclear medicine diagnostic scans worldwide. The ^{99m}Tc is produced via β-decay from ^{99}Mo ($T_{1/2}$ = 65.94 h). The amount of ^{99}Mo consumed for radiopharmaceuticals increases year by year [3].

*e-mail: jumabiala@pvamu.edu

The present work reports on the production routes of ^{99}Mo and other medical radioisotopes using inverse-kinematics nuclear reactions.

2 Experimental method

The experiment was performed at the Cyclotron Institute at Texas A&M University. The beam of ^{100}Mo at 12 MeV/nucleon was delivered by the K500 superconducting cyclotron and transported to the target chamber of the Momentum Achromat Recoil Spectrometer (MARS) [4]. The beam impinged on a cryogenic gas cell filled with ^{4}He gas. The experimental setup is schematically shown in Fig. 1. The cryogenic gas cell had a length of 10 cm with 4 μm Havar entrance and exit windows of 19.0 mm diameter. The target temperature was kept at T = 77 K during all the irradiations. The gas-cell target, the aluminum catcher foils and the Faraday cup to monitor the beam current, were set up in the MARS target chamber. The beam current was periodically monitored and was nearly constant (within 15%). The aluminum catcher foil with 127 μm thickness, mounted in a target frame with 12.7 mm diameter hole and positioned after the gas-cell target, was used to collect the produced ^{99}Mo nuclei along with other coproduced nuclides. In order to measure the neutron yield produced during the irradiation, ten neutron detectors were positioned around the chamber. After an irradiation period lasting a few hours (See Table 1), we turned the beam off, then removed the catcher foil from the frame and eventually placed it in front of a 70% efficient HPGe detector. We then measured the foil's gamma activity to determine the production efficiency for ^{99}Mo, as well as that of other coproduced radioisotopes. Also, the produced isotopes could be detected directly with the MARS spectrometer in coincidence with produced neutrons. Three aluminum catcher

https://doi.org/10.1051/epjconf/202125208003

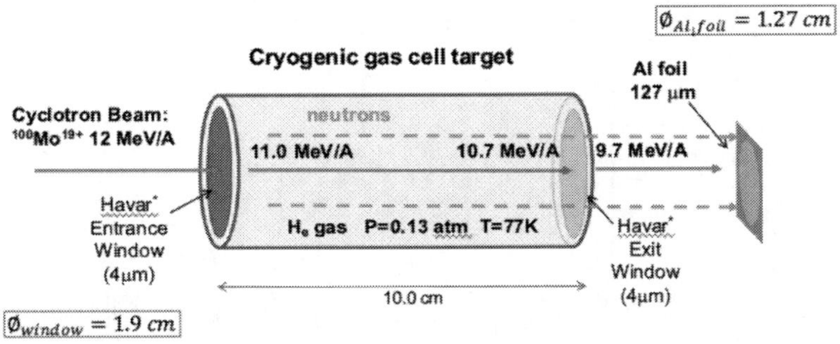

Figure 1. Schematic diagram of the irradiation setup. A ^{100}Mo beam at 12 MeV/nucleon enters the gas cell and interacts with the helium gas. The heavy reaction products, including ^{99}Mo, after exiting the gas cell are implanted in the Al catcher. The energies of the beam are listed as it passes through the entrance window, the gas target and the exit window.

foils (referred to as source 1, source 2, and source 3) were irradiated for different irridiation times, at different beam currents and gas pressures. The irradiation conditions for each foil are shown in Table 1. The beam energy degradation in the gas target and the Mylar windows was determined using LISE++ calculations [5, 6]. We estimated that the beam energy degreaded more when using gas target at higher pressure (source 3). We therefore expected ^{99}Mo and other isotopes of interest to be predominantly produced in source 3.

Table 1. Irradiation conditions for each foil.

Source	Irradiation time (hour)	Beam current (nA)	Gas pressure (Torr)	Target thickness (mg/cm^3)
Source 1	11.6	2.0	102	0.085
Source 2	10.5	7.0	213	0.178
Source 3	7.9	3.5	1008	0.841

A particle-identification plot of the total energy deposited in the Si detectors ΔE–E (55 μm –500 μm) versus the Y-position in the MARS focal plane is shown in Fig. 2. Loci representing different isotopes including mass 99 are clearly visible and well separated.

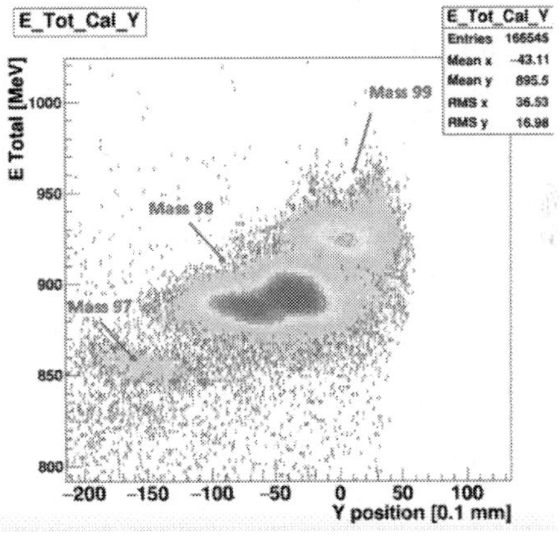

Figure 2. Particle identification spectrum of the total energy deposited in the Si detectors versus the Y-position on the MARS focal plane. Different isotopes including mass 99 are clearly visible and well separated.

3 Data analysis and results

After the irradiations and an appropriate cooling time, the foils were removed, and the gamma activities emitted from the activated foils were measured. The foils were placed at a distance of 50 mm from the end cap of the detector. Under this condition, the dead time of the counting system was around 2–3%, thus avoiding the pile-up effect. The energy resolution of the detector system was 1.0–2.0 keV full width at half maximum (FWHM) for the peaks of interest. The energy calibration was performed using known γ-rays obtained in the spectra. The absolute photopeak efficiencies for the actual source-detector geometry were obtained with the Monte Carlo codes GEANT4 [7] and EGSnrc [8, 9]. The radioactivity levels of the isotopes were determined by the quantification of the photopeaks of the γ-rays taking into account the branching ratios and the absolute efficiencies of the detector. A detailed description of the γ-ray analysis of all the observed γ-ray peaks will be presented in [10]. Room background spectra of good statistics were recorded before the run and carefully subtracted from the spectra acquired with all three sources in order not to affect the uncertainty in determination of the ^{99}Mo yields. Figure 3 displays a background-subtracted γ-decay spectrum for foil 3, obtained during an accumulation period of 16 h, starting 10 h after the end of the irradiation when most of the irrelevant short-lived intense activities of the source (shorter than 1 h) were exhausted. The peaks associated with ^{99}Mo and the most intense impurities are indicated. The most prominent γ peaks are those of ^{99}Mo γ–decay (140.5 keV, 89.4% relative intensity) and

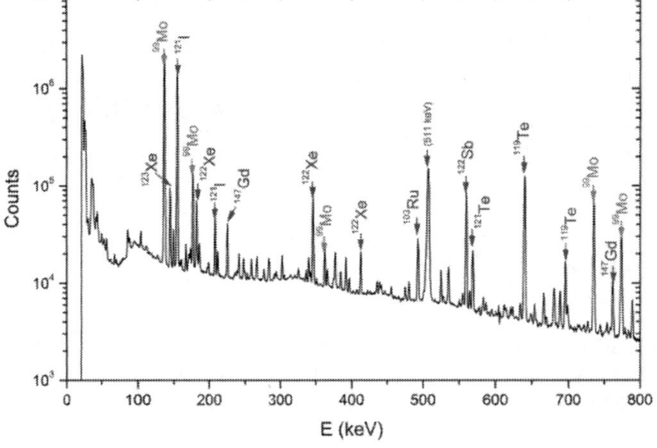

Figure 3. Typical background-subtracted γ-ray spectrum of the Al catcher foil following the interaction of a 12 MeV/nucleon ^{100}Mo beam with the helium gas cell. Different γ-rays coming from the decay of ^{99}Mo along with other radionuclides are indicated.

the other impurities have smaller activities. The reaction presented a breakthrough of ^{99}Mo production. The analysis is underway, the radioactivities for ^{99}Mo and the other impurities are being extracted. The data analysis of radionuclides detected in coincidence with neutrons is underway.

4 Summary

Production routes of ^{99}Mo and other isotopes by means of an inverse kinematics have been demonstrated in this work. This production scheme allows us to use essentially two different targets, a gas and a solid catcher. Furthermore, the neutrons produced in the forward direction could be used to irradiate other targets and produce a variety of radioisotopes of interest. With the neutron detectors, the radioisotope production could also be exploited in coincidence with heavier fragments measured in the MARS detector. We are currently working on the extraction of the relevant cross-sections to compare to data if available and to theoretical models. The present data call for further investigations and the findings will be discussed in the forthcoming papers.

Acknowledgements

We are grateful to the support staff of the Cyclotron Institute for providing the primary beam. This work was performed with financial support from the U.S. Department of Energy under Grant No. DE-FG03-93ER40773 and NNSA Grant No. DENA0003841 (CENTAUR).

References

[1] G.A. Souliotis, M.R.D. Rodrigues, K. Wang, V. Iacob, N. Nica, B. Roeder, G. Tabacaru, M. Yu, P. Zanotti-Fregonara, A. Bonasera, Appl. Radiat. Isotopes 149, 89 (2019)

[2] S. Takács, Z. Szűcs, F. Tárkányi et al., J. Radioanal. Nucl. Chem. 257, 195 (2003)

[3] M.S. Uddina,, M. Hagiwara, F. Tarkanyi, F. Ditroi, M. Baba, Appl. Radiat. Isotopes 60, 911 (2004)

[4] R.E. Tribble, R.H. Burch, C.A. Gagliardi, Nucl. Instrum. Methods A 285, 441 (1989)

[5] O.B. Tarasov, D. Bazin , Nucl. Phys. A 746, 411c (2004)

[6] O.B. Tarasov, D. Bazin, Nucl. Instr. and Meth. B 204, 174 (2003)

[7] S. Agostinelli et al., Nucl. Instrum. Methods Phys. Res. 506, 250 (2003)

[8] I. Kawrakow, Med. Phys. 27, 485 (2000)

[9] I. Kawrakow, D.W.O. Rogers, *NRCC Report PIRS-701*, NRC, Ottawa, 2003. accessible at. https://nrc-cnrc.github.io/EGSnrc/

[10] M.R.D. Rodrigues et al., *Production of Medical Radionuclide ^{99}Mo in Inverse Kinematics* (in preparation)

AUTHOR INDEX

AUTHOR INDEX

AUTHOR INDEX